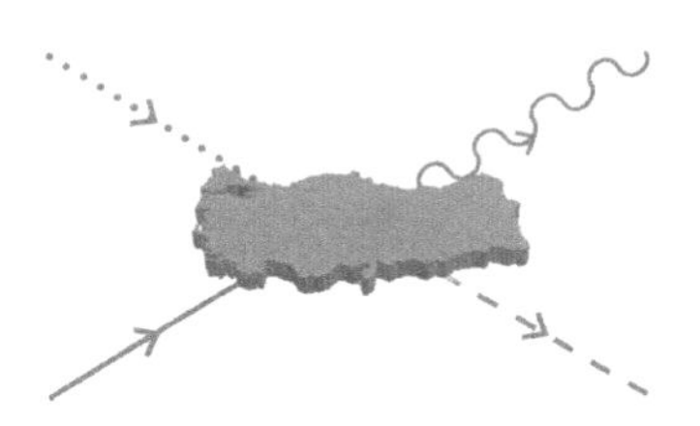

PROCEEDINGS OF THE 13TH REGIONAL CONFERENCE

MATHEMATICAL PHYSICS

PROCEEDINGS OF THE 13TH REGIONAL CONFERENCE

MATHEMATICAL PHYSICS

Antalya, Turkey, 27–31 October 2010

Editors

Uğur Camcı (Akdeniz University, Turkey)

İbrahim Semiz (Boğaziçi University, Turkey)

NEW JERSEY • LONDON • SINGAPORE • BEIJING • SHANGHAI • HONG KONG • TAIPEI • CHENNAI

Published by

World Scientific Publishing Co. Pte. Ltd.
5 Toh Tuck Link, Singapore 596224
USA office: 27 Warren Street, Suite 401-402, Hackensack, NJ 07601
UK office: 57 Shelton Street, Covent Garden, London WC2H 9HE

British Library Cataloguing-in-Publication Data
A catalogue record for this book is available from the British Library.

Cover image of Antalya, Turkey, courtesy of Mehmet Demirci.

MATHEMATICAL PHYSICS
Proceedings of the 13th Regional Conference

ISBN 978-981-4417-52-5

Printed in Singapore by World Scientific Printers.

ORGANIZING COMMITTEES

INTERNATIONAL ADVISORY COMMITTEE

LOCAL ORGANIZING COMMITTEE

U. Camcı	– Akdeniz University & TUG, Antalya
Z. Eker	– Akdeniz University & TUG, Antalya
N. Ünal	– Akdeniz University, Antalya
İ. Boztosun	– Akdeniz University, Antalya
M. Helvacı	– Akdeniz University, Antalya
Y. Sucu	– Akdeniz University, Antalya
S. Helhel	– Akdeniz University, Antalya
M. Dernek	– Akdeniz University, Antalya
B.C. Lütfüoğlu	– Akdeniz University, Antalya
Y. Küçükakça	– Onsekiz Mart University, Çanakkale
İ. Semiz	– Boğaziçi University, İstanbul
E. Abadoğlu	– Yeditepe University, İstanbul
E. Güdekli	– İstanbul University, İstanbul

LIST OF PARTICIPANTS

: Oral presentation
: Work included in the proceedings

Muhammad Afzal	Quaid-i-Azam University, Islamabad, Pakistan
Bobomurat Ahmedov 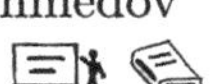	Institute of Nuclear Physics, Tashkent, Uzbekistan
Farhad Ardalan	Sharif University of Technology & IPM, Tehran, Iran
Emrumiye Arlı	Çukurova University, Adana, Turkey
Orhan Bayrak	Akdeniz University, Antalya, Turkey
İsmail Boztosun	Akdeniz University, Antalya, Turkey
Çetin Camcı	Çanakkale Onsekiz Mart University, Çanakkale, Turkey
Uğur Camcı	Akdeniz University & TUG, Antalya, Turkey
Diyadin Can	İstanbul University, Istanbul, Turkey
Ali Kazım Çamlıbel	Boğaziçi University, Istanbul, Turkey
Tarık Çelik	Turkish Academy of Sciences, Ankara, Turkey
Hakan Çiftçi	Gazi University, Ankara, Turkey
Naresh Dadhich 	Inter-University Center for Astronomy and Astrophysics, Pune, India
Durmuş Ali Demir	İzmir Institute of Technology, İzmir, Turkey
Mustafa Dernek	Akdeniz University, Antalya, Turkey
Koray Düztaş	Boğaziçi University, Istanbul, Turkey
Mohammad Reza Ejtehadi	Sharif University of Technology, Tehran, Iran
Zeki Eker	Akdeniz University & TUG, Antalya, Turkey
Ganim Geçim	Akdeniz University, Antalya, Turkey

Cristiano Germani	Ludwig-Maximilians-University, München, Germany
Ertan Güdekli	İstanbul University, Istanbul, Turkey
Semra Gürtaş	Akdeniz University, Antalya, Turkey
Richard Lewis Hall	Concordia University, Montreal, Canada
Mahmut Hortaçsu	İstanbul Technical University, Istanbul, Turkey
Viqar Husain	University of New Brunswick & AARMS, Canada
Ibrar Hussain	National University of Science & Technology, Rawalpindi, Pakistan
George Jorjadze	Razmadze Mathematical Institute, Tiblisi, Georgia
Mesut Karakoç	Akdeniz University, Antalya, Turkey
Vahid Karimipour	Sharif University of Technology, Tehran, Iran
Mehmet Koca	Sultan Qaboos University, Oman
Burak Kurt	Akdeniz University, Antalya, Turkey
Veli Kurt	Akdeniz University, Antalya, Turkey
Yusuf Küçükakça	Akdeniz University, Antalya, Turkey
C. S. Lim	Kobe University, Kobe, Japan
Bekir Can Lütfüoğlu	Akdeniz University, Antalya, Turkey
Fazal Mahomed	University of the Witwatersrand & Centre for Differential Equations, Johannesburg, South Africa
Alexei Morozov	Institute of Theoretical and Experimental Physics, Moscow, Russia
Ali Mostafazadeh	Koç University, Istanbul, Turkey
Işıl Başaran Öz	Akdeniz University, Antalya, Turkey
Sacit Özdemir	Ankara University, Ankara, Turkey
Asghar Qadir	National University of Science & Technology, Rawalpindi, Pakistan
Seifallah Randjbar-Daemi	International Centre for Theoretical Physics, Trieste, Italy

Muneer Ahmad Rashid	National University of Science and Technology & CAMP, Rawalpindi, Pakistan
Shahin Rouhani	Sharif University of Technology, Tehran, Iran
Makoto Sakamoto	Kobe University, Kobe, Japan
Özgür Sevinç	İstanbul University, Istanbul, Turkey
Ghulam Shabbir	Ghulam Ishaq Khan Institute of Engineering Sciences and Technology, N.W.F.P, Pakistan
Muhammad Sharif	Punjab University, Lahore, Pakistan
Azad Akhter Siddiqui	National University of Science & Technology, Rawalpindi, Pakistan
Yusuf Sucu	Akdeniz University, Antalya, Turkey
Ferhat Taşkın	Erciyes University, Kayseri, Turkey
Francesco Toppan	Brazilian Center for Physics Research (CBPF), Rio de Janeiro, Brazil
Gülçin Uluyazı	İstanbul University, Istanbul, Turkey
Nuri Ünal	Akdeniz University, Antalya, Turkey
Maria A. H. Vozmediano	Instituto de Ciencia de Materiales & CSIC, Madrid, Spain
Fevziye Yasuk	Erciyes University, Kayseri, Turkey
Özlem Yeşiltaş	Gazi University, Ankara, Turkey
Mustafa Kürşat Yıldız	Erciyes University, Kayseri, Turkey
İhsan Yılmaz	Çanakkale Onsekiz Mart University, Çanakkale, Turkey

PREFACE

by Asghar Qadir

The Regional Conferences started in one sense in Adana, Turkey, in 1985. However, at that time these Conferences had not been given that name and the "Region" had not been identified. It was clear that it included Turkey, Iran and Pakistan. Pakistan was represented by Faheem Hussain and had its base in the International Centre for Theoretical Physics (ICTP). In the second one, also held in Turkey, it was agreed that it should be formalized as a regular series and Faheem Hussain undertook to hold the third one in Islamabad, Pakistan. It was at this stage that the "Region" was defined and the series acquired the name of the Regional Conferences. Faheem came back to Quaid-i-Azam University full of plans for the Third Regional Conference in Mathematical Physics and associated me in the organization. I got in touch with World Scientific to publish our Proceedings (also the first one in the series). The theme of the Conference was Superstrings and Relativity and it was held in 1989. It turned out to be a superb international conference. The Proceedings came out the same year and has been well cited. In a very real sense, then, the Third was the start of the series.

We tried to keep the series circulating between Turkey (with Mehmet Koca as the main organizer there), Iran (with Farhad Ardalan) and Pakistan, to be held every other year. Regularity was not always maintained and the series did not always get the Proceedings published when the Conference was held. Later, the experiment of holding the Conference in some other place (Georgia) was tried but each time it led to the series suffering a break. It has finally been realized that the series has its roots in the Region and should stay there. Unfortunately, the "young Turks" who started the series became old fogies and felt that the baton needed to be handed over to younger people. However, this has not yet settled into a self-sustaining mode. We are still looking into how this could be achieved.

This (Thirteenth) Conference is dedicated to Faheem Hussain as he died due to cancer some time after the Twelfth Conference. The next one was to be held in Turkey and I tried to arrange that it should be held there in memory of Faheem, so I "leaned" on Uğur Camcı to organize it. I would like to share some memories of Faheem. He left the University soon after the Third Conference and went to Germany. He finally ended up at the ICTP, first with charge of developing their Diploma course in High Energy Physics. That was so successful that the Mathematics section of the ICTP developed one as well. When Galeno Denardo retired as Director of External Activities at the ICTP, Faheem took that charge up and continued with it till he retired from the ICTP. He has left fond memories with the High Energy Physics group, the Diploma programme and with External Activities. All feel that he was immensely successful in each of his capacities.

Begging further indulgence, I would like to give some personal reminiscences about Faheem as well. In some sense I "knew" Faheem since 1964. I was in the undergraduate first year of Physics at the Imperial College, London and Faheem was a PhD student there. I did not meet him, but, as it usually happens, the junior students know about the senior students and the seniors do not deign to notice the juniors. This unequal relationship changed when I attended my first "conference" in Pakistan, soon after joining the Institute of Mathematics of what was then called the Islamabad University (IU). The conference was a "National Science Conference" (more populary known as *the mela* or carnival). Recognizing Faheem in the audience, I approached him and found that he was in the Institute of Physics of IU. With Faheem, to know him was to become friends. I doubt if any young person could have been found who was immune to his charm. There was a total lack of pretension or self-importance. He would talk about *issues* not people or petty politics or complain about how much more he deserved than he got. Not Faheem!

Faheem was very much a product of intellectual England of the 1960's. There was a general feeling then that "the millennium was at hand". All that was required was the correct political system and the will of the workers to bring it about. The only real danger was the sword of Damocles, of nuclear bombs, hanging over our heads. As such there was a concerted effort to change from the aggressive mode of the 1940's and 1950's to a mode of love and friendship. Among the most effective in this direction were the folk singers whose songs would be sung at the protest marches for nuclear disarmament. Faheem's PhD supervisor, P.T. Matthews, had a life-size poster of Bob Dylan (one of the leaders among the folk singers) on the wall facing the door as one entered his office. However, while the general mood was of some form of socialism, Faheem was a Marxist in every sense. People often associate Marxism with belligerence. Faheem was a living counter-example to this misconception. While he was vehement in his beliefs I did not see him display *malice* towards anyone. His vehement opposition never carried ill-will.

Very often Marxists, especially in Pakistan, pay lip service to the equality of all people. For them "all people are equal *but some are more equal than others*". Not Faheem! He lived his belief in a way I have not witnessed in others. His Marxism cost him in Pakistan but he never wavered in following his honest opinions.

It would be wrong to give the impression that social awareness and activism were all that Faheem stood for. When he joined Islamabad University, he was generally regarded as having an exceptional grasp of the fundamentals of Physics, despite the galaxy of stars in the Institute of Physics. He was no less emotional or less devoted to his Physics than to his Marxism. He was one of those who exuded real joy in the subject. To me, the late Arif-uz-Zaman and Faheem Hussain epitomized the quintessential physicist, in their pure joy in the subject. For them, Physics was not just a job, or a career, but a way of life. That enthusiasm infected those around them — not only their students but also their colleagues. Faheem would take on duties that furthered the cause of Physics because it needed to be done, not because he was asked to do it or would get credit for it.

We all felt that the declining standards of education were adversely affecting the level of the incoming students. Characteristically, Faheem was not ready to just sit and complain about it. We started a series of popular level lectures delivered at the various Colleges and Schools in the region. There would be general seminars where teachers from all Schools and Colleges, not just those where the lecture was held, would be invited. I also started a similar one in Mathematics. They were very successful and we then started giving such lectures for the particular School or College students and teachers and then devoted the day to that institution to resolve problems that the teachers were facing in teaching or understanding the courses. This made a major difference to the interest and enthusiasm of the students that came to the University.

Faheem had married an American, Jane, and had two children, Nadeem and Amina. His casual style carried through to his dealing with his family. His children called him by his first name rather than by the relationship. With that approach, he was not put on a pedestal by his children but was their *friend*. Though I have seen others actually adopting this way of managing family relations, in Pakistan there are very few such examples.

His wife retained her separate identity and was not regarded by Faheem or herself as Mrs. Faheem Hussain, but as Jane Steinfels. He never tried to dominate anyone — his wife, his children, his students or his colleagues. The same approach characterized his period as Chairman of his Department. He never tried to project himself or to present himself as other than he was.

When Zia-ul-Haq set himself up as the *Khalifah* of Pakistan there was a movement to try to restore democracy. Among other attempts people started distributing literature against the current regime. The regime reacted by picking up people and locking them away. In these raids, they also picked up (and tortured) three faculty members of the Quaid-i-Azam University (QAU), as it was now called. They then started interrogating others. Faheem and a handful of others took it on themselves to try to protect the others and to go to visit the locked-up colleagues. This despite the fact that Faheem stood in severe danger of being picked up himself. I remember going with him to the FIA office when two of our colleagues had been called there. One of them had written an article entitled "The New Great Game and Pakistan", read by the worthy FIA as "The Great Game of Pakistan". What could be more treasonous than calling Pakistan a game? This was not a trivial matter. They had already locked away someone for "being in possession of a library". The implication was that this was a library of seditious material. The evidence displayed was a picture published in the newspaper showing a pile of books topped by a copy of Solzhenitsyn's *Cancer Ward*, which had a picture of a hammer and sickle on it!

Faheem had not been appointed as full Professor in the Physics Department because of the lack of a position. In 1983, I became Chairman of the Department of Mathematics at QAU. In view of Faheem's competence in various branches of mathematics, especially Group Theory, I wanted to get him to join the Mathematics Department. Unfortunately, the Vice Chancellor at the time refused to allow it, saying that he did not want Mathematics to be developed at the expense of Physics. I argued that if Physics wanted to retain him they should arrange for an extra position of Professor of Physics instead of keeping him from a higher appointment, but my argument was ignored.

I had long been saying that there should be joint courses at the University to utilize available faculty more efficiently. A new Dean of the Faculty of Science accepted my point and asked me to set up a three-person committee for joint courses in Physics and Mathematics, to start with. Faheem's casual and humble way of managing his department as Chairman had influenced me strongly. I asked for him and Kamaluddin Ahmed. We presented our proposal for combined courses, which was approved and successfully implemented. Unfortunately, it was later discontinued due to a desire of many Chairmen to maintain control over the teachers and students, as if they were possessions.

In the mid 1980's, Jane developed a disease that required expensive treatment, which Faheem could not afford on his salary as an Associate Professor at QAU. This problem forced him to take leave from the University and accept a job in Libya. I tried arguing with him about leaving the University, using his Marxist opposition to religious fundamentalism to dissuade him from going to Qaddafi's land. However, when he told me the reason I could no longer argue. His time in Libya reduced his extreme opposition to people following religion and generally made his views milder. That did not mean that he ceased to be a Marxist but that he realized that other forces need to be considered.

The Libyan interlude caused a hiatus in Faheem's research. His Marxist interest in Philosophy was extended to the philosophy of science, partly because of me. In Libya, he tried to work on the philosophical problems of Quantum Theory. That led to one paper that other physicists did not take so seriously. On his return, we tried to work further in that area with no success and then published a joint paper on the problem of accelerated frames in Special Relativity. Before he could get back to major research in his field he needed to get back in touch with mainstream research in it. He got an opportunity to go to Germany for a spell and picked up the recent excitement of Superstring Theory in which he worked there. His earlier interest in Particle Phenomenology was also pursued.

During his Sabbatical, he had also visited the ICTP, from where he got involved in the Regional Conferences in Mathematical Physics mentioned earlier. The break-up of Faheem's marriage to Jane led to his leaving the country again. He had now got back into the swing of research and was able to go again on a year's leave. He then got married to a Triestino, Sara, and accepted a position at ICTP. Faheem attended the Eleventh Conference as he was at the point of retiring from AS-CITP (as it was now known). He discussed with me about returning to Pakistan. I had just taken early retirement from QAU (due to severe differences with the administration) and accepted a position at the COMSATS Institute of Information Technology (CIIT), Islamabad. I asked him to apply for a Higher Education Commission (HEC) *foreign* faculty Professor and get himself placed at CIIT. He had doubts about the ethics of such an arrangement but I persuaded him that it was *not* unethical as the whole idea was to reverse the brain-drain.

I found that CIIT was not going to be appropriate for me, so I asked to be released from my (verbal) commitment there and joined the National University of Sciences and Technology (NUST). I asked Faheem to change his choice and join me there. NUST is run by the military and Faheem (like so many others) did not want to have anything to do with a military-run institution. He wanted to join the National Centre for Physics (NCP) but the HEC did not allow placement there as it is not a university. As such, when he came back he joined CIIT but was allowed to also work at the NCP. It had been agreed that the next Regional Conference, which would be the Twelfth, would be in Pakistan. Prof. Riazuddin was the Convener of the Conference and Faheem the Secretary. The Conference was held in 2006 and was again a very successful one, with G. 't Hooft, the Nobel laureate, speaking as well. The Proceedings were, once again, published by World Scientific in 2007. He and I were both among the editors. This was the last time that we worked together.

Faheem did not feel comfortable with the HEC position and CIIT (perhaps because the Physics Department had shifted to the other end of Islamabad — and beyond). He got an offer at the Lahore University of Management Sciences (LUMS) in their newly started Faculty of Sciences and Engineering. He worked there till the end.

Faheem was a great nationalist and, oddly enough, a great *inter*nationalist. Many would regard the two as incompatible. Not Faheem! His nationalism was not of the "us against them" type. It was a deep commitment to bring Pakistan into line with what he thought it should be, especially as regards equality of social status. He felt that Pakistani society needed to be open to influences from other societies and should learn tolerance. He was particularly keen to oppose the anti-Indian sentiment prevalent in the country. To this end, he always tried to involve Indians in the activities he arranged and made special arrangements that they should get especially hospitable treatment. I believe that all the Indians had especially warm feelings for him. Of course, his internationalism was not limited to Indians. He went out of his way for many others and I have heard extreme praise for this aspect of his character from any number of other people. I think it made him especially appropriate for his job as Director of External Activities at AS-ICTP. In that capacity, he urged me to set up a regional Network for Cosmology, Astrophysics and Relativity, which I did. Long after he left AS-ICTP, he continued to take active interest in the Network and to try to ensure that it functioned at the best possible level.

He developed cancer of the prostrate while in Italy but it seemed to have been treated successfully. When it recurred, it was again (apparently) successfully treated. This happened a third time. However, the recurrences soon swamped the periods of remission and he spent much of the last year in Italy. He never admitted to me the pain he suffered or talked of his illness in anything other than the most hopeful terms. However, towards the end he stopped answering my questions about his health and would write about other things. That seemed to me to indicate that his ailment was seriously deteriorating. I still did not expect that the end was so close. I received an e-mail about extreme deterioration of his condition with an empty feeling in the pit of my stomach, knowing that this was the end. I suppose I was better prepared when, about half an hour later, I got an SMS that he had passed away. I hope that enough in Pakistan carry something of his spirit in them.

Faheem was a very dear friend. We repeatedly re-discovered that we shared the same birthday and would then wish each other on it. He was one of the few men of principles around. We had a lot in common in our views but had our differences of opinion as well. Never do I recall any time when our differences led to the least ill-will. That is extremely rare and the credit must go to Faheem for it. He had the most charming way of holding very strong opinions of his own without that leading to a condemnation of the other person's opinions. For example, he did not think much of my work in Economics as he felt that Mathematics could not be effectively applied to those problems. He expressed this view but it was never said in a way that made me defensive about *my* work. Instead, it was an academic criticism of *the field in which I was working.* I had not expected to mourn his loss as I had not expected to survive him. Though there were long periods when we did not meet, it is a loss that is not easy to bear, knowing that there will be no further meetings.

Asghar Qadir Antalya, Turkey
National University of Science & Technology, Pakistan October 2010

CONTENTS

Part B General Relativity and Cosmology 101

Part C Quantum Gravity 151

Part D Quantum Field Theory 183

PART A

Formal Aspects

GEOMETRIC SPECTRAL INVERSION

RICHARD L. HALL

Department of Mathematics and Statistics, Concordia University, 1455 de Maisonneuve Boulevard West, Montreal, Quebec, Canada H3G 1M8
E-mail: rhall@mathstat.concordia.ca
www.mathstat.concordia.ca/faculty/rhall

A discrete eigenvalue E_n of a Schrödinger operator $H = -\Delta + vf(r)$ is given, as a function $F_n(v)$ of the coupling parameter $v \geq v_c$. It is shown how the potential shape $f(x)$ can be reconstructed from $F_n(v)$. A constructive inversion algorithm and a functional inversion sequence are both discussed.

Keywords: Schrödinger operator; Discrete spectrum; Envelope theory; Kinetic potentials; Spectral inversion.

1. Introduction

We suppose that a discrete eigenvalue $E_n = F_n(v)$ of the Schrödinger Hamiltonian

$$H = -\Delta + vf(x)$$

is known for all sufficiently large values of the coupling parameter $v \geq v_c$ and we use this data to reconstruct the potential shape $f(x)$. The usual 'forward' problem would be: given the potential (shape) $f(x)$, find the corresponding energy trajectories $F_n(v)$. For example, if the potential shape is $f(x) = -\text{sech}^2(x)$, then the eigenvalues as functions of the coupling v are given[1] by the formula

$$F_n(v) = -\left[\left(v + \frac{1}{4}\right)^{\frac{1}{2}} - \left(n + \frac{1}{2}\right)\right]^2, \quad n = 0, 1, 2, \ldots, \quad v \geq n(n+1). \tag{1.1}$$

These energy graphs are illustrated in Fig. 1.

The problem we now consider here is the inverse of this, namely $F \to f$. We call this problem 'geometric spectral inversion'. It must at once be distinguished both from inverse scattering theory[2–6] and, more specifically,

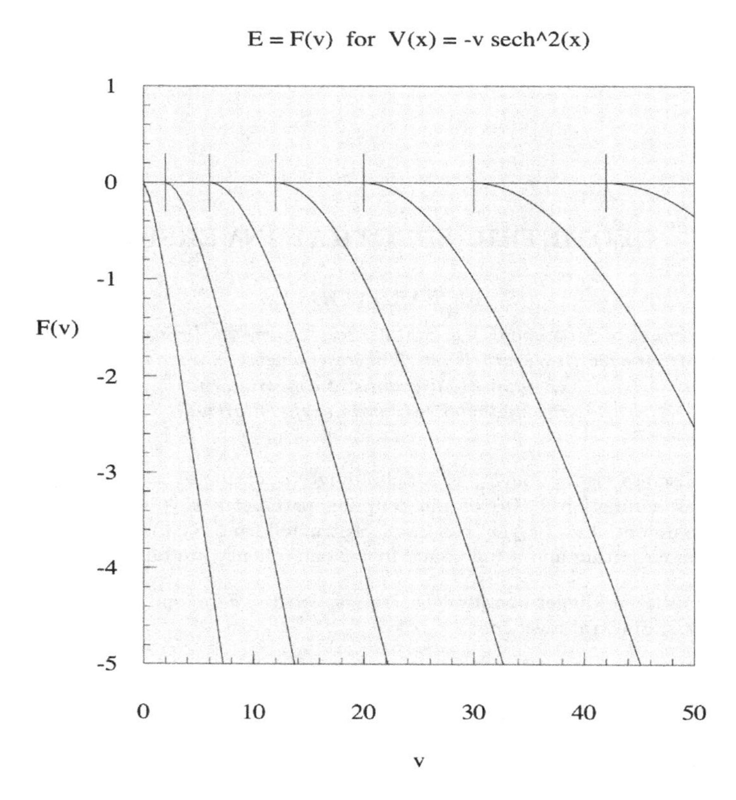

Fig. 1. Discrete eigenvalues of $H = -\Delta - v\ \mathrm{sech}^2(x)$ as functions of the coupling $v > 0$.

from the 'inverse problem in the coupling constant' discussed, for example, by Chadan *et al.*[2,7–10] In this latter problem, the discrete part of the 'input data' is a set $\{v_i\}$ of values of the coupling constant that all yield the identical energy eigenvalue E. The index i might typically represent the number of nodes in the corresponding eigenfunction. In contrast, for the problem discussed in the present paper, i is kept fixed and the input data is the graph $(F(v), v)$, where the coupling parameter has any value $v > v_c$, and v_c is the critical value of v for the support of a discrete eigenvalue with i nodes. Geometric spectral inversion has been discussed in a series of earlier papers[11–16]. Here we shall mainly discuss the bottom of the spectrum $i = 0$. However, on the basis of results we have obtained for the inversion IWKB of the WKB approximation[13], there is good reason to expect that inversion is possible starting from any discrete eigenvalue trajectory $F_n(v)$,

$n > 0$. In fact, perhaps not surprisingly, IWKB yields better results starting from higher trajectories; moreover, they become asymptotically exact as the eigenvalue index is increased without limit.

After recalling some basic results in the remainder of this introduction, in sections 2-4 we discuss a constructive algorithm for generating $f(x)$ from $F(v)$. In section 5 we apply this method to some examples. In sections 6 and 7, we briefly describe an established geometric spectral theory[17–22] that enables us in section 8 to effect 'functional inversion'[16]. This latter method allows us to start with a seed potential $f^{[0]}(x)$ and from this reconstruct $f(x)$ by means of a sequence of functional operations. This latter method is applied to some examples in sections 9 and 10.

By making suitable assumptions concerning the class of potential shapes, general theoretical progress has already been made with this inversion problem[11,12]. The assumptions that we retain for most of the present paper are that $f(x)$ is symmetric, monotone increasing for $x > 0$, and bounded below: consequently the minimum value is $f(0)$. We assume that our spectral data, the energy trajectory $F(v)$, derives from a potential shape $f(x)$ with these features. We have discussed[23] how two potential shapes f_1 and f_2 can cross over and still preserve spectral ordering $F_1 < F_2$. It is known[12] that lowest point $f(0)$ of f is given by the limit

$$f(0) = \lim_{v \to \infty} \frac{F(v)}{v}. \tag{1.2}$$

We have proved[11] that a potential shape f has a finite flat portion ($f'(x) = 0$) in its graph starting at $x = 0$ if and only if the mean kinetic energy is bounded. That is to say, $s = F(v) - vF'(v) \le K$, for some positive number K. More specifically, the size b of this patch can be estimated from F by means of the inequality:

$$s \le K \quad \Rightarrow \quad f(x) = f(0), \quad |x| \le b, \quad \text{and} \quad b = \frac{\pi}{2} K^{-\frac{1}{2}}. \tag{1.3}$$

The monotonicity of the potential, which allows us to prove results like this, also yields the

Concentration Lemma[11]

$$q(v) = \int_{-a}^{a} \psi^2(x, v)dx > \frac{f(a) - F'(v)}{f(a) - f(0)} \quad \to \quad 1, \quad \text{as } v \to \infty, \tag{1.4}$$

where $\psi(x, v)$ is the normalized eigenfunction satisfying $H\psi = F(v)\psi$. More importantly, perhaps, if $F(v)$ derives from a symmetric monotone potential shape f which is bounded below, then f is *uniquely* determined[12]. The significance of this result can be appreciated more clearly upon consideration of an example. Suppose the bottom of the spectrum of H is given by $F(v) = \sqrt{v}$, what is $f(x)$? It is well known, of course, that $f(x) = x^2 \rightarrow F_o(v) = \sqrt{v}$; but are there any other potential pre-images of this spectral function $F_o(v)$? Are scaling arguments reversible? A possible source of initial disquiet for anyone who ponders such questions is the uncountable number of (unsymmetric) perturbations[24] of the harmonic oscillator $f(x) = x^2$ there are, all of which have the identical spectrum $E_n = (2n+1), n = 0, 1, 2, \ldots,$ to that of the unperturbed oscillator (with coupling $v = 1$).

If, in addition to symmetry and monotonicity, we also assume that a potential shape $f_a(x)$ vanishes at infinity and that $f_a(x)$ has area, then a given trajectory function $F_a(v)$ corresponding to $f_a(x)$ can be 'scaled'[12] to a standard form in which the new function $F(v) = \alpha F_a(\beta v)$ corresponds to a potential shape $f(x)$ with area -2 and minimum value $f(0) = -1$. Thus square-well potentials, which of course are completely determined by depth and area, are immediately invertible; moreover it is known that, amongst all standard potentials, the square-well is 'extremal' for it has the lowest possible energy trajectory. In Ref. 12 an approximate variational inversion method is developed; it is also demonstrated constructively that all separable potentials are invertible. However, these results and additional constraints are not used in the present paper. When a potential has area $2A$, we first assumed, during our early attempts at numerical inversion, that it would be very useful to determine A from $F(v)$ and then appropriately constrain the inversion process. However, the area constraint did not turn out to be helpful. Thus the numerical method we have established for constructing $f(x)$ from $F(v)$ does not depend on use of this constraint, and is therefore not limited to the reconstruction of potentials which vanish at infinity and have area.

2. Constructive inversion

Much of numerical analysis assumes that errors arising from arithmetic computations or from the computation of elementary functions is negligibly small. The errors usually studied in depth are those that arise from the discrete representation of continuous objects such as functions, or from operations on them, such as derivatives or integrals. In this paper we shall take this separation of numerical problems to a higher level. We shall assume that we have a numerical method for solving the eigenvalue problem in the *forward* direction $f(x) \rightarrow F(v)$ that is reliable and may be considered for our purposes to be essentially error free. Our main emphasis will be on the design of an effective algorithm for the inverse problem *assuming* that the forward problem is numerically soluble. The forward problem is essential to our methods because we shall need to know not only the given exact energy trajectory $F(v)$ but also, at each stage of the reconstruction, what eigenvalue a partly reconstructed potential generates. This line of thought immediately indicates that we shall also need a way of temporarily extrapolating a partly reconstructed potential to all x.

Our constructive inversion algorithm hinges on the assumed symmetry and monotonicity of $f(x)$. This allows us to start the reconstruction of $f(x)$ at $x = 0$, and sequentially increase x. In Section 2 it is shown how numerical estimates can be made for the shape of the potential near $x = 0$, that is for $x < b$, where b is a parameter of the algorithm. In Section 3 we explore the implications of the potential's monotonicity for the 'tail' of the wave function. In Section 4 we establish a numerical representation for the form of the unknown potential for $x > b$ and construct our inversion algorithm. In Section 5 the algorithm is applied to three test problems.

2.1. *The reconstruction of $f(x)$ near $x = 0$*

Since the energy trajectory $F(v)$ which we are given is assumed to arise from a symmetric monotone potential, and since the spectrum generated by the potential is invariant under shifts along the x-axis, we may assume without loss of generality that the minimum value of the potential occurs at $x = 0$. We now investigate the behaviour of $F(v)$, either analytically or numerically, for large values of v. The purpose is to establish a value for the starting point $x = b > 0$ of our inversion algorithm and the shape of the potential in the interval $x \in [0, b]$. First of all, the minimum value $f(0)$ of the potential is provided by the limit (1.2). Now, if the mean kinetic energy $s = (\psi, -\Delta\psi) = F(v) - vF'(v)$ is found to be bounded above by a positive number K, then we know[11] that the potential shape $f(x)$ satisfies $f(x) = f(0)$, $x \in [0, b]$, where b is given by (1.3). In this case we have a value for b and also the shape $f(x)$ inside the interval $[0, b]$.

If the mean potential energy s is (or appears numerically to be) unbounded, then we adopt another strategy: we model $f(x)$ as a shifted power potential near $x = 0$. Since we never know $f(x)$ *exactly,* we shall need another symbol for the approximation we are currently using for $f(x)$. We choose this to be $g(x)$ and we suppose that the bottom of the spectrum of $-\Delta + vg(x)$ is given by $G(v)$. The goal is to adjust $g(x)$ until $G(v)$ is close to the given $F(v)$. Thus we write

$$f(x) \approx g(x) = f(0) + Ax^q, \quad x \in [0, b]. \tag{2.1}$$

Therefore we have three positive parameters to determine, b, A, and q. We first suppose that $g(x)$ has the form (2.1) for *all* $x \geq 0$. We now choose a 'large' value v_1 of v. This is related to the later choice of b by a bootstrap argument: the idea is that we choose v_1 so large that the turning point determined by

$$\psi_{xx}(x, v_1)/\psi(x, v_1) = v_1 f(x) - F(v) = 0 \tag{2.2}$$

is equal to b. The concentration lemma guarantees that this is possible. By scaling arguments we have

$$G(v) = f(0)v + E(q)(vA)^{\frac{2}{2+q}}, \tag{2.3}$$

where $E(q)$ is the bottom of the spectrum of the pure-power Hamiltonian $-\Delta + |x|^q$. We now 'fit' $G(v)$ to $F(v)$ by the equations $G(v_1) = F(v_1)$ and $G(2v_1) = F(2v_1)$ which yield the estimate for q given by

$$\eta = \frac{2}{2+q} = \frac{\log(F(2v_1) - 2v_1 f(0)) - \log(F(v_1) - v_1 f(0))}{\log(2)}. \tag{2.4}$$

Thus A is given by

$$A = ((F(v_1) - v_1 f(0))/E(q))^{\frac{1}{\eta}}/v_1. \tag{2.5}$$

We choose b to be equal to the turning point corresponding to the *model* potential $g(x)$ with the smaller value of v, that is to say so that $f(0)+Ab^q = F(v_1)/v_1$, or

$$b = \left(\frac{F(v_1) - v_1 f(0)}{Av_1}\right)^{\frac{1}{q}}. \tag{2.6}$$

Thus we have determined the three parameters which define the potential model $g(x)$ for $x \in [-b, b]$.

3. The tail of the wavefunction

Let us suppose that the ground-state wave function is $\psi(x, v)$. Thus the turning point $\psi_{xx}(x, v) = 0$ occurs for a given v when

$$x = x_t(v) = f^{-1}(R(v)), \quad R(v) = \left(\frac{F(v)}{v}\right). \tag{3.1}$$

The concentration lemma (1.4) quantifies the tendency of the wave function to become, as the coupling v is increased, progressively more concentrated on the patch $[-c, c]$, where $x = c$ is the point (perhaps zero) where $f(x)$ first starts to increase. This allows us to think in terms of the wave function having a 'tail'. We think of a symmetric potential as having been determined from $x = 0$ up to the current point x. The question we now ask is: what value of v should we use to determine how $f(x)$, or, more particularly, our *approximation* $g(x)$ for $f(x)$, continues beyond the current point. We have found that a good choice is to choose v so that the turning point $x_t(v) = x/2$, or some other similar fixed fraction $\sigma < 1$ of the current x value. The algorithm seems to be insensitive to this choice. Since $g(x)$ has been constructed up to the current point, and $F(v)$ is known, the value of v required follows by inverting (3.1). It has been proved[12] that $R(v)$ is monotone and therefore invertible. Hence we have the following general recipe for v:

$$v = R^{-1}(g(\sigma x)), \quad \sigma = \frac{1}{2}. \tag{3.2}$$

Since we can only determine Schrödinger eigenvalues of $H = -\Delta + vg(x)$ if the potential is defined for all x, we must have a policy about temporarily extending $g(x)$. We have tried many possibilities and found the simplest and most effective method is to extend $g(x)$ in a straight line, with slope to be determined.

4. Constructive inversion algorithm

We must first define the 'current' approximation $g(x)$ for the potential $f(x)$ sought. For values of x less than b, $g(x)$ is defined either as the horizontal line $f(x) = f(0)$ or as the shifted power potential (2.1). For values of x greater than b, the x-axis is divided into steps of length h. Thus the 'current' value of x would be of the form $x = x_k = b + kh$, where k is a positive integer. The idea is that $g(x_k)$ is determined sequentially and $g(x)$ is interpolated linearly between the x_k points. We suppose that $\{g(x_k)\}$ have already been determined up to k and we need to find $y = g(x_{k+1})$. For $x \geq x_k$ we let

$$g(x) = g(x_k) + (y - g(x_k))\frac{x - x_k}{h}. \tag{4.1}$$

If, from a study of $F(v)$, the underlying potential $f(x)$ has been shown[12] to be bounded above, it is convenient to rescale $F(v)$ so that it corresponds to a potential shape $f(x)$ which vanishes at infinity. In this case it is slightly more efficient to modify (4.1) so that for large x the straight-line extrapolation of $g(x)$ is 'cut' to zero instead of becoming positive. In either case we now have for the current point x_k an approximate potential $g(x)$ parameterized by the 'next' value $y = g(x_{k+1})$. The task of the inversion algorithm is simply to choose this value of y.

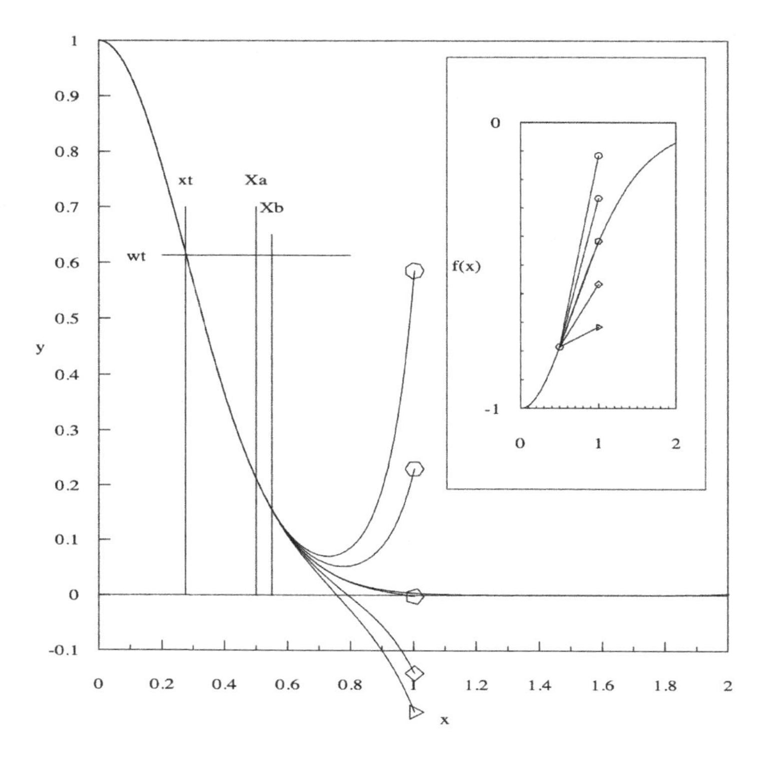

Fig. 2. We illustrate the ideas discussed in the text for the case of the sech-squared potential. The inset graph shows the sech-squared potential perturbed from $x = x_a$ by five straight line extensions; meanwhile the main graph shows the corresponding set of five wave functions which agree for $0 \le x \le x_a$ and then continue with different 'tails' dictated by the corresponding potential extensions. The value of the coupling v is the value that makes the turning point of the wave function occur at $x = x_a/2$. This figure illustrates the sort of graphical study that has lead to the algorithm described in this paper.

Let us suppose that, for given values of k and y, the bottom of the spectrum of $H = -\Delta + vg(x)$ is given by $G(v, k, y)$, then the inversion algorithm may be stated in the following succinct form in which $\sigma < 1$ is a fixed parameter. Find y such that

$$vg(\sigma x_k) = F(v) = G(v, k, y); \quad \text{then} \quad g(x_{k+1}) = y. \tag{4.2}$$

The value of v is first chosen so that the turning point of the wave function generated by g occurs at σx_k; after this, the value of y is chosen so that G 'fits' F for this value of v. The value of the parameter σ chosen for the examples discussed in section 5 below is $\sigma = \frac{1}{2}$. The idea behind this choice can best be understood from a study of Figure 1: the value of the coupling v must be such that the current value of x for which y is sought is in the 'tail' of the corresponding wave function; that is to say, the turning point σx should be before x, but not too far away. Fortunately the inversion algorithm seems to be insensitive to the choice of σ.

5. Some examples

The first example we consider is the unbounded potential whose shape $f(x)$ and corresponding exact energy trajectory $F(v)$ are given by the $\{f, F\}$ pair

$$f(x) = -1 + |x|^{\frac{3}{2}} \quad \longleftrightarrow \quad F(v) = -v + E(3/2)v^{\frac{4}{7}}, \tag{5.1}$$

where $E(3/2)$ is the bottom of the spectrum of $H = -\Delta + |x|^{\frac{3}{2}}$ and has the approximate value $E(3/2) \approx 1.001184$. Applying the inversion algorithm to $F(v)$ we obtain the reconstructed potential shown in Figure 3. We first

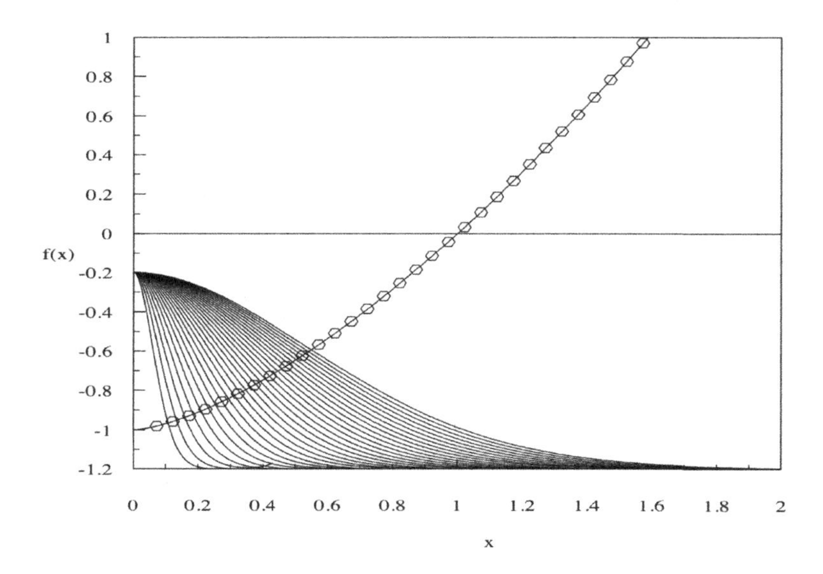

Fig. 3. Constructive inversion of the energy trajectory $F(v)$ for the shifted power potential $f(x) = -1 + |x|^{\frac{3}{2}}$. For $x \leq b = 0.072$, the algorithm correctly generates the model $f(x)$; for larger values of x, in steps of size $h = 0.05$, the hexagons indicate the reconstructed values for the potential $f(x)$, shown exactly as a smooth curve. The unnormalized wave functions are also shown.

set $v_1 = 10^4$ and find that the initial shape is determined (as described in Section 2) to be $-1 + x^{1.5}$ for $x < b = 0.072$. For larger values of x the step size is chosen to be $h = 0.05$ and 40 iterations are performed by the inversion algorithm. The results are plotted as hexagons on top of the exact potential shape shown as a smooth curve.

The following two examples are bounded potentials both having large-x limit zero, lowest point $f(0) = -1$, and 'area' -2. The exponential potential[1,25] has the $\{f, F\}$ pair

$$f(x) = -e^{-|x|} \quad \longleftrightarrow \quad J'_{2|E|^{\frac{1}{2}}}(2v^{\frac{1}{2}}) = 0 \quad \equiv \quad E = F(v), \tag{5.2}$$

where $J'_\nu(x)$ is the derivative of the Bessel function of the first kind of order ν.

For the sech-squared potential[1] we have

$$f(x) = -\mathrm{sech}^2(x) \quad \longleftrightarrow \quad F_0(v) = -\left[\left(v + \frac{1}{4}\right)^{\frac{1}{2}} - \frac{1}{2}\right]^2. \tag{5.3}$$

In Figure 4 the two energy trajectories are plotted. Since the two *potentials*

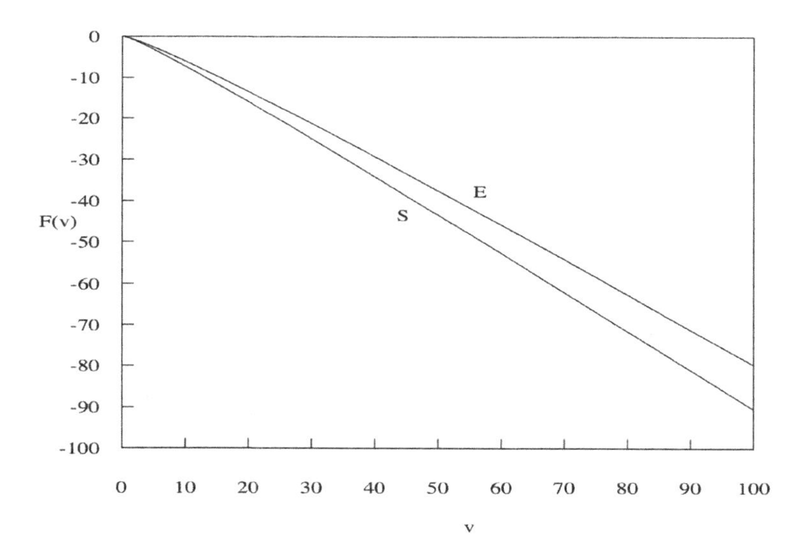

Fig. 4. The ground-state energy trajectories $F(v)$ for the exponential potential (E) and the sech-squared potential (S). For small v, $F(v) \approx -v^2$; for large v, $\lim_{v\to\infty}(F(v)/v) = -1$. The shapes of the underlying potentials are buried in the details of $F(v)$ for intermediate values of v.

have lowest value -1 and 'area' -2 it follows[12] that the corresponding trajectories both have the form $F(v) \approx -v^2$ for small v and they both satisfy the large-v limit $\lim_{v\to\infty}(F(v)/v) = -1$. Thus the differences between the potential shapes is somehow encoded in the fine differences between these two similar energy curves for intermediate values of v : it is the task of our inversion theory to decode this information and reveal the underlying potential shape. If we apply the inversion algorithm to these two problems we obtain the results shown in Figures 5 and 6. The parameters used are

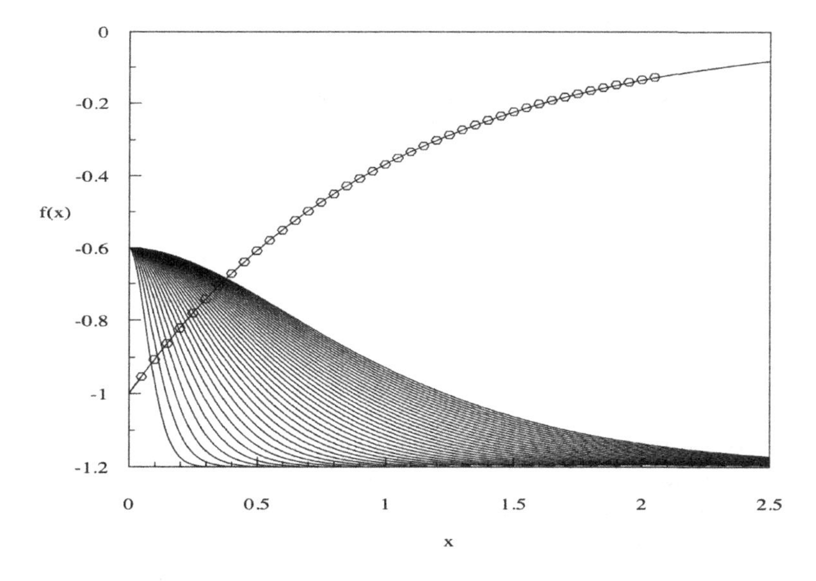

Fig. 5. Constructive inversion of the energy trajectory $F(v)$ for the exponential potential $f(x) = -\exp(x)$. For $x \leq b = 0.048$, the algorithm correctly generates the model $f(x) = -1 + |x|$; for larger values of x, in steps of size $h = 0.05$, the hexagons indicate the reconstructed values for the potential $f(x)$, shown exactly as a smooth curve. The unnormalized wave functions are also shown.

exactly the same as for the first problem described above. The time taken to perform the inversions was less than 20 seconds if we discount, in the case of the exponential potential, the extra time taken to compute $F(v)$ itself.

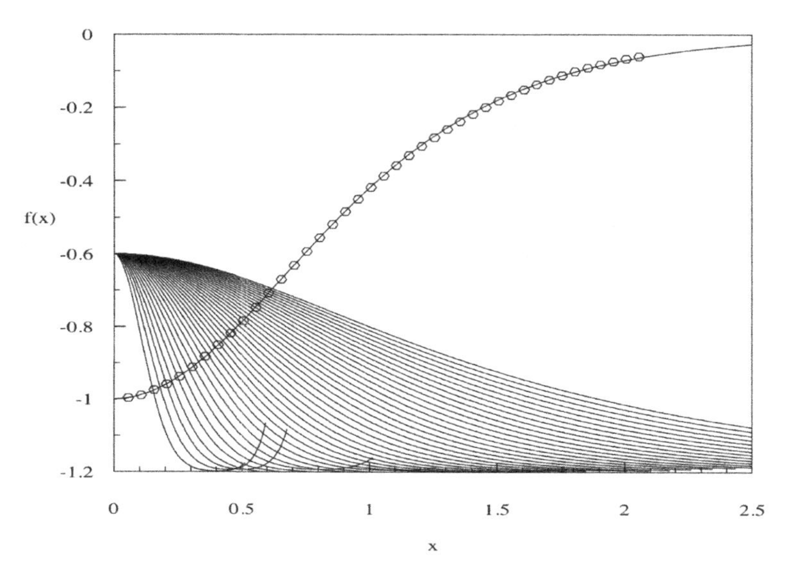

Fig. 6. Constructive inversion of the energy trajectory $F(v)$ for the sech-squared potential $f(x) = -\mathrm{sech}^2(x)$. For $x \leq b = 0.1$, the algorithm correctly generates the model $f(x) = -1 + x^2$; for larger values of x, in steps of size $h = 0.05$, the hexagons indicate the reconstructed values for the potential $f(x)$, shown exactly as a smooth curve. The unnormalized wave functions are also shown.

6. Kinetic potentials

Geometry is involved with this problem because we deal with a family of operators depending on a continuous parameter v. This immediately leads to a family of spectral manifolds, and, more particularly, to the consideration of smooth transformations of potentials, and to the transformations which they in turn induce on the spectral manifolds. This is the environment in which we are able to construct the following functional inversion sequence:

$$f^{[n+1]} = \bar{f} \circ \bar{f}^{[n]^{-1}} \circ f^{[n]} \equiv \bar{f} \circ K^{[n]}. \tag{6.1}$$

A *kinetic potential* is the constrained mean value of the potential shape $\bar{f}(s) = \langle f \rangle$, where the corresponding mean kinetic energy $s = \langle -\Delta \rangle$ is held constant. It turns out that kinetic potentials may be obtained from the corresponding energy trajectory F by what is essentially a Legendre transformation[26] $\bar{f} \leftrightarrow F$ given[19] by

$$\bar{f}(s) = F'(v), \quad s = F(v) - vF'(v),$$

and

$$F(v)/v = \bar{f}(s) - s\bar{f}'(s), \quad 1/v = -\bar{f}'(s). \tag{6.2}$$

These transformations are well defined because of the definite convexities of F and $\bar{f}$; they complete the definition of the inversion sequence (1.2), up to the choice of a starting seed potential $f^{[0]}(x)$. They differ from Legendre transformations only because of our choice of signs. The choice has been made so that the eigenvalue can be written (exactly) in the semi-classical forms

$$E = F(v) = \min_{s>0}\left\{s + v\bar{f}(s)\right\} = \min_{x>0}\left\{K^{[f]}(x) + vf(x)\right\} \tag{6.3}$$

where the kinetic- and potential-energy terms have the 'usual' signs.

7. Envelope theory

The term 'kinetic potential' is short for 'minimum mean iso-kinetic potential'. If the Hamiltonian is $H = -\Delta + vf(x)$, where $f(x)$ is potential *shape,* and $\mathcal{D}(H) \subset L^2(\Re)$ is the domain of H, then the ground-state kinetic potential $\bar{f}(s) = \bar{f}_0(s)$ is defined[18,19] by the expression

$$\bar{f}(s) = \inf_{\substack{\psi\in\mathcal{D}(H)\\ (\psi,\psi)=1\\ (\psi,-\Delta\psi)=s}} (\psi, f\psi). \tag{7.1}$$

The extension of this definition to the higher discrete eigenvalues (for v sufficiently large) is straightforward[19] but not explicitly needed in the present paper. The idea is that the min-max computation of the discrete eigenvalues is carried out in two stages: in the first stage (7.1) the mean potential shape is found for each fixed value of the mean kinetic energy s; in the second and final stage we minimize over s. Thus we arrive at the semi-classical expression which is the first equality of Eq. (1.4). It is well known that $F(v)$ is concave ($F''(v) < 0$) and it follows immediately that $\bar{f}(s)$ is convex. More particularly, we have[11]

$$F''(v)\bar{f}''(s) = -\frac{1}{v^3}. \tag{7.2}$$

Thus, although kinetic potentials are *defined* by (7.1), the transformations (6.2) may be used in practice to go back and forth between F and $\bar{f}$.

Kinetic potentials have been used to study smooth transformations of potentials and also linear combinations. The present work is an application of the first kind. Our goal is to devise a method of searching for a transformation g, which would convert the initial seed potential $f^{[0]}(x)$ into the (unknown) goal $f(x) = g(f^{[0]})$. We shall summarize briefly how one proceeds in the forward direction, to approximate F, if we know $f(x)$. The K functions are then introduced, by a change of variable, so that the potential $f(x)$ is exposed and can be extracted in a sequential inversion process.

In the forward direction we assume that the lowest eigenvalue $F^{[0]}(v)$ of $H^{[0]} = -\Delta + vf^{[0]}(x)$ is known for all $v > 0$ and we assume that $f(x)$ is given; hence, since the potentials are symmetric and monotone for $x > 0$, we have defined the transformation function g. 'Tangential potentials' to $g(f^{[0]})$ have the form $a+bf^{[0]}(x)$, where the coefficients $a(t)$ and $b(t)$ depend on the point of contact $x = t$ of the tangential potential to the graph of $f(x)$. Each one of these tangential potentials generates an energy trajectory of the form $\mathcal{F}(v) = av + F^{[0]}(bv)$, and the *envelope* of this family (with respect to t) forms an approximation $F^A(v)$ to $F(v)$. If the transformation g has definite convexity, then $F^A(v)$ will be either an upper or lower bound to $F(v)$. It turns out[19] that all the calculations implied by this envelope approximation can be summarized nicely by kinetic potentials. Thus the whole procedure just described corresponds exactly to the expression:

$$\bar{f} \approx \bar{f}^A = g \circ \bar{f}^{[0]}, \tag{7.3}$$

with $\approx$ being replaced by an inequality in case g has definite convexity. Once we have an approximation $\bar{f}^A$, we immediately recover the corresponding energy trajectory F^A from the general minimization formula (6.3).

The formulation that reveals the potential shape is obtained when we use x instead of s as the minimization parameter. We achieve this by the following general definition of x and of the K function associated with f :

$$f(x) = \bar{f}(s), \quad K^{[f]}(x) = \bar{f}^{-1}(f(x)). \tag{7.4}$$

The monotonicity of $f(x)$ and of $\bar{f}$ guarantee that x and K are well defined. Since $\bar{f}^{-1}(f)$ is a convex function of f, the second equality in (1.4) immediately follows[18]. In terms of K the envelope approximation (2.3) becomes simply

$$K^{[f]} \approx K^{[f^{[0]}]}. \tag{7.5}$$

Thus the envelope approximation involves the use of an approximate K function that no longer depends on f, and there is now the possibility that we can invert (1.4) to extract an approximation for the potential shape.

We end this summary by listing some specific results that we shall need. First of all, the kinetic potentials and K functions obey[17,18] the following elementary shift and scaling laws:

$$f(x) \to a + bf(x/t) \Rightarrow \left\{ \bar{f}(s) \to a + b\bar{f}(st^2), \quad K^{[f]}(x) \to \frac{1}{t^2} K^{[f]}\left(\frac{x}{t}\right) \right\}. \tag{7.6}$$

Pure power potentials are important examples which have the following formulas:

$$f(x) = |x|^q \Rightarrow \left\{ \bar{f}(s) = \left(\frac{P}{s^{\frac{1}{2}}}\right)^q, \quad K(x) = \left(\frac{P}{x}\right)^2 \right\}, \tag{7.7}$$

where, if the bottom of the spectrum of $-\Delta + |x|^q$ is $E(q)$, then the P numbers are given[18] by the following expressions with $n = 0$:

$$P_n(q) = |E_n(q)|^{\frac{(2+q)}{2q}} \left[\frac{2}{2+q}\right]^{\frac{1}{q}} \left[\frac{|q|}{2+q}\right]^{\frac{1}{2}}, \quad q \neq 0. \tag{7.8}$$

We have allowed for $q < 0$ and for higher eigenvalues since the formulas are essentially the same. The $P_n(q)$ as functions of q are interesting in themselves[18]: they have been proven to be monotone increasing, they are probably concave, and $P_n(0)$ corresponds exactly to the log potential. By contrast the $E_n(q)$ are not so smooth: for example, they have infinite slopes at $q = 0$. But this is another story. An important observation is that the K functions for the pure powers are *all* of the form $(P(q)/x)^2$ and they are invariant with respect to both potential shifts and multipliers: thus $a + b|x|^q$ has the same K function as does $|x|^q$. For the harmonic oscillator $P_n(2) = (n + \frac{1}{2})^2, \quad n = 0, 1, 2, \ldots$. Other specific examples may be found in the references cited.

The last formulas we shall need are those for the ground state of the sech-squared potential:

$$f(x) = -\text{sech}^2(x) \Rightarrow \left\{ \bar{f}(s) = -\frac{2s}{(s + s^2)^{\frac{1}{2}} + s}, \quad K(x) = \sinh^{-2}(2x) \right\}. \tag{7.9}$$

8. Functional inversion

The inversion sequence (6.1) is based on the following idea. The goal is to find a transformation g so that $f = g \circ f^{[0]}$. We choose a seed $f^{[0]}$, but, of course, f is unknown. In so far as the envelope approximation with $f^{[0]}$ as a basis is 'good', then an approximation $g^{[1]}$ for g would be given by $\bar{f} = g^{[1]} \circ \bar{f}^{[0]}$. Thus we have

$$g \approx g^{[1]} = \bar{f} \circ \bar{f}^{[0]^{-1}}. \tag{8.1}$$

Applying this approximate transformation to the seed we find:

$$f \approx f^{[1]} = g^{[1]} \circ f^{[0]} = \bar{f} \circ \bar{f}^{[0]^{-1}} \circ f^{[0]} = \bar{f} \circ K^{[0]}. \tag{8.2}$$

We now use $f^{[1]}$ as the basis for another envelope approximation, and, by repetition, we have the ansatz (1.2), that is to say

$$f^{[n+1]} = \bar{f} \circ \bar{f}^{[n]^{-1}} \circ f^{[n]} = \bar{f} \circ K^{[n]}. \tag{8.3}$$

A useful practical device is to invert the second expression for F given in (1.4) to obtain

$$K^{[f]}(x) = \max_{v>0} \{F(v) - vf(x)\}. \tag{8.4}$$

The concavity of $F(v)$ explains the max in this inversion, which, as it stands, is exact. In a situation where f is unknown, we have f on both sides and nothing can be done with this formal result. However, in the inversion sequence which we are considering, (8.4) is extremely useful. If we re-write (8.4) for stage $[n]$ of the inversion sequence it becomes:

$$K^{[n]}(x) = \max_{v>0} \left\{F^{[n]}(v) - vf^{[n]}(x)\right\}. \tag{8.5}$$

In this application, the current potential shape $f^{[n]}$ and consequently $F^{[n]}(v)$ can be found (by shooting methods) for each value of v. The minimization can then be performed even without differentiation (for example, by using a Fibonacci search) and this is a much more effective method for $K^{[n]} = \bar{f}^{[n]^{-1}} \circ f^{[n]}$ than finding $\bar{f}^{[n]}(s)$, finding the functional inverse, and applying the result to $f^{[n]}$.

9. Functional inversion for pure powers

We now treat the case of pure-power potentials given by

$$f(x) = A + B|x|^q, \quad q > 0, \tag{9.1}$$

where A and $B > 0$ are arbitrary and fixed. We shall prove that, starting from another pure power as a seed, the inversion sequence converges in just two steps. The exact energy trajectory $F(v)$ for the potential (9.1) is assumed known. Hence, so is the exact kinetic potential given by (7.7) and the general scaling rule (7.6), that is to say

$$\bar{f}(s) = A + B\left(\frac{P(q)}{s^{\frac{1}{2}}}\right)^{q}. \tag{9.2}$$

We now suppose that a pure power is also used as a seed, thus we have

$$f^{[0]}(x) = a + b|x|^{p} \quad \Rightarrow \quad K^{[0]}(x) = \left(\frac{P(p)}{x}\right)^{2}, \tag{9.3}$$

where the parameters a, $b > 0$, $p > 0$ are arbitrary and fixed. The first step of the inversion (6.3) therefore yields

$$f^{[1]}(x) = \left(\bar{f} \circ K^{[0]}\right)(x) = A + B\left(\frac{P(q)|x|}{P(p)}\right)^{q}. \tag{9.4}$$

The approximate potential $f^{[1]}(x)$ now has the correct x power dependence but has the wrong multiplying factor. Because of the invariance of the K functions to multipliers, this error is completely corrected at the next step, yielding:

$$K^{[1]}(x) = \left(\frac{P(q)}{x}\right)^{2} \quad \Rightarrow \quad f^{[2]}(x) = \left(\bar{f} \circ K^{[1]}\right)(x) = A + B|x|^{q}. \tag{9.5}$$

This establishes our claim that power potentials are inverted without error in exactly two steps.

The implications of this result are a little wider than one might first suspect. If the potential that is being reconstructed has the asymptotic form of a pure power for small or large x, say, then we know that the inversion sequence will very quickly produce an accurate approximation for that part of the potential shape. More generally, since the first step of the inversion process involves the construction of $K^{[0]}$, the general invariance property $K^{[a+bf]} = K^{[f]}$ given in (7.6) means that the seed potential $f^{[0]}$ may be chosen without special consideration to gross features of f already arrived at by other methods. For example, the area (if the potential has area), or the starting value $f(0)$ need not be incorporated in $f^{[0]}$, say, by adjusting a and b.

10. A more general example

We consider the problem of reconstructing the sech-squared potential $f(x) = -\text{sech}^2(x)$. We assume that the corresponding exact energy trajectory $F(v)$ and, consequently, the kinetic potential $\bar{f}(s)$ are known. Thus[18] from the potential shape $f(x) = -\text{sech}^2(x)$ we have:

$$\left\{ F(v) = -\left[\left(v + \frac{1}{4}\right)^{\frac{1}{2}} - \frac{1}{2}\right]^2, \quad \bar{f}(s) = -\frac{2s}{(s + s^2)^{\frac{1}{2}} + s} \right\}. \tag{10.1}$$

The seed is essentially x^2, but we use a scaled version of this for the purpose of illustration in Fig. 7. Thus we have

$$f^{[0]} = -1 + \frac{x^2}{20} \quad \Rightarrow \quad K^{[0]}(x) = \frac{1}{4x^2} \tag{10.2}$$

This potential generates the exact eigenvalue

$$F^{[0]}(v) = -v + \left(\frac{v}{20}\right)^{\frac{1}{2}}, \tag{10.3}$$

which, like the potential itself, is very different from that of the goal. After the first iteration we obtain

$$f^{[1]}(x) = \bar{f}\left(K^{[0]}(x)\right) = -\frac{2}{1 + (1 + 4x^2)^{\frac{1}{2}}}. \tag{10.4}$$

A graph of this potential is shown as $f1$ in Fig. 7. In order to continue analytically we would need to solve the problem with Hamiltonian $H^{[1]} = -\Delta + vf^{[1]}(x)$ exactly to find an expression for $F^{[1]}(v)$. We know no way of doing this. However, it can be done numerically, with the aid of the inversion formula (8.5) for K. The first 5 iterations shown in Fig. 7 suggest convergence of the functional sequence.

11. Conclusion

We have discussed two methods that may be used to reconstruct the potential $f(r)$ in a Schrödinger operator $H = -\Delta + vf(r)$ if an eigenvalue curve $E = F_n(v)$ is known as a function of the coupling v. Such functions as $F(v)$ are met, for example, as one-body approximations to certain N-body systems, such as atoms. In addition to the inversion of physical data to infer the structure of the underlying system, the situation also presents a rather fascinating mathematical problem. Work is under way extending these results and methods to a wider class of problems, including those with singular potentials.

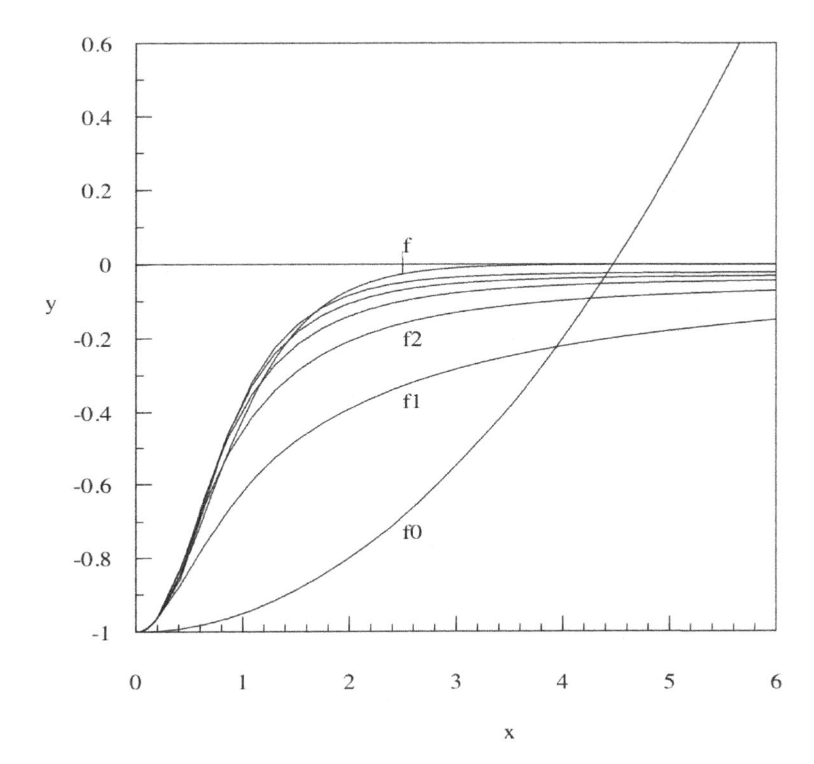

Fig. 7. The energy trajectory F for the sech-squared potential $f(x) = -\text{sech}^2(x)$ is approximately inverted starting from the seed $f^{[0]}(x) = -1 + x^2/20$. The first step can be completed analytically yielding $f1 = f^{[1]}(x) = -2/\{1 + \sqrt{1 + 4x^2}\}$. Four more steps $\{fk = f^{[k]}\}_{k=2}^{5}$ of the inversion sequence approaching f are performed numerically.

Acknowledgements

Partial financial support of this work under Grant No. GP3438 from the Natural Sciences and Engineering Research Council of Canada, and the hospitality of the organizers of the 13th Regional Conference on Mathematical Physics, Antalya, Turkey, are gratefully acknowledged.

References

1. S. Flügge, *Practical Quantum Mechanics*, (Springer, New York, 1974. The exponential potential is discussed on p 196 and the sech-squared potential on p 94.
2. K. Chadan and P. C. Sabatier, *Inverse Problems in Quantum Scattering*

Theory, (Springer, New York, 1989). The 'inverse problem in the coupling constant' is discussed on p 406.
3. R. G. Newton, *Scattering Theory of Waves and Particles*, (Springer, New York, 1982).
4. B. N. Zakhariev and A.A.Suzko, *Direct and Inverse Problems: Potentials in Quantum Scattering Theory*, (Springer, Berlin, 1990).
5. G. Eilenberger, *Solitons*, (Springer, Berlin, 1983).
6. M. Tod, *Nonlinear Waves and Solitons*, (Kluwer, Dodrecht, 1989).
7. K. Chadan, *C. R. Acad. Sci. Paris Sèr. II*, **299**, 271 (1984).
8. K. Chadan and H. Grosse, *C. R. Acad. Sci. Paris Sèr. II*, **299**, 1305 (1984).
9. K. Chadan and R. Kobayashi, *C. R. Acad. Sci. Paris Sèr. II*, **303**, 329 (1986).
10. K. Chadan and M. Musette, *C. R. Acad. Sci. Paris Sèr. II*, **305**, 1409 (1987).
11. R. L. Hall, *Phys. Rev.* A **50**, 2876 (1995).
12. R. L. Hall, *J. Phys. A: Math. Gen.* **28**, 1771 (1995).
13. R. L. Hall, *Phys. Rev.* A **51**, 1787 (1995).
14. R. L. Hall, *J. Math. Phys.* **40**, 699 (1999).
15. R. L. Hall, *J. Math. Phys.* **40**, 2254 (1999).
16. R. L. Hall, *Phys. Lett.* A **265**, 28 (2000).
17. R. L. Hall, *Phys. Rev.* D **22**, 2062 (1980).
18. R. L. Hall, *J. Math. Phys.* **24**, 324 (1983).
19. R. L. Hall, *J. Math. Phys.* **25**, 2708 (1984).
20. R. L. Hall, *Phys. Rev.* A **39**, 550 (1989).
21. R. L. Hall, *J. Math. Phys.* **33**, 1710 (1992).
22. R. L. Hall, *J. Math. Phys.* **34**, 2779 (1993).
23. R. L. Hall, *J. Phys. A: Math. Gen.* **25**, 4459 (1992).
24. O. L. De Lange and R. E. Raab, *Operator Methods in Quantum Mechanics*, (Oxford University Press, Oxford, 1991). Perturbed harmonic oscillators with identical spectra to that generated by $f(x) = x^2$ are given on p 71.
25. H. S. W. Massey and C. B. O. Mohr, *Proc. Roy. Soc.* **148**, 206 (1934).
26. I. M. Gelfand and S. V. Fomin, *Calculus of Variations*, (Prentice-Hall, Englewood Cliffs, 1963). Legendre transformations are discussed on p 72.
27. *Handbook of Mathematical Functions*, edited by M. Abramowitz and I. A. Stegun (Dover, New York, 1972).

HEUN FUNCTIONS AND THEIR USES IN PHYSICS

M. HORTAÇSU
Istanbul Technical University, Department of Physics, Istanbul, Turkey

Most of the theoretical physics known today is described by using a small number of differential equations. If we study only linear systems, different forms of the hypergeometric or the confluent hypergeometric equations often suffice to describe this problem. These equations have power series solutions with simple relations between consecutive coefficients and can be generally represented in terms of simple integral transforms. If the problem is nonlinear, one often uses one form of the Painlevé equation. There are important examples, however, where one has to use more complicated equations. An example often encountered in quantum mechanics is the hydrogen atom in an external electric field, the Stark effect. One often bypasses this difficulty by studying this problem using perturbation methods. If one studies certain problems in astronomy or general relativity, encounter with Heun equation is inevitable. This is a general equation whose special forms take names as Mathieu, Lamé and Coulomb spheroidal equations. Here the coefficients in a power series expansions do not have two way recursion relations. We have a relation at least between three or four different coefficients. A simple integral transform solution also is not obtainable. Here I will try to introduce this equation and give some examples where the result can be expressed in terms of solutions of this equation. Although this equation was discovered more than hundred years ago, there is not a vast amount of literature on this topic and only advanced mathematical packages can identify it. Its popularity, however, increased recently, mostly among theoretical physicists, with ninety four papers in SCI in the last twenty five years. More than two thirds of the papers which use these functions in physical problems were written in the last decade.

Keywords: Heun equation.

1. Introduction

Most of the theoretical physics known today is described by using a small number of differential equations. If we study only linear systems, different forms of the hypergeometric or the confluent hypergeometric equations often suffice to describe this problem. These equations have power series solutions with simple relations between consecutive coefficients and can be generally represented in terms of simple integral transforms. If the problem is nonlinear, one often uses one form of the Painlevé equation.

Let us review some well known facts about linear second order differential equations. Differential equations are classified according to their singularity structure[1,2]. If a differential equation has no singularities over the full complex plane, it can only be a constant. Singularities are classified as regular or irregular singular points. If the coefficient of the first derivative has at most single poles, and the coefficient of the term without a derivative has at most double poles when the coefficient of the second derivative is unity, this second order differential equation has regular singularities. Then we have one regular solution while expanding around these singular points. The second solution around a regular singular point has a branch cut. If the poles of these coefficients are higher, we have irregular singularities and the general solution has essential singularities around these points[3].

As stated in Morse and Feshbach[1], an example of a second order differential equation with one regular singular point is

$$\frac{d^2w}{dz^2} = 0. \tag{1}$$

This equation has one solution which is constant. The second solution blows up at infinity. The differential equation

$$\frac{d^2w}{dz^2} + k^2 w = 0 \tag{2}$$

has one irregular singularity at infinity which gives an essential singularity at this point. The equation

$$z\frac{d^2w}{dz^2} + (1+a)\frac{dw}{dz} = 0 \tag{3}$$

has two regular singular points, at zero and at infinity. In physics an often used equation is the hypergeometric equation

$$z(1-z)\frac{d^2w}{dz^2} + [c-(1+a+b)z]\frac{dw}{dz} - abw = 0. \tag{4}$$

This equation has three regular singular points, at zero, one and infinity. Jacobi, Legendre, Gegenbauer, Tchebycheff equations are special forms of this equation. When the singular points at $z = 1$ and z equals infinity are "coalesced"[4] at infinity, we get the confluent hypergeometric equation

$$z\frac{d^2w}{dz^2} + (c-z)\frac{dw}{dz} - aw = 0, \tag{5}$$

with an essential singularity at infinity and a regular singularity at zero. Bessel, Laguerre, Hermite equations can be reduced to this form.

An important property of all these equations is that they allow infinite series solutions about one of their regular singular points and a recursion relation can be found between two consecutive coefficients of the series. This fact allows us having an idea about the general properties of the solution, as asymptotic behaviour at distant points, the radius of convergence of the series, etc.

A new equation was introduced in 1889 by Karl M. W. L. Heun[5]. This is an equation with four regular singular points. This equation is discussed in the book edited by Ronveaux[6]. All the general information we give below is taken from this book.

As discussed there, any equation with four regular singular points can be transformed to the equation given below:

$$\frac{d^2w}{dz^2} + [\frac{c}{z} + \frac{d}{z-1} + \frac{e}{z-f}]\frac{dw}{dz} - \frac{abz-q}{z(z-1)(z-f)}w = 0. \tag{6}$$

There is a relation between the constants given as $a + b + 1 = c + d + e$. Then this equation has regular singularities at zero, one, f and infinity. If we try to obtain a solution in terms of a power series, one can not get a recursion relation between two consecutive coefficients. Such a relation exists between at least three coefficients. A simple solution as an integral transform also can not be found[1].

One can obtain different confluent forms of this equation. When we "coalesce" two regular singular points at zero and infinity, we get the confluent Heun equation:

$$\frac{d}{dz}((z^2-1)\frac{dw}{dz}) + [-p^2(z^2-1) + 2p\beta z - \lambda - \frac{m^2+s^2+2msz}{(z^2-1)}]w = 0. \tag{7}$$

Special forms of this equation are obtained in problems with two Coulombic centers,

$$\frac{d}{dz}((z^2-1)\frac{dw}{dz}) + [-p^2(z^2-1) + 2p\beta z - \lambda - \frac{m^2}{(z^2-1)}]w = 0, \tag{8}$$

whose special form is the spheroidal equation,

$$\frac{d}{dz}((z^2-1)\frac{dw}{dz}) + [-p^2(z^2-1) - \lambda - \frac{m^2}{(z^2-1)}]w = 0. \tag{9}$$

Another form is the algebraic form of the Mathieu equation:

$$\frac{d}{dz}((z^2-1)\frac{dw}{dz}) + [-p^2(z^2-1) - \lambda - \frac{1}{4(z^2-1)}]w = 0. \tag{10}$$

If we coalesce two regular singular points pairwise, we obtain the double confluent form:

$$D^2w+(\alpha_1 z+\frac{\alpha_{-1}}{z})Dw+[(B_1+\frac{\alpha_1}{2})z+(B_0+\frac{\alpha_1\alpha_{-1}}{2})+(B_{-1}-\frac{\alpha_{-1}}{2})\frac{1}{z})]w=0. \quad (11)$$

Here $D = z\frac{d}{dz}$. If we equate the coefficient $\alpha_i, i = -1, 1$ to zero, we can reduce the new equation to the Mathieu equation, an equation with two irregular singularities at zero and at infinity. Another form is the biconfluent form, where three regular singularities are coalesced. The result is an equation with a regular singularity at zero and an irregular singularity at infinity of higher order:

$$z^2\frac{d^2w}{dz^2}+z\frac{dw}{dz}w+(A_0+A_1z+A_2z^2+A_3z^3-z^4)w=0. \quad (12)$$

The anharmonic equation in three dimensions can be reduced to this equation:

$$\frac{d^2w}{dr^2}+(E-\frac{\nu}{r^2}-\mu r^2-\lambda r^4-\eta r^6)w=0. \quad (13)$$

In the triconfluent case, all regular singular points are "coalesced" at infinity which gives the equation below:

$$\frac{d^2w}{dz^2}+(A_0+A_1z+A_2z^2-\frac{9}{4}z^4)w=0. \quad (14)$$

These different forms are used in different physics problems.

In SCI I found 94 papers when I searched for Heun functions. More than two thirds of these papers were published in the last ten years. The rest of the papers were published between 1990 and 1999, except a single paper in 1986 on a mathematical problem[7]. This shows that although the Heun equation was found in 1889, it was largely neglected in the physics literature until recently. Earlier papers on this topic are mostly articles in mathematics journals.

The list of books on this topic also is not very long. There is a book edited by A. Ronveaux, which is a collection of papers presented in the "Centennial Workshop on Heun's Equations: Theory and Application. Sept. 3-8 1989, Schloss Ringberg". It was published by the Oxford University Press in 1995 by the title *Heun's Differential Equations*[6]. There is an extensive bibliography at the end of this book. I also found frequent references to two other books: *Mathieusche Funktionen und Sphaeroidfunktionen mit Anwendungen auf physikalische und technische Probleme* by Joseph

Meixner and Friedrich Wilhelm Schaefke, published by Springer Verlag in 1954[8] and a 1963 Dover reprint of a book first published in 1946, *Theory and Applications of Mathieu Functions* by N.W. McLachlan[9]. These are on functions which are special cases of the Heun equation. Some well known papers on different mathematical properties of these functions can be found in references 10–14.

A reason why more physicists are interested in the Heun equation recently may be, perhaps, a demonstration of the fact that we do not have simple problems especially in "the theoretical theory" section of particle physics anymore and people doing research on this field have to tackle more difficult problems, either with more difficult metrics or in higher dimensions. Both of these extensions may necessitate the use of the Heun functions among the solutions. We can give the Eguchi-Hanson case as an example. The wave equation in the background of the Eguchi-Hanson metric[15] in four dimensions has hypergeometric functions as solutions[16] whereas the Nutku helicoid[17,18] metric, another instanton metric which is little bit more complicated than the Eguchi-Hanson one, gives us Mathieu function solutions[19], a member of the Heun function set. We also find that the same equation in the background of the Eguchi-Hanson metric, trivially extended to five dimensions, gives Heun type solutions[20].

Note that the problem need not be very complicated to end up with these equations. We encounter Mathieu functions if we consider two dimensional problems with elliptical boundaries[21] . Let us use $x = \frac{1}{2}a\cosh\mu\cos\theta, y = \frac{1}{2}a\sinh\mu\sin\theta$, where a is the distance from the origin to the focal point. Then the Helmholtz equation can be written as

$$\partial_{\mu\mu}\psi + \partial_{\theta\theta}\psi + \frac{1}{4}a^2k^2[\cosh^2\mu - \cos^2\theta]\psi = 0 \tag{15}$$

which separates into two equations

$$\frac{d^2H}{d\theta^2} + (b - h^2\cos^2\theta)H = 0, \tag{16}$$

$$-\frac{d^2M}{d\mu^2} + (b - h^2\cosh^2\mu)M = 0. \tag{17}$$

a, k, h are constants given in Ref. 21. The solutions to these two equations can be represented as Mathieu and modified Mathieu functions.

If we combine different inverse powers of r, starting from first up to the fourth, or if we combine the quadratic potentials with inverse even powers of two, four and six, we see that the solution of the Schrodinger equation involves Heun functions[22]. Solution of the Schrödinger equation to symmetric double Morse potentials, like $V(x) = \frac{B^2}{4}\sinh 2x - (s+\frac{1}{2})B\cosh x$, where $s = (0, 1/2, 1, 3/2, \ldots)$,[23] also needs these functions. Similar problems are treated in references 24–26.

o In atomic physics further problems such as separated double wells, Stark effect, hydrogen molecule ion use these functions. Physics problems which end up in these equations are given in the book by S.Y. Slavyanov and S. Lay[27]. Here we see that even the Stark effect, hydrogen atom in the presence of an external electric field, gives rise to this equation. As described in page 166 of Slavyanov's book, cited above (original reference is Epstein[28], also treated by S.Yu Slavyanov[29]), when all the relevant constants, namely Planck constant over 2π, electron mass and electron charge are set to unity, the Schrodinger equation for the hydrogen atom in a constant electric field of magnitude F in the z direction is given by

$$\left(\Delta + 2[E - (Fz - \frac{1}{r})]\right)\Psi = 0. \tag{18}$$

Here Δ is the laplacian operator.

Using parabolic coordinates, where the cartesian ones are given in terms of the new coordinates by $x = \sqrt{\xi\eta}\cos\phi, y = \sqrt{\xi\eta}\sin\phi, z = \frac{\xi-\eta}{2}$ and writing the wave function in the product form

$$\Psi = \sqrt{\xi\eta}V(\xi)U(\eta)\exp(im\phi), \tag{19}$$

we get two separated equations:

$$\frac{d^2V}{d\xi^2} + \left(\frac{E}{2} + \frac{\beta_1}{\xi} + \frac{F}{4}\xi + \frac{1-m^2}{4\xi^2}\right)V(\xi) = 0, \tag{20}$$

$$\frac{d^2U}{d\eta^2} + \left(\frac{E}{2} + \frac{\beta_2}{\eta} + \frac{F}{4}\eta + \frac{1-m^2}{4\eta^2}\right)U(\eta) = 0. \tag{21}$$

Here β_1 and β_2 are separation constants that must add to one. We note that these equations are of the biconfluent Heun form.

The hydrogen molecule also is treated in reference 27, (original reference is 30). When the hydrogen-molecule ion is studied in the Born-Oppenheimer approximation, where the ratio of the electron mass to the proton mass is very small, one gets two singly confluent Heun equations if the prolate spheroidal coordinates $\xi = \frac{r_1+r_2}{2c}, \eta = \frac{r_1+r_2}{2c}$ are used. Here c is the distance between the two centers. Assuming

$$\psi = \sqrt{\xi\eta}V(\xi)U(\eta)\exp(im\phi), \tag{22}$$

we achieve separation into two equations:

$$\frac{d}{d\xi}\Big((1-\xi^2)\frac{dV}{d\xi}\Big) + \Big(\lambda^2\xi^2 - \kappa\xi - \frac{m^2}{1-\xi^2} + \mu\Big)V = 0, \tag{23}$$

$$\frac{d}{d\eta}\Big((1-\eta^2)\frac{dU}{d\eta}\Big) + \Big(\lambda^2\eta^2 - \frac{m^2}{1-\eta^2} + \mu\Big)U = 0. \tag{24}$$

Both equations are of the confluent Heun type.

If we mention some recent papers on atomic physics with Heun type solutions we find three relatively recent papers which treat atoms in magnetic fields:

o Exact low-lying states of two interacting equally charged particles in a magnetic field are studied in by Truong and Bazzali[31].

o The energy spectrum of a charged particle on a sphere under a magnetic field and Coulomb force is studied by Ralko and Truong[32]. Both papers reduce the problem to a biconfluent Heun equation.

o B.S. Kandemir presented an analytical analysis of the two-dimensional Schrödinger equation for two interacting electrons subjected to a homogeneous magnetic field and confined by a two-dimensional external parabolic potential. Here also a biconfluent Heun equation is used[33].

o Dislocation movement in crystalline materials, quantum diffusion of kinks along dislocations are some solid state applications of this equation. The book by S.Y. Slavyanov and S. Lay[27] is a general reference on problems solved before 2000.

o In a relatively recent work P. Dorey, J. Suzuki, R. Tateo[34] show that equations in finite lattice systems also reduce to Heun equations.

In the rest of this paper I will comment only on papers on particle physics and general relativity.

o In a relatively early work, Teukolsky studied the perturbations of the Kerr metric and found out that they were described by two coupled singly confluent Heun equations[35].

o Quasi-normal modes of rotational gravitational singularities were also studied by E.W. Leaver[36] with Heun type equations.

In recent applications in general relativity, these equations become indispensible when one studies phenomena in higher dimensions, or in different geometries. Some references are:

o D. Batic, H. Schmid, M. Winklmeier studied the Dirac equation in the Kerr-Newman metric and static perturbations of the non-extremal Reisner-Nordstrom solution and ended up with Heun type solutions[37]. D. Batic and

H. Schmid also studied the Dirac equation for the Kerr-Newman metric and computed its propagator[38]. They found that the equation satisfied is a form of a generalized Heun equation described in Reference 37. In later work Batic, with collaborators, continued studying Heun equations and their generalizations[39,40].

Prof. P.P. Fiziev studied problems whose solutions are Heun equations extensively:

o In a paper published in gr-qc/0603003, he studied the exact solutions of the Regge-Wheeler equation in the Schwarschild black hole interior[41]. In another paper he obtained exact solutions of the same equation and quasi-normal modes of compact objects[42].

o He presented a novel derivation of the Teukolsky-Starobinsky identities, based on properties of the confluent Heun functions[43]. These functions define analytically all exact solutions to the Teukolsky master equation, as well as to the Regge-Wheeler and Zerilli ones.

o In a talk given at 29th Spanish Relativity Meeting (ERE 2006), he depicted in more detail the exact solutions of Regge-Wheeler equation, which described the axial perturbations of Schwarzschild metric in linear approximation, in the Schwarzschild black hole interior and on Kruskal-Szekeres manifold in terms of the confluent Heun functions[44].

o All classes of exact solutions to the Teukolsky master equation were described in terms of confluent Heun functions in Reference 45.

o In reference 46 he reveals important properties of the confluent Heun's functions by deriving a set of novel relations for confluent Heun's functions and their derivatives of arbitrary order. Specific new subclasses of confluent Heun's functions are introduced and studied. A new alternative derivation of confluent Heun's polynomials is presented.

o In another paper[47] he, with a collaborator, noted that weak gravitational, electromagnetic, neutrino and scalar fields, considered as perturbations on Kerr background satisfied the Teukolsky Master Equation. Two non-trivial equations were obtained after separating the variables, one equation only with the polar angle and another using only the radial variable. These were solved by transforming each one into the form of a confluent Heun equation.

Among other papers on this subject one may cite the following papers:

o In reference 48, R. Manvelyan, H.J.W. Müller Kirsten, J.Q. Liang and Y. Zhang, calculated the absorption rate of a scalar field by a D3 brane in ten dimensions in terms of modified Mathieu functions, and obtained the S-matrix.

o In reference 49, T. Oota and Y. Yasui studied the scalar laplacian on a wide class of five dimensional toric Sasaki-Einstein manifolds, ending in two Heun's differential equations.

o In reference 50, S. Musiri and G. Siopsis found out that the wave equation obtained in calculating the asymptotic form of the quasi-normal frequencies for large AdS black holes in five dimensions reduced to a Heun equation.

o A. Al-Badawi and I. Sakallı studied the Dirac equation in the rotating Bertotti-Robinson spacetime[51], ending up with a confluent Heun equation for the angular part of the equation for a massless particle.

o Mirjam Cvetic, and Finn Larsen studied grey body factors and event horizons for rotating black holes with two rotation parameters and five charges in five dimensions. When the Klein Gordon equation for a scalar particle in this background is written, one gets a confluent Heun equation. In the asymptotic region this equation turns into the hypergeometric form[52]. When they studied the similar problem for the rotating black hole with four $U(1)$ charges, they again obtained a confluent Heun equation for the radial component of the Klein Gordon equation, which they reduced to the hypergeometric form by making approximations[53]. These two papers are partly repeated in Ref. 54. Same equations were obtained which were reduced to approximate forms which gave solutions in the hypergeometric form. Other relevant references I could find are listed as references 55–62.

I first "encountered" this type of equation when we tried to solve the scalar wave equation in the background of the Nutku helicoid instanton[17]. In this case one gets the Mathieu equation which is a special case of the Heun class[19].

o The helicoid instanton is a double-centered solution. As remarked above, for the simpler instanton solution of Eguchi-Hanson[15], hypergeometric solutions are sufficient[16]. Here one must remark that another paper using the Eguchi-Hanson metric ends up with the confluent Heun equation[63]. These two papers show that sometimes judicious choice of the coordinate system and separation ansatz matters.

o Sucu and Ünal obtained closed solutions for the spinor particle written in the background of the Nutku helicoid instanton metric[16]. One can show that these solutions can be expanded in terms of Mathieu functions if one attempts to use the separation of variables method, as described by

L.Chaos-Cador and E. Ley-Koo[64].

In the subsequent sections I will summarize some work Tolga Birkandan and I have done on this topic[20,65,66].

2. Dirac equation in the background of the Nutku helicoid metric[a]

The Nutku helicoid metric is given as

$$ds^2 = \frac{1}{\sqrt{1+\frac{a^2}{r^2}}}[dr^2 + (r^2+a^2)d\theta^2 + \left(1+\frac{a^2}{r^2}\sin^2\theta\right)dy^2 - \frac{a^2}{r^2}\sin 2\theta dydz + \left(1+\frac{a^2}{r^2}\cos^2\theta\right)dz^2] \tag{25}$$

where $0 < r < \infty$, $0 \leq \theta \leq 2\pi$, y and z are along the Killing directions and will be taken to be periodic coordinates on a 2-torus[19].

If we make the following transformation

$$r = a\sinh x, \tag{26}$$

the metric is written as

$$\begin{aligned} ds^2 &= \frac{a^2}{2}\sinh 2x(dx^2 + d\theta^2) \\ &+ \frac{2}{\sinh 2x}[(\sinh^2 x + \sin^2\theta)dy^2 \\ &- \sin 2\theta dydz + (\sinh^2 x + \cos^2\theta)dz^2]. \end{aligned} \tag{27}$$

We write the system in the form $L\psi = \Lambda\psi$, where L is the Dirac operator, Λ is the eigenvalue, and try to obtain the solutions for the different components.

We use the Newman-Penrose formalism[67,68] to write the Dirac equation, with four components, in this metric. We use a solution of the type $\psi_i = e^{i(k_y y + k_z z)}\Psi_i(x,\theta)$. We also use $k_y = k\cos\phi$, $k_z = k\sin\phi$. The transformation $\Psi_{1,2} = \frac{1}{\sqrt{\sinh 2x}} f_{1,2}$ is used for the upper components to have similar equations for all components. Then these equations read:

[a]The material of this section is based on Refs. 65,66.

$$(\partial_x + i\partial_\theta)\Psi_3 \ + iak[\cos(\theta-\phi+ix)]\Psi_4 = \Lambda\frac{a}{\sqrt{2}}f_1 \tag{28}$$

$$(\partial_x - i\partial_\theta)\Psi_4 \ - iak[\cos(\theta-\phi-ix)]\Psi_3 = \Lambda\frac{a}{\sqrt{2}}f_2, \tag{29}$$

$$(-\partial_x + i\partial_\theta)f_1 \ + iak[\cos(\theta-\phi+ix)]f_2 = -\Lambda\frac{a\sinh 2x}{\sqrt{2}}\Psi_3, \tag{30}$$

$$(-\partial_x - i\partial_\theta)f_2 \ - iak[\cos(\theta-\phi-ix)]f_1 = -\Lambda\frac{a\sinh 2x}{\sqrt{2}}\Psi_4. \tag{31}$$

We solve our equations in terms of $f_{1,2}$ and substitute these expressions in equations, given above. This substitution gives us second order, but uncoupled equations for the lower components:

$$\left(\partial_{xx}+\partial_{\theta\theta} + \frac{a^2}{2}[k^2(-\cos[2(\theta-\phi)] - \cosh 2x) - \Lambda^2\sinh 2x]\right)\Psi_{3,4} = 0. \tag{32}$$

We can separate this equation into two ordinary differential equations by the ansatz $\Psi_{3,4} = R(x)S(\theta-\phi)$. Using this ansatz gives us two ordinary differential equations. The equation for S reads

$$\partial_{\Theta\Theta}S(\Theta) - \left(\frac{a^2}{2}k^2\cos(2\Theta) - n\right)S(\Theta) = 0, \tag{33}$$

where $\Theta = (\theta-\phi)$. This equation is of the Mathieu type and the solution can be written immediately.

$$S(\theta) = C_1 Se(n, \frac{a^2k^2}{4}, \theta-\phi) + C_2 So(n, \frac{a^2k^2}{4}, \theta-\phi), \tag{34}$$

where Se, So are the even and odd Mathieu functions. The solutions should be periodic in the angular variable Θ. This fact forces n, the separation constant, to take discrete values. It is known that the angular Mathieu functions satisfy an orthogonality relation such that functions with different n values are perpendicular to each other. Here, we integrate the angular variable from zero to 2π.

The equation for $R(x)$ reads

$$\left\{\partial_{xx} - [\frac{a^2}{2}(k^2\cosh 2x) + \Lambda^2\sinh 2x] + n\right\}R(x) = 0. \tag{35}$$

This solution is of the double confluent form which can be reduced to the form

$$R(x) = D_1 Se(n, A_6, i(x+b)) + D_2 So(n, A_6, i(x+b)) \tag{36}$$

by several transformations. Here C_1, C_2, D_1, D_2 are arbitrary constants. A_6, b are given in the original reference[65].

As a result of this analysis we see that the solutions of the Dirac equation, written in the background of the Nutku helicoid metric, can be expressed in terms of Mathieu functions, which is a special form of Heun function[65,66]. Mathieu function is a related but much more studied function with singularity structure same as the double confluent Heun equation.

o One can also show[65,66] that one can use the similar metric in five dimensions and obtain, in general, double confluent Heun functions which can be reduced to Mathieu functions.

3. Scalar field in the background of the extended Eguchi-Hanson solution[b]

To go to five dimensions, we can add a time component to the Eguchi-Hanson metric[15] so that we have

$$ds^2 = -dt^2 + \frac{1}{1-\frac{a^4}{r^4}}dr^2 + r^2(\sigma_x^2 + \sigma_y^2) + r^2(1 - \frac{a^4}{r^4})\sigma_z^2 \tag{37}$$

where

$$\sigma_x = \frac{1}{2}(-\cos\xi d\theta - \sin\theta \sin\xi d\phi) \tag{38}$$

$$\sigma_y = \frac{1}{2}(\sin\xi d\theta - \sin\theta\cos\xi d\phi) \tag{39}$$

$$\sigma_z = \frac{1}{2}(-d\xi - \cos\theta d\phi). \tag{40}$$

This is a vacuum solution.

If we take the solution to the wave equation Φ as

$$\Phi = e^{ikt}e^{in\phi}e^{i(m+\frac{1}{2})\xi}\varphi(r,\theta), \tag{41}$$

we can write the reduced wave equation as

$$\begin{aligned}\Delta\varphi(r,\theta) = (&\frac{r^4-a^4}{r^2}\partial_{rr} + \frac{3r^4+a^4}{r^3}\partial_r + k^2r^2 + \frac{4a^4m^2}{a^4-r^4} \\ &+ 4\partial_{\theta\theta} + 4\cot\theta\partial_\theta + \frac{8mn\cos\theta - 4(m^2+n^2)}{\sin^2\theta})\varphi(r,\theta) = 0.\end{aligned} \tag{42}$$

[b]The material of this section is based on Ref. 20.

If we take $\varphi(r,\theta) = f(r)g(\theta)$, the solution of the radial part is expressed in terms of confluent Heun (H_C) functions.

$$f(r) = \left(-a^4 + r^4\right)^{\frac{m}{2}} H_C\left(0, m, m, \frac{k^2a^2}{2}, \frac{m^2}{2} - \frac{\lambda}{4} - \frac{k^2a^2}{4}, \frac{a^2+r^2}{2a^2}\right)$$
$$+ \left(a^2+r^2\right)^{-\frac{m}{2}} \left(r^2 - a^2\right)^{\frac{m}{2}}$$
$$\times H_C\left(0, -m, m, \frac{k^2a^2}{2}, \frac{m^2}{2} - \frac{\lambda}{4} - \frac{k^2a^2}{4}, \frac{a^2+r^2}{2a^2}\right). \quad (43)$$

If the variable transformation $r = a\sqrt{\cosh x}$ is made, the solution can be expressed as

$$f(x) = [\sinh^m(x)]\, H_C\left(0, m, m, \frac{k^2a^2}{2}, \frac{m^2}{2} - \frac{\lambda}{4} - \frac{k^2a^2}{4}, \frac{\cosh(x)+1}{2}\right)$$
$$+2\left(\cosh x + 1\right)^{-\frac{m}{2}} \left(\cosh x - 1\right)^{\frac{m}{2}}$$
$$\times H_C\left(0, -m, m, \frac{k^2a^2}{2}, \frac{m^2}{2} - \frac{\lambda}{4} - \frac{k^2a^2}{4}, \frac{\cosh(x)+1}{2}\right)\Big\}. \quad (44)$$

We tried to express the equation for the radial part in terms of $u = \frac{a^2+r^2}{2a^2}$ to see the singularity structure more clearly. Then the radial differential operator reads

$$4\frac{d^2}{du^2} + 4\left(\frac{1}{u-1} + \frac{1}{u}\right)\frac{d}{du} + k^2a^2\left(\frac{1}{u-1} + \frac{1}{u}\right) + \frac{m^2}{u^2(1-u)^2}. \quad (45)$$

This operator has two regular singularities at zero and one, and an irregular singularity at infinity, the singularity structure of the confluent Heun equation. This is different from the hypergeometric equation, which has regular singularities at zero, one and infinity.

The angular solution is in terms of hypergeometric functions:

$$g(\theta) = 2\left(\cos\theta - 1\right)^{\frac{m-n}{2}} \Big[2^{-m}\left(\cos\theta + 1\right)^{-\frac{n+m}{2}}$$
$$\times {}_2F_1([-n + \frac{\sqrt{\lambda+1}}{2} + \frac{1}{2}, -n - \frac{\sqrt{\lambda+1}}{2} + \frac{1}{2}], [1-n-m], \frac{\cos\theta+1}{2})$$
$$+ 2^n\left(\cos\theta + 1\right)^{\frac{n+m}{2}}$$
$$\times {}_2F_1([m + \frac{\sqrt{\lambda+1}}{2} + \frac{1}{2}, m - \frac{\sqrt{\lambda+1}}{2} + \frac{1}{2}], [1+m+n], \frac{\cos\theta+1}{2})\Big]. \quad (46)$$

4. Derivation of a new generalized Heun equation[c]

The Dirac equation, with $\Lambda = 0$, written in the background of the Nutku helicoid metric, reads as follows:

$$(\partial_x + i\partial_\theta)\Psi_3 \ + iak[\cos(\theta - \phi + ix)]\Psi_4 = 0, \tag{47}$$

$$(\partial_x - i\partial_\theta)\Psi_4 \ - iak[\cos(\theta - \phi - ix)]\Psi_3 = 0, \tag{48}$$

$$(-\partial_x + i\partial_\theta)f_1 \ + iak[\cos(\theta - \phi + ix)]f_2 = 0, \tag{49}$$

$$(-\partial_x - i\partial_\theta)f_2 \ - iak[\cos(\theta - \phi - ix)]f_1 = 0. \tag{50}$$

These equations have simple solutions[16] which can also be expanded in terms of products of radial and angular Mathieu functions[64,65]. Problem arises when these solutions are restricted to boundary[20,66].

To impose these boundary conditions we need to write the little Dirac equation, the Dirac equation restricted to the boundary, where the variable x takes a fixed value x_0. We choose to write the equations in the form

$$\frac{\sqrt{2}}{a}\{i\frac{d}{d\theta}\Psi_3 \ + ika\cos(\theta - \phi + ix_0)\Psi_4\} = \lambda f_1, \tag{51}$$

$$\frac{\sqrt{2}}{a}\{-i\frac{d}{d\theta}\Psi_4 \ - iak\cos(\theta - \phi - ix_0)\Psi_3\} = \lambda f_2, \tag{52}$$

$$\frac{\sqrt{2}}{a}\{-i\frac{d}{d\theta}f_1 \ - iak\cos(\theta - \phi + ix_0)f_2\} = \lambda \Psi_3, \tag{53}$$

$$\frac{\sqrt{2}}{a}\{i\frac{d}{d\theta}f_2 \ + iak\cos(\theta - \phi - ix_0)f_1\} = \lambda \Psi_4. \tag{54}$$

Here λ is the eigenvalue of the little Dirac equation. We take $\lambda = 0$ as the simplest case. The transformation

$$\Theta = \theta - \phi - ix_0 \tag{55}$$

can be used. Then we solve f_1 in the latter two equations in terms of f_2:

$$\begin{aligned} &- \frac{d^2}{d\Theta^2}f_2 - \tan\Theta\frac{d}{d\Theta}f_2 \\ &+ \frac{(ak)^2}{2}[\cos(2\Theta)\cosh(2x_0) - i\sin(2\Theta)\sinh(2x_0) + \cosh(2x_0)]f_2 = 0. \end{aligned} \tag{56}$$

When we make the transformation

$$u = e^{2i\Theta}, \tag{57}$$

[c]The material of this section is based on Ref.20.

the equation reads,

$$\Big\{4(u+1)u[u\frac{d^2}{du^2}+\frac{d}{du}]-2iu(u-1)\frac{d}{du} \tag{58}$$

$$+\frac{(ak)^2}{2}(u+1)[ue^{-2x_0}+\frac{1}{u}e^{2x_0}+\cosh(2x_0)]\Big\}f_2=0. \tag{59}$$

This equation has irregular singularities at $u = 0$ and ∞ and a regular singularity at $u = -1$. This is still another generalized Heun equation, different from the one given by reference 37.

Acknowledgments

I thank the organizers of 13th Regional Conference on Mathematical Physics for inviting me to this meeting and dedicate this work to the Memory of my dear friend Faheem Hussain. I thank Tolga Birkandan for collaboration. This work is supported by TÜBİTAK, the Scientific and Technological Council of Turkey and by TÜBA, the Academy of Sciences of Turkey.

References

1. P.M. Morse and H. Feshbach, *Methods of Theoretical Physics*, (McGraw Hill Company, New York, 1953).
2. E.L. Ince, *Ordinary Differential Equations*, (Dover Publications 1926, 1956).
3. P.M. Morse and H. Feshbach, *Methods of Theoretical Physics*, (McGraw Hill Company, New York 1953), p.532.
4. P. Dennery, A. Krzywicki, *Mathematics for Physicists* , (Dover Publications, 1967).
5. K. Heun, *Mathematische Annalen* **33**, 161 (1889).
6. A. Ronveaux (ed.), *Heun's Differential Equations*, (Oxford University Press, 1995).
7. G. Valent, *Siam Journal on Mathematical Analysis* **17**, 688 (1986).
8. J. Meixner and F. W. Schaefke, *Mathieusche Funktionen und Sphaeroidfunktionen mit Anwendungen auf physikalische und technische Probleme*, (Springer Verlag, 1954).
9. N.W. McLachlan, *Theory and Applications of Mathieu Functions*, (Dover reprint from 1946 edition, 1963).
10. R. Schafke, D. Schmidt, *SIAM Journal of Mathematical Analysis* **11**, 848 (1980).
11. R.S. Maier, math.CA/0408317
12. R.S. Maier, *Journal of Differential Equations* **213**, 171 (2005); math.CA/0203264.
13. N. Gurappa, P.K. Panigrahi, *Journal of Physics A: Mathematical and General* **37**, L605 (2004).

14. K. Kuiken, *SIAM Journal of Mathematical Analysis* **10**, 655 (1979).
15. T. Eguchi, A.J. Hanson, *Physics Letters* **74B**, 249 (1978).
16. Y. Sucu, N. Ünal, *Classical and Quantum Gravity* **21**, 1443 (2004).
17. Y. Nutku, *Physical Review Letters* **77**, 4702 (1996).
18. D. Lorenz-Petzold, *Journal of Mathematical Physics* **24**, 2632 (1983).
19. A.N. Aliev, M. Hortaçsu, J. Kalaycı, Y. Nutku, *Classical and Quantum Gravity* **16**, 631 (1999).
20. T. Birkandan, M. Hortaçsu, *Journal of Mathematical Physics* **49**, 0054101 (2008).
21. P.M. Morse and H. Feshbach, *Methods of Theoretical Physics* (McGraw Hill Company, New York, 1953), p.1407.
22. B.D.B. Figueiredo, *Journal of Mathematical Physics* **46**, 113503 (2005); math-phys/0509013.
23. B.D.B. Figueiredo, *Journal of Mathematical Physics* **48**, 013503 (2007).
24. B.D.B. Figueiredo, math-ph 10402071.
25. L.J. El-Jaick, B.D.B. Figueiredo, *Journal of Mathematical Physics* **49**, 083508 (2008).
26. L.J. El-Jaick, B.D.B. Figueiredo, *Journal of Mathematical Physics* **50**, 123511 (2008).
27. S.Y. Slavyanov, W. Lay, *Special Functions, A Unified Theory Based on Singularities*, (Oxford University Press, 2000).
28. P.S. Epstein, *Physical Review* **2**, 695 (1926).
29. S.Y. Slavyanov, *Asymptotic Solutions of the One-dimensional Schrodinger Equation* (Leningrad University Press, 1991) (in Russian). Translation into English: S.Y. Slavyanov, *Asymptotic Solutions of the One-dimensional Schrodinger Equation* (Amer. Math. Soc. Trans. of Math. Monographs) **151** (1996).
30. A.H. Wilson, *Proceedings of Royal Society* **A118**, 617 (1928). Also Ref. 27, p.167.
31. T.T. Truong, D. Bazzali, *Physics Letters A* **269**, 186 (2000).
32. A. Ralko, T.T. Truong, *Journal of Physics A-Mathematical and General* **35**, 9573 (2002).
33. B.S. Kandemir, *Journal of Mathematical Physics* **46**, 032110 (2005).
34. P.Dorey, J. Suzuki and R. Tateo, *Journal of Physics A: Mathematical and General* **37**, 2047 (2004).
35. S.A. Teukolsky, *Physical Review Letters* **29**, 1114 (1972). Also Ref.27, p.170.
36. E.W. Leaver, *Proceedings of Royal Society London A* **402**, 285 (1985). Also Ref.27, p.171.
37. D. Batic, H. Schmid, M. Winklmeier, *Journal of Physics A: Mathematical and General* **39**, 12559 (2006); gr-qc/0607017.
38. D. Batic, H. Schmid, *Progress in Theoretical Physics* **116**, 517 (2006).
39. D. Batic, H. Schmid, *Journal of Mathematical Physics* **48**, 042502 (2007).
40. D. Batic, M. Sandoval, e-Print: arXiv:0805.4399 [gr-qc].
41. P.P. Fiziev, e-print gr-qc/0603003.
42. P.P. Fiziev, *Classical and Quantum Gravity* **23**, 2447 (2006); gr-qc/0509123.
43. P.P. Fiziev, *Physical Review D* **80**, 124001 (2009).

44. P.P. Fiziev, *Journal of Physics Conference Series* **66**, 012016, (2007); gr-qc/0702014.
45. P.P. Fiziev, *Classical and Quantum Gravity* **27**, 135001 2010.
46. P.P. Fiziev, *Journal of Physics-Mathematical and Theoretical* **43**, 035203 (2010).
47. Roumen S. Borissov, Plamen P. Fiziev, e-print: arXiv:0903.3617 [gr-qc].
48. R. Manvelyan, H.J.W. Müller-Kirsten, J.Q. Liang and Y. Zhang, *Nuclear Physics B* **579**, 177-208 (2000); hep-th/0001179.
49. T. Oota, Y. Yasui, *Nuclear Physics B* **742**, 275 (2006).
50. S. Musiri, G. Siopsis, *Physics Letters B* **576**, 309 (2003); hep-th/0308196,
51. A. Al-Badawi, I. Sakallı, *Journal of Mathematical Physics* **49**, 052501 (2008).
52. M. Cvetic and F. Larsen, *Physical Review D* **56**, 4994 (1997); hep-th/9705192.
53. M. Cvetic and F. Larsen, *Nuclear Physics B* **506**, 107 (1997); hep-th/9706071.
54. M.Cvetic and F. Larsen, *Journal of High Energy Physics*, 0909:088 (2009); arXiv:0908.1136 [hep-th].
55. G. Siopsis, *Nuclear Physics B* **715**, 483 (2005); hep-th/0407157.
56. L. Anguelva, P. Langfelder, *Journal of High Energy Physics*, Issue: 3 Article 057, (2003).
57. S.R. Lau, *Classical and Quantum Gravity* **21**, 4147 (2004); also *Journal of Computational Physics* **199**, 376 (2004).
58. H. Suzuki, E. Takasugi and H. Umetsu, *Progress of Theoretical Physics* **100**, 491 (1998).
59. A. Zecca, *Nuovo Cimento B* **125**, 191 (2010).
60. A. Ensico and N. Kamran, *Communications in Mathematical Physics* **290**, 105 (2009).
61. G. Esposito and R. Roychowdhury, *General Relativity and Gravitation* **42**, 1221 (2010).
62. S. Yoshida, N. Uchikata and T. Fumatase, *Physical Review D* **81**, 044005 (2010).
63. A. Malmendier, *Journal of Mathematical Physics* **44**, 4308 (2003).
64. L. Chaos-Cador and E. Ley-Koo, *Revista Mexicana de Fisica* **48**, 67 (2002).
65. T. Birkandan, M. Hortaçsu, *Journal of Physics A: Mathematical and Theoretical* **40**, 1105 (2007).
66. T. Birkandan, M. Hortaçsu, *Journal of Mathematical Physics* **48**, 092301 (2007).
67. E.T. Newman and R. Penrose, *Journal of Mathematical Physics* **3**, 566 (1962).
68. E.T. Newman and R. Penrose, *Journal of Mathematical Physics* **4**, 998 (1962).

COXETER GROUPS, QUATERNIONS, SYMMETRIES OF POLYHEDRA AND 4D POLYTOPES

MEHMET KOCA* and N. ÖZDEŞ KOCA†

Sultan Qaboos University, College of Science,
Physics Department, P.O Box 36, Al-Khoudh, 123 Muscat, Sultanate of Oman
**E-mail: kocam@squ.edu.om*
†E-mail: nazife@squ.edu.om

Emergence of the experimental evidence of E_8 in the analysis of one dimensional Ising-model invokes further studies of the Coxeter-Weyl groups generated by reflections regarding their applications to polytopes. The Coxeter group $W(H_4)$ which describes the Platonic Polytopes 600-cell and 120-cell in 4D singles out in the mass relations of the bound states of the Ising model for it is a maximal subgroup of the Coxeter-Weyl group $W(E_8)$. There exists a one-to-one correspondence between the finite subgroups of quaternions and the Coxeter-Weyl groups of rank 4 which facilitates the study of the rank-4 Coxeter-Weyl groups. In this paper we study the systematic classifications of the 3D-polyhedra and 4D-polytopes through their symmetries described by the rank-3 and rank-4 Coxeter-Weyl groups represented by finite groups of quaternions. We also develop a technique on the constructions of the duals of the polyhedra and the polytopes and give a number of examples. Applications of the rank-2 Coxeter groups have been briefly mentioned.

Keywords: Ising-model; Coxeter group; Polytopes; Quaternions.

1. Introduction

Zamolodchikov[1] has proved that the one-dimensional Ising model at critical temperature perturbed by an external magnetic field leads to eight spinless bosons with masses

$$\begin{aligned} &m_1, \quad m_3 = 2m_1 \cos\frac{\pi}{30}, \quad m_4 = 2m_2 \cos\frac{7\pi}{30}, \quad m_5 = 2m_2 \cos\frac{4\pi}{30}, \\ &m_2 = \tau m_1, \quad m_6 = \tau m_3, \quad m_7 = \tau m_4, m_8 = \tau m_5 \end{aligned} \tag{1}$$

where $\tau = (1+\sqrt{5})/2$ is the golden ratio. It can be described by an affine Toda Field Theory of E_8 (Ref. 2). The masses can be related to the radii of the Gosset's polytope of E_8 (Ref. 3). In the derivation of the mass relations in (1) the maximal subgroup $W(H_4)$ (Ref. 4) of the Coxeter-Weyl group

$W(E_8)$ plays a crucial role. This motivates us to study the structure of the group $W(H_4)$ in terms of quaternions and its 4D polytopes.

Radu Coldea of the University of Oxford and his colleagues, from Oxford, Bristol University, the ISIS Rutherford Laboratory and the Helmholtz Zentrum Berlin[5] performed a neutron scattering experiment on $CoNb_2O_6$ (cobalt niobate) which describes quantum Ising chain. They have determined masses of the five emerging particles, first two are obeying the relation $m_2 = \tau m_1$. A modification of the Hamiltonian describing the Ising model has been suggested[6] which confirms the mass relations in (1). It seems that E_8 displays itself as a Coxeter-Weyl group $W(E_8)$ at absolute zero temperature (the coldest regime) and perhaps as a Lie group in the form of $E_8 \times E_8$ describing the heterotic superstring theory at very hot regime (Planck scale). The Coxeter-Weyl group $W(E_8)$ includes the crystallographic (tetrahedral and octahedral symmetries in 3D and 4D) as well as the quasi crystallographic symmetries (icosahedral symmetry in 3D and its generalization to 4D). The icosahedral group of rank-3 describes fully the structures of the fullerenes such as the C_{60} molecule which is represented by a truncated icosahedrons, the icosahedral quasicrystals and the viral structures displaying the icosahedral symmetry.

The Platonic solids, tetrahedron, cube, octahedron, icosahedron and dodecahedron have been discovered by the people lived in Scotland nearly 1000 years earlier than the ancient Greeks. Their models carved on the stones are now kept in the Ashmolean Museum at Oxford[7]. They were used in old Greek as models to describe fundamental matter associating tetrahedron with fire, cube with earth, air with octahedron, and water with icosahedron. Archimedes discovered the semi-regular convex solids now known as Archimedean solids, and several centuries later they were rediscovered by the renaissance mathematicians. The prisms and anti-prisms as well as four regular non-convex polyhedra have been discovered in 1620 by Kepler completing the classifications of the regular polyhedra. Nearly two centuries later, in 1865, Catalan constructed the dual solids of the Archimedean solids, now known as Catalan solids[8] . Extensions of the platonic solids to 4D dimensions have been made in 1855 by L. Schlaffli[9] and their generalizations to higher dimensions in 1900 by T. Gosset[10]. Further important contributions are made by W. A. Wythoff[11] among many others and in particular by the contemporary mathematicians H.S.M. Coxeter[12] and J.H. Conway[13].

The Coxeter symmetries in 3D and 4D are richer than those in higher dimensions since the analogous structures to the Coxeter groups $W(H_3)$

and $W(H_4)$ do not exist in higher dimensions. For example, the number of Platonic solids is five in 3D and six in 4D contrary to the higher dimensional cases where there exist only three platonic polytopes which are the generalizations of tetrahedron, octahedron and cube. The Platonic and Archimedean solids[14] as well as the Catalan solids[15] can be described with the rank-3 Coxeter groups $W(A_3), W(B_3)$ and $W(H_3)$. The 4D polytopes are described by the rank-4 Coxeter groups $W(A_4), W(B_4)$ and $W(H_4)$ and the group $W(F_4)$.

In what follows we study the regular and semi regular 4D polytopes as the orbits of the rank-4 Coxeter groups expressed in terms of quaternions. There is a one-to-one correspondence between the finite subgroups of quaternions and the symmetries of the regular and semi regular 4D polytopes. In Section 2 we introduce the finite subgroups of quaternions. The rank-4 Coxeter groups are constructed in terms of quaternions in Section 3. The analysis of the 4D polytopes requires a study of the 3D polyhedra which will be introduced in Section 4. Examples of the 4D polytopes and their dual polytopes will be given in Section 5. Some concluding remarks are made in Section 7.

2. Finite subgroups of quaternions

This section deals with the introduction of quaternions and their relevance to the orthogonal transformations in four dimensions and gives the list of finite subgroups of quaternions.

2.1. *Quaternions and O(4) transformations*

Let $q = q_0 + q_i e_i$, $(i = 1, 2, 3)$ be a real unit quaternion with its conjugate defined by $\bar{q} = q_0 - q_i e_i$ and the norm $q\bar{q} = \bar{q}q = 1$. The imaginary units satisfy

$$e_i e_j = -\delta_{ij} + \varepsilon_{ijk} e_k, \quad (i, j, k = 1, 2, 3). \tag{2}$$

Let p, q be unit quaternions and r represents an arbitrary quaternion. Then the transformations[16]

$$r \to prq : [p, q]; \qquad r \to p\bar{r}q : [p, q]^* \tag{3}$$

define the orthogonal group $O(4)$ which preserves the norm $r\bar{r} = \bar{r}r$. The first term in (3) represents a proper rotation and the second includes also the reflection, generally called rotary reflection. In particular, the group element

$$r \to p\bar{r}p : [p, -p]^* \tag{4}$$

represents the reflection with respect to the hyperplane orthogonal to the quaternion p. The orthogonal transformations in 3D can be simply written as $r \rightarrow \pm pr\bar{p} : \pm[p, \bar{p}]$.

2.2. *Finite subgroups of quaternions*

The finite subgroups of quaternions are well known and their classification can be found in the reference 17. They are given as follows.

(a) Cyclic group of order n with n an odd number.
(b) Cyclic group of order $2n$ is generated by $\langle p = \exp\left(e_1 \frac{\pi}{n}\right)\rangle$ and dicyclic group of order $4n$ is generated by the generators $\langle p = \exp\left(e_1 \frac{\pi}{n}\right), e_2\rangle$.
(c) The binary tetrahedral group can be represented by the set of 24 unit quaternions:

$$T = \left\{\pm 1, \pm e_1, \pm e_2, \pm e_3, \frac{1}{2}\left(\pm 1, \pm e_1, \pm e_2, \pm e_3\right)\right\}. \tag{5}$$

(d) The binary octahedral group consists of 48 unit quaternions. Let the set $T' = (V_1 \oplus V_2 \oplus V_3)$ represent the 24 unit quaternions, where

$$V_1 = \left\{\frac{1}{\sqrt{2}}(\pm 1, \pm e_1), \frac{1}{\sqrt{2}}(\pm e_2, \pm e_3)\right\},$$
$$V_2 = \left\{\frac{1}{\sqrt{2}}(\pm 1, \pm e_2), \frac{1}{\sqrt{2}}(\pm e_3, \pm e_1)\right\},$$
$$V_3 = \left\{\frac{1}{\sqrt{2}}(\pm 1, \pm e_3), \frac{1}{\sqrt{2}}(\pm e_1, \pm e_2)\right\}. \tag{6}$$

Then the union of the set $O' = T \oplus T'$ represents the binary octahedral group.

(e) The binary icosahedral group $I = \langle b, c\rangle$ of order 120 can be generated by two unit quaternions $b = \frac{1}{2}(\tau + \sigma e_1 + e_2) \in 12(1)_+$ and $c = \frac{1}{2}(\tau - \sigma e_1 + e_2) \in 12(1)_+$. We display its elements in Table 1 as the sets of conjugacy classes which represent a number of icosahedra, dodecahedra and one icosidodecahedron.

In Table 1 the golden ratio $\tau = \frac{1+\sqrt{5}}{2}$ and $\sigma = \frac{1-\sqrt{5}}{2}$ satisfy the relations

$$\tau + \sigma = 1, \quad \tau\sigma = -1, \quad \tau^2 = \tau + 1, \quad \sigma^2 = \sigma + 1.$$

Table 1. Conjugacy classes of the binary icosahedral group I represented by quaternions.

Order of the elements	Conjugacy classes denoted by the number of elements and order of the elements
1	1
2	-1
10	$12_+ : \frac{1}{2}(\tau \pm e_1 \pm \sigma e_3), \frac{1}{2}(\tau \pm e_2 \pm \sigma e_1), \frac{1}{2}(\tau \pm e_3 \pm \sigma e_2)$
5	$12_- : \frac{1}{2}(-\tau \pm e_1 \pm \sigma e_3), \frac{1}{2}(-\tau \pm e_2 \pm \sigma e_1), \frac{1}{2}(-\tau \pm e_3 \pm \sigma e_2)$
10	$12'_+ : \frac{1}{2}(\sigma \pm e_1 \pm \tau e_2), \frac{1}{2}(\sigma \pm e_2 \pm \tau e_3), \frac{1}{2}(\sigma \pm e_3 \pm \tau e_1)$
5	$12'_- : \frac{1}{2}(-\sigma \pm e_1 \pm \tau e_2), \frac{1}{2}(-\sigma \pm e_2 \pm \tau e_3), \frac{1}{2}(-\sigma \pm e_3 \pm \tau e_1)$
6	$20_+ : \frac{1}{2}(1 \pm e_1 \pm e_2 \pm e_3), \frac{1}{2}(1 \pm \tau e_1 \pm \sigma e_2), \frac{1}{2}(1 \pm \tau e_2 \pm \sigma e_3),$ $\frac{1}{2}(1 \pm \tau e_3 \pm \sigma e_1)$
3	$20_- : \frac{1}{2}(-1 \pm e_1 \pm e_2 \pm e_3), \frac{1}{2}(-1 \pm \tau e_1 \pm \sigma e_2),$ $\frac{1}{2}(-1 \pm \tau e_2 \pm \sigma e_3), \frac{1}{2}(-1 \pm \tau e_3 \pm \sigma e_1)$
4	$30 : \pm e_1, \pm e_2, \pm e_3, \frac{1}{2}(\pm\sigma e_1 \pm \tau e_2 \pm e_3), \frac{1}{2}(\pm\sigma e_2 \pm \tau e_3 \pm e_1),$ $\frac{1}{2}(\pm\sigma e_3 \pm \tau e_1 \pm e_2)$

3. Correspondence between the rank-4 Coxeter-Weyl groups and the finite subgroups of quaternions

Let $I_2(n)$ denotes the Coxeter diagram representing two vectors with the angle $\frac{2\pi}{n}$ between them. Then the Coxeter diagram $I_2(n) \oplus I_2(n)$ of rank-4 is represented by Figure 1.

Fig. 1. The Coxeter diagram $I_2(n) \oplus I_2(n)$.

Let $p = \exp\left(e_1 \frac{\pi}{n}\right)$ and $q = \exp\left(e_1 \frac{\pi}{n}\right) e_2$ be two orthogonal unit quaternions generating the dicyclic group of order $4n$. Then the following set of quaternions describes the root system of the diagram of Figure 1:

$$I_2(n) \oplus I_2(n) = \left\{p^k, q^k\right\}; \ k = 1, 2, ..., 2n. \tag{7}$$

If $s, t \in (I_2(n) \oplus I_2(n))$ are arbitrary elements of the root system then the group $Aut\,(I_2(n) \oplus I_2(n)) = \{[s,t] \oplus [s,t]^*\}$ of order $4n \times 4n$ is represented by the elements of the dicyclic group[18].

The F_4 diagram shown in Figure 2 with its quaternionic roots[19] leads to its automorphism group $Aut(F_4) = W(F_4) : \gamma = \{[O,O] \oplus [O,O]^*\}$ which is of the order $48 \times 48 = 2304$. Here γ is the Dynkin diagram symmetry generator of Figure 2.

The automorphism group of H_4 with its quaternionic simple roots[20] is

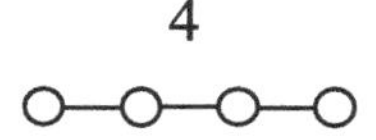

Fig. 2. The Coxeter diagram of F_4.

given as $W(H_4) = \{[I, I] \oplus [I, I]^*\}$ with an order of $120 \times 120 = 14,400$. The rank-4 Coxeter-Weyl groups $W(B_4)$ of order 384 and $W(D_4)$ of order 192 are the subgroups of the group $W(F_4)$. On the other hand the Coxeter-Weyl group $W(A_4) \approx S_5$ of order 120 is a subgroup of the group $W(H_4)$.[21]

4. Rank-3 Coxeter-Weyl groups and polyhedra

These groups are discussed extensively in the references 14 and 15. Quaternionic representations of these groups can be classified as follows. The Coxeter-Dynkin diagram of the group $W(A_3)$ is given in Figure 3.

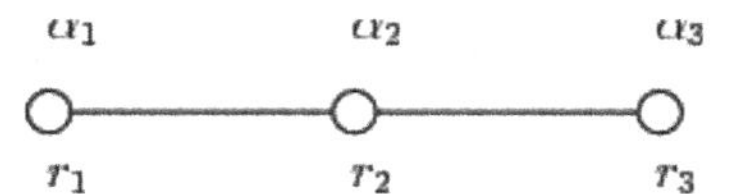

Fig. 3. The Coxeter-Dynkin diagram of the group $W(A_3)$.

If the roots α_i are chosen to be $\alpha_1 = e_1 + e_2$, $\alpha_2 = e_3 - e_2$, $\alpha_3 = e_2 - e_1$ then the group elements are given by the set $W(A_3) = \{[T, \bar{T}] \oplus [T', \bar{T}']^*\} \approx S_4$.

The Coxeter diagram of the group $W(B_3)$ is represented by the quaternionic roots as in Figure 4.

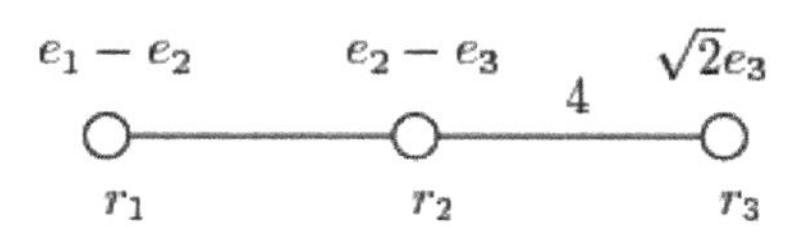

Fig. 4. The Coxeter diagram of the group $W(B_3)$.

Note that all the roots in the Coxeter diagram in Figure 4 have the same norm contrary to the Dynkin diagrams where some of the roots are represented as short roots. The group elements are represented by the set

$$W(B_3) \approx Aut(A_3) = \{[T, \bar{T}] \oplus [T', \bar{T}'] \oplus [T, \bar{T}]^* \oplus [T', \bar{T}']^*\} \approx S_4 \times C_2$$

which is isomorphic to the octahedral group O_h.

The Coxeter diagram of the group $W(H_3)$ is represented by the unit quaternions as shown in Figure 5.

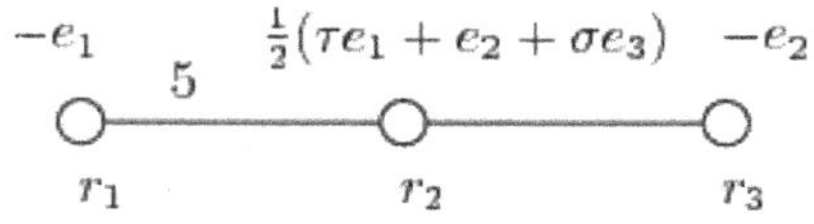

$-e_1$ — 5 — $\frac{1}{2}(\tau e_1 + e_2 + \sigma e_3)$ — $-e_2$

r_1 r_2 r_3

Fig. 5. The Coxeter diagram of the group $W(H_3)$.

The group elements can be written in terms of the elements of the binary icosahedral group as $W(H_3) = \{[I, \bar{I}] \oplus [I, \bar{I}]^*\} \approx A_5 \times C_2$ which is isomorphic to the icosahedral group I_h. The symmetry of the prisms are represented by the group elements given by the set of quaternions $W(I_2(n) \oplus A_1) \approx D_n \times C_2$.

Below we display some polyhedra as the orbits of these Coxeter groups. Let α_i and ω_i denote respectively the set of simple roots and the basis vectors of the dual space satisfying the relations

$$(\alpha_i, \omega_j) = \delta_{ij}, \ (\alpha_i, \alpha_j) = C_{ij}, \ (\omega_i, \omega_j) = (C^{-1})_{ij}. \tag{8}$$

Note that the roots are normalized to $\sqrt{2}$. Define an arbitrary vector in the dual space by $\Lambda = a_1\omega_1 + a_2\omega_2 + a_3\omega_3 \equiv (a_1 a_2 a_3)$ and the $\Lambda - orbit$ by $O(\Lambda) = W(G)\Lambda$.

The Platonic and Archimedean solids as well as Catalan solids can be described as the orbits of the rank-3 Coxeter groups. Some examples are given below.

4.1. *The tetrahedral group $W(A_3) \approx S_4$*

Let us consider the group $W(A_3)$ where the Cartan matrix C **and its inverse** C^{-1} are given by

$$C = \begin{bmatrix} 2 & -1 & 0 \\ -1 & 2 & -1 \\ 0 & -1 & 2 \end{bmatrix}, \qquad C^{-1} = \frac{1}{4}\begin{bmatrix} 3 & 2 & 1 \\ 2 & 4 & 2 \\ 1 & 2 & 3 \end{bmatrix}. \tag{9}$$

The orbit $O(100) = \frac{1}{2}(\pm e_1 \pm e_2 \pm e_3)$ with even number of $(-)$ sign and the orbit $O(001) = \frac{1}{2}(\pm e_1 \pm e_2 \pm e_3)$ with odd number of $(-)$ sign represent two tetrahedra dual to each other.

The union of two dual tetrahedra constitutes a cube. Similarly the orbits $O(110)$ and $O(011)$ represent two truncated tetrahedra.

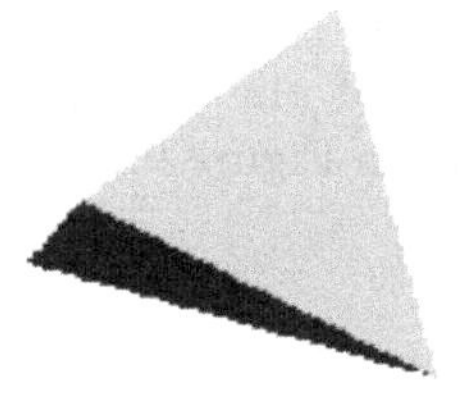

Fig. 6. Tetrahedron represented by the orbit $O(100)$.

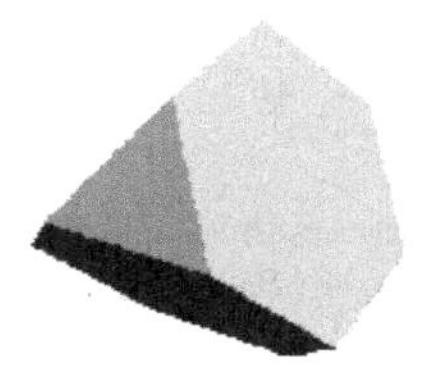

Fig. 7. The truncated tetrahedron represented by the orbit $O(110)$.

4.2. *The octahedral group* $W(B_3) \approx S_4 \times C_2$

The Cartan matrix and its inverse are given by the matrices

$$C = \begin{bmatrix} 2 & -1 & 0 \\ -1 & 2 & -\sqrt{2} \\ 0 & -\sqrt{2} & 2 \end{bmatrix}, \qquad C^{-1} = \begin{bmatrix} 1 & 1 & \frac{1}{\sqrt{2}} \\ 1 & 2 & \sqrt{2} \\ \frac{1}{\sqrt{2}} & \sqrt{2} & \frac{3}{2} \end{bmatrix}. \tag{10}$$

The orbit $O(100) = (\pm e_1, \pm e_2, \pm e_3)$ represents an octahedron and the orbit $O(001) = \frac{1}{2}(\pm e_1 \pm e_2 \pm e_3)$ represents a cube which is the dual of octahedron.

(a) (b)

Fig. 8. The octahedron (a) and its dual cube (b) as the respective orbits $W(B_3)(100)$ and $W(B_3)(001)$.

All non-chiral Archimedean solids with octahedral symmetry can be constructed as the orbits of the octahedral group. The non-chiral Archimedean solids are shown in Figure 9.

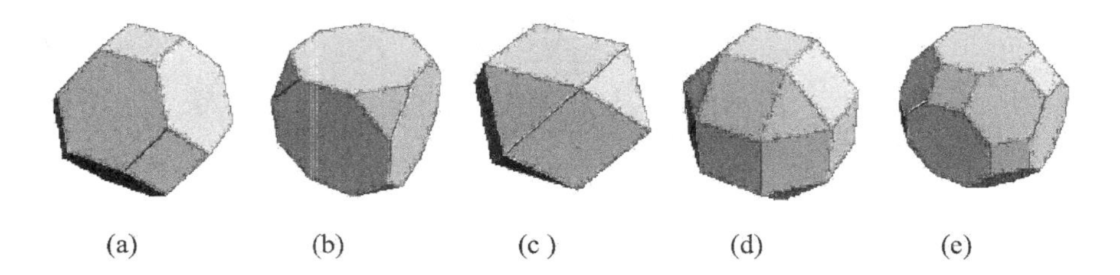

(a) (b) (c) (d) (e)

Fig. 9. The Archimedean solids with octahedral symmetry (a) truncated octahedron as $W(B_3)(110)$, (b) truncated cube as $W(B_3)(011)$, (c) cuboctahedron as $W(B_3)(010)$, (d) small rhombicuboctahedron as $W(B_3)(101)$, (e) great rhombicuboctahedron as $W(B_3)(111)$.

4.3. *The icosahedral symmetry* $W(H_3) \approx A_5 \times C_2$

The Cartan matrix and its inverse are given by the matrices

$$C = \begin{bmatrix} 2 & -\tau & 0 \\ -\tau & 2 & -1 \\ 0 & -1 & 2 \end{bmatrix}, \qquad C^{-1} = \frac{1}{2}\begin{bmatrix} 3\tau^2 & 2\tau^3 & \tau^3 \\ 2\tau^3 & 4\tau^2 & 2\tau^2 \\ \tau^3 & 2\tau^2 & \tau+2 \end{bmatrix}. \tag{11}$$

The platonic as well as Archimedean solids possessing the icosahedral symmetry can be constructed as the orbits of the Coxeter group $W(H_3)$. For example, two dual platonic solids, the icosahedron and dodecahedron, are represented by the orbits $O(100)$ and $O(001)$ respectively as shown in Figure 10.

The non-chiral Archimedean solids with the icosahedral symmetry are represented by the orbits $W(H_3)(010)$, $W(H_3)(110)$, $W(H_3)(011)$, $W(H_3)(101)$ and $W(H_3)(111)$. They are respectively (a) icosidodecahedron, (b) truncated dodecahedron, (c) truncated icosahedron, (d) small rhombicosidodecahedron, and (e) great rhombicosidodecahedron as shown in the Figure 11.

There are two more Archimedean solids. They are the chiral solids: the snub cube and the snub dodecahedron. They are generated by the proper subgroups of the octahedral group and the icosahedral group respectively. Construction of their vertices using the Coxeter diagrams of the groups

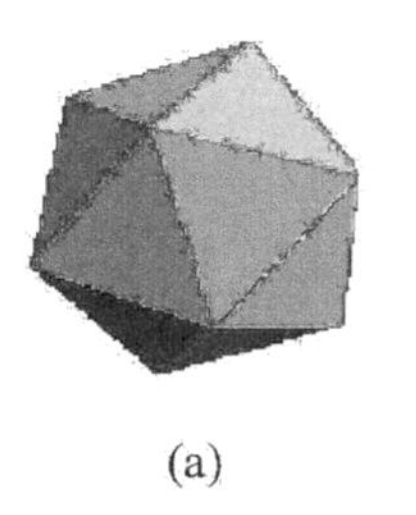
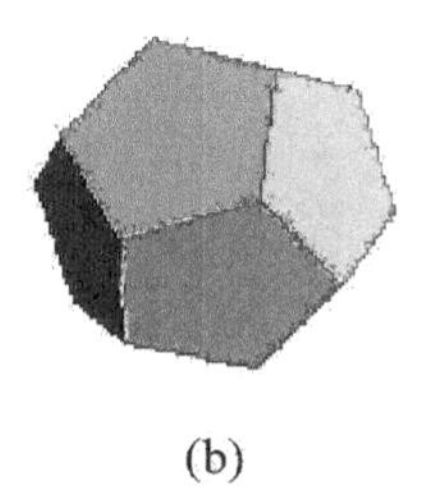

(a) (b)

Fig. 10. The iocahedron (a) and its dual dodecahedron (b) as the respective orbits $W(H_3)(001)$ and $W(H_3)(100)$.

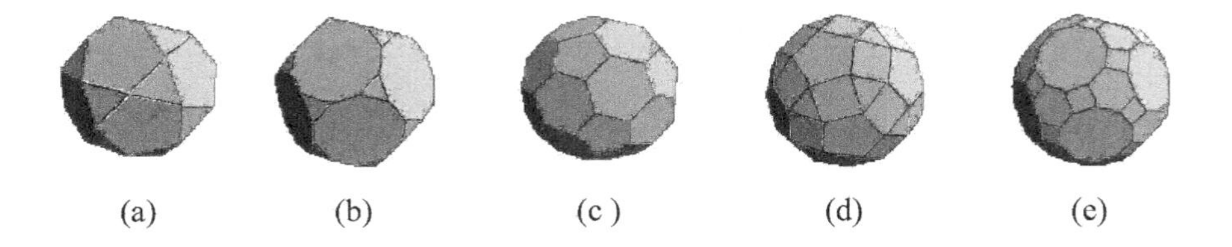

(a) (b) (c) (d) (e)

Fig. 11. The non-chiral Archimedean solids with icosahedral symmetry.

$W(B_3)$ and $W(H_3)$ has been obtained in reference 22. The vertices of the snub cube can be obtained as follows

$$\begin{aligned}(W(B_3))/C_2)\Lambda_I &= \Big\{ \left(\pm xe_1 \pm e_2 \pm x^{-1}e_3\right), \left(\pm xe_2 \pm e_3 \pm x^{-1}e_1\right), \\ &\qquad \left(\pm xe_3 \pm e_1 \pm x^{-1}e_2\right) \Big\} \\ (W(B_3))/C_2)\Lambda_{II} &= \Big\{ \left(\pm e_1 \pm xe_2 \pm x^{-1}e_3\right), \left(\pm e_2 \pm xe_3 \pm x^{-1}e_1\right), \\ &\qquad \left(\pm e_3 \pm xe_1 \pm x^{-1}e_2\right) \Big\}\end{aligned} \tag{12}$$

with $W(B_3)/C_2 = \{[T, \bar{T}] \oplus [T', \bar{T}']\}$ and in terms of quaternionic units two vectors are given by $\Lambda_I = (xe_1 + e_2 + x^{-1}e_3)$ and $\Lambda_{II} = (e_1 + xe_2 + x^{-1}e_3)$ where $x^3 - x^2 - x - 1 = 0, x \approx 1.8393$. The left and the right oriented snub cubes are shown in Figure 12.

Snub dodecahedron can be obtained by the action of the proper subgroup of the icosahedral group to two vectors expressed in terms of quater-

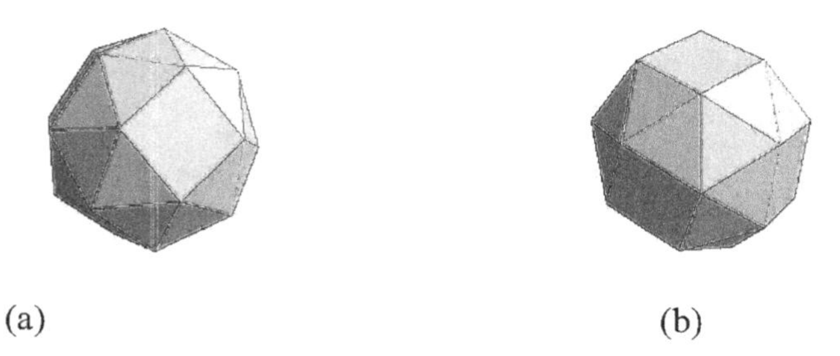

Fig. 12. The snub cube (a) as the orbit $(W(B_3)/C_2)\Lambda_I$ and (b) $(W(B_3)/C_2)\Lambda_{II}$.

nions as follows:

$$\Lambda_I = \frac{1}{\sqrt{2}}\left[\sigma(x^2-1)e_1 - xe_2 + (1-\tau x^3)e_3\right],$$
$$\Lambda_{II} = \frac{1}{\sqrt{2}}\left[-\sigma(x^2-1)e_1 - xe_2 + (1-\tau x^3)e_3\right],$$
$$x^3 - x^2 - x - \tau = 0, \quad x \approx 1.94315. \tag{13}$$

Now the orbits generated by these vectors $(W(H_3)/C_2)\Lambda_I$ and $(W(H_3)/C_2)\Lambda_{II}$ are represented by two snub dodecahedra as shown in Figure 13.

Fig. 13. The snub dodecahedron (a) as the orbit of Λ_I and its mirror image (b) as the orbit of ΛII.

4.4. *Catalan solids*

The Catalan solids are the duals of the Archimedean solids and are extensively studied in reference 15. Here we give a few examples which are quite useful in the crystallography or in the classification of viral structures.

They are the unions of the orbits of the relevant Coxeter groups. The dual of the cuboctahedron possessing the octahedral symmetry is the rhombic dodecahedron. Its 14 vertices are the union of the orbits $W(B_3)(100) \oplus \frac{1}{\sqrt{2}}W(B_3)(001)$ which is represented in Figure 14.

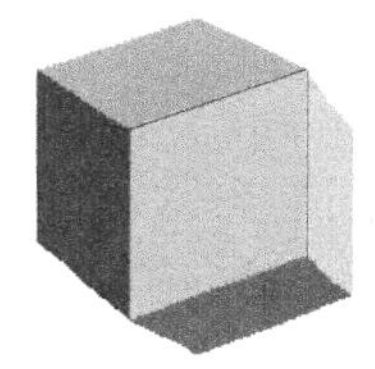

Fig. 14. The rhombic dodecahedron.

The dual of the icosidodecahedron is the rhombic triacontahedron. Its 32 vertices is the union of the orbits $W(H_3)(100) \oplus \tau W(H_3)(001)$ which is displayed in Figure 15.

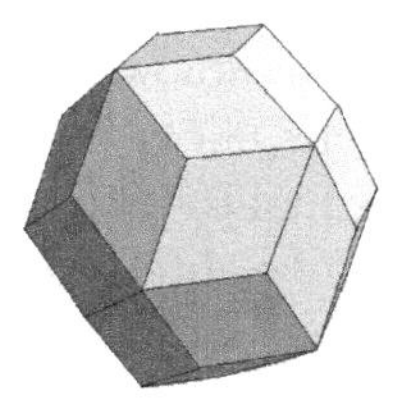

Fig. 15. The rhombic triacontahedron.

The duals of the snub cube and snub dodecahedron have been constructed in the reference 22 using the quaternionic representation of the chiral groups and the vertices of the solids of concern. They are displayed in the Figure 16.

5. Rank-4 Coxeter groups and 4D polytopes

As we have mentioned in the introduction there are five rank-4 groups $W(D_4), W(B_4), W(F_4), W(A_4)$, and $W(H_4)$ describing the regular and semiregular polytopes. We will discuss the group $W(D_4)$ with its relevance to the snub 24-cell later. Now let us start first with the group $W(B_4)$.

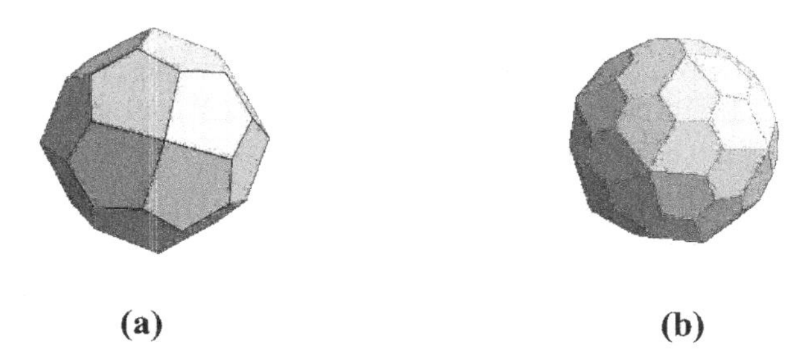

Fig. 16. Dual of the left oriented snub cube (a) and the dual of left oriented snub dodecahedron (b).

5.1. *The Coxeter group $W(B_4)$ of order 384*

This group is an extension of the octahedral group $W(B_3)$ to four dimensions[23]. As it is expected the generalizations of the octahedron and the cube in terms of quaternions are straightforward. The hyper octahedron, also known as (16-cell), is given by the orbit $W(B_4)(1000) = \{\pm 1, \pm e_1, \pm e_2 \pm e_3\}$. It consists of 16 tetrahedra whose vertices are 16 different combinations of the unit quaternions $\{1, e_1, e_2, e_3\}$ including the sign changes. Similarly the hyper cube, known as 8-cell, is represented by the orbit $W(B_4)(1000) = \frac{1}{2}(\pm 1 \pm e_1 \pm e_2 \pm e_3)$. It consists of 8 cubic cells. Hyperoctahedron and hyper cube are dual to each other as expected. The vertices of any quasi regular polytope can be obtained as the orbit $W(B_4)(a_1a_2a_3a_4)$ where $a_i \geq 0, (i = 1, 2, 3, 4)$, is a real number. For the values $a_i = 0$ or 1 many semi regular polytopes have been discussed in the reference 24.

5.2. *The Coxeter group $W(F_4)$ of order 1152*

It is a unique Coxeter group in the sense that it has correspondences neither in three dimensions nor in any other higher dimensions. It has a self dual polytope with 24 vertices with 24 cells made of octahedra. It is simply called 24-cell which is represented in terms of quaternions either by the orbit

$$W(F_4)(1000) = T = \{(\pm 1, \pm e_1, \pm e_2, \pm e_3), \frac{1}{2}(\pm 1 \pm e_1 \pm e_2 \pm e_3)\}$$

or by

$$W(F_4)(0001) = T' = (V_1 \oplus V_2 \oplus V_3).$$

The Dynkin diagram symmetry of F_4 transforms these two sets of vertices to each other. Thus the polytope 24-cell is said to be self dual. The quasi regular and semi regular polytopes can be constructed[25] by the same technique explained as above for the group $W(B_4)$.

5.3. *The Coxeter group* $W(H_4) \subset W(E_8)$

The Coxeter group $W(H_4)$ is a maximal subgroup of the group $W(E_8)$. Its two dual platonic polytopes are called the 120-cell and 600-cell represented by quaternions as follows:

120-cell: $W(H_4)(1000) = \sum_{i,j=0}^{4} \oplus p^i T' p^j$. It has 600 vertices. Here $p \in I$, $p^5 = \pm 1$.

Each cell is a dodecahedron. At each vertex there are 4 dodecahedra.

The 600-cell is represented by the set of quaternions I which also represents the binary icosahedral group:

600-cell: $W(H_4)(0001) = I = \sum_{i=0}^{4} \oplus p^i T$.

Each cell is a tetrahedron. At each vertex there are 20 tetrahedra. When the sphere S^3 representing the vertices of the 600-cell is sliced by parallel hyperplanes one obtains $3D$ polyhedra represented by the conjugacy classes of the binary icosahedral group I in Table 1.

They consist of two points ± 1 corresponding to the poles of the sphere S^3, 4 icosahedra, 2 dodecahedra and one icosidodecahedron which can be readily seen from the Table 1. The 120-cell and 600-cell are dual to each other. In addition to these two platonic solids we will study two more polytopes possessing the symmetry $W(H_4)$. Actually any polytope represented by the orbit $W(H_4)(a_1 a_2 a_3 a_4)$ can be worked out and their branching under its maximal subgroups can be studied[26]. Here we will give the following examples.

720-cell = 120 + 600: This is the orbit represented by $W(H_4)(0100)$ consisting of 1200 vertices. Its cells are made of 120 icosidodecahedra and 600 tetrahedra. At any vertex there join 3 icosidodecahedra and 2 tetrahedra whose centers are represented respectively by the vectors $\omega_4, r_3 r_4 \omega_4, (r_3 r_4)^2 \omega_4$ up to a scale factor and by the vectors $\omega_1, r_1 \omega_1$.

The dual polytope of the 720-cell consists of 720 vertices which is the union of the orbits $W(H_4)(0001) \oplus \frac{2}{3\tau} W(H_4)(1000)$.[26] Each cell of the dual polytope is a dipyramid with an equilateral triangular base. The dual polytope is the projection of the Voronoi cell of E_8 (see Ref. 27) with 19,440 vertices into 4D space. Another semi regular polytope of the Coxeter group $W(H_4)$ has 720 vertices and 720 cells which we denote by $720' - cell = 120 + 600$. The $720' - cell$ is represented by the orbit $W(H_4)(0010)$. It consists of the cells of 120 icosahedra and 600 octahedra and with this property it is the only Archimedean solid with the $W(H_4)$

symmetry. At each vertex there are 2 icosahedra and 5 octahedra. Their centers can be represented respectively by $\omega_4, r_4\omega_4$ up to a scale factor $\frac{2\tau}{2+\sigma}$ and $\omega_1, r_1r_2\omega_1, (r_1r_2)^2\omega_1, (r_1r_2)^3\omega_1, (r_1r_2)^4\omega_1$. These 7 vertices form a dipyramid with a pentagonal base. The 720 vertices of the dual polytope consist of the set $\frac{2\tau}{2+\sigma}W(H_4)(0001) \oplus W(H_4)(1000)$.

It is also interesting to note that the symmetries of the semi regular 4D polytopes, the grand antiprism and the snub 24-cell, are represented by two maximal subgroups of the Coxeter group $W(H_4)$. We display the maximal subgroups of the Coxeter group $W(H_4)$ in Table 2.[18,21]

Table 2. Maximal subgroups of the Coxeter group $W(H_4)$.

Maximal subgroup	Order	Polytope
$Aut(A_2 \oplus A_2)$	144	
$Aut(H_2 \oplus H_2)$	400	Grand antiprism with 100 vertices
$W(H_3 \oplus A_1)$	240	
$Aut(A_4)$	240	
$(W(D_4)/C_2) : S_3$	576	Snub 24-cell with 96 vertices

One of these semi regular polytope is the Grand Antiprism (GA) which was first determined by a computer calculation by Conway and Guy in 1965[28]. This semi regular polytope has 100 vertices which can be obtained as a subset of the set of quaternions I (600-cell) under the decomposition of its maximal subgroup $Aut(H_4 \oplus H_2)$.[18] Let us take the quaternions $b = \frac{1}{2}(\tau + \sigma e_1 + e_2) \in 12(1)_+$ and $c = \frac{1}{2}(\tau - \sigma e_1 + e_2) \in 12(1)_+$. Then the vertices of the GA are given by the set of quaternions

$$GA = \{b^m c b^n, b^m e_3 b^n\}, \quad m, n = 0, 1, ..., 9 \tag{14}$$

which consists of 100 elements of the set of quaternions I from which the set of 20 quaternionic roots of the Coxeter diagram $H_2 \oplus H_2' = \{b^m, e_3 b^m\}; m = 0, 1, ..., 9$ are removed. The GA has 100 vertices, 320 cells (20 pentagonal antiprisms and 300 tetrahedra), 500 edges and 720 faces (700 triangles and 20 pentagons). The dual polytope of the GA consists of 320 vertices which are given as the orbits of the group $Aut(H_2 \oplus H_2)$ as follows:

$$\text{200 vertices like } J_1 = \sum_{i.j=0}^{4} \oplus b^i V_1 b^j;$$

$$\text{100 vertices like } J_3' = \sum_{i.j=0}^{4} \oplus b^i \frac{1}{\sqrt{2}}(\pm 1 \pm e_3) b^j \quad \text{and}$$

$$\text{20 vertices like} \left\{ \frac{\tau}{\sqrt{2}} b^m, \frac{\tau}{\sqrt{2}} e_3 b^m \right\};\ m = 0, 1, ..., 9. \tag{15}$$

The typical cell of the dual polytope has 14 vertices with faces of 4 pentagons, 4 kites and 2 isosceles trapezoids as shown in Figure 17.

Fig. 17. A typical cell of the dual polytope of the grand antiprism.

Another interesting semi regular polytope is the snub 24-cell. The snub 24-cell[29] consists of 96 vertices $S = I - T = \sum_{i=1}^{4} p^i T$ with the symmetry group $(W(D_4)/C_2) : S_3 = \{[T,T] \oplus [T,T]^*\}$ of order 576. Its 144 cells consist of 24 icosahedra and 120 tetrahedra. The dual of the snub 24-cell has the vertices $T \oplus T' \oplus S'$ with $S' = \sum_{i=1}^{4} p^i \bar{p}^{\dagger i} T'$ where $p^\dagger$ is obtained from p by exchanging $\tau \leftrightarrow \sigma$. One cell of the dual polytope is a solid with 8 vertices as shown in Figure 18.

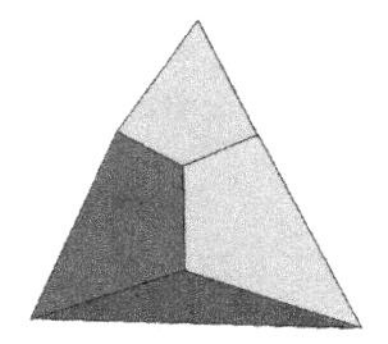

Fig. 18. A typical cell of the dual polytope of the snub 24-cell.

5.4. *The Coxeter group* $W(A_4)$ *of order 120*

The automorphism group $Aut(A_4) \approx W(A_4) : \gamma$ is an extension of the Coxeter group $W(A_4)$ by the Dynkin diagram symmetry γ. It is a maximal subgroup of the group $W(H_4)$ and has its own Platonic and Archimedean polytopes. We discuss the following examples. The 5-cell is an extension of tetrahedron to 4D and can be represented either by the orbit $W(A_4)(1000)$ or $W(A_4)(0001)$. It has 5 vertices consisting of 5 tetrahedra, 4 of which are meeting at one vertex. Similarly the orbit $W(A_4)(0100)$ represents the rectified 5-cell consisting of 5 tetrahedral and 5 octahedral cells. The rectified cell has 10 vertices. At each vertex there are 2 tetrahedra and 3 octahedra whose centers respectively consist of the vertices $\omega_1, r_1\omega_1$ and $\omega_4, r_3r_4\omega_4, (r_3r_4)^2\omega_4$ up to a scale factor. These vertices form a dipyramid which is a typical cell of the dual polytope[30,31]. The vertices of the dual polytope consist of the union of the orbits $W(A_4)(1000) \oplus \frac{3}{2}W(A_4)O(0001)$. Another polytope with the symmetry group $Aut(A_4)$ is called runcinated 5-cell consisting of 20 vertices and 30 cells composed of 10 tetrahedra and 20 triangular prisms. It is represented by the orbit $W(A_4)(1001)$. 2 tetrahedra and 6 triangular prisms join to each vertex. The dual polytope has 30 vertices consisting of the union of the orbits

$$W(A_4)(1000) \oplus W(A_4)(0100) \oplus W(A_4)(0010) \oplus W(A_4)(0001).$$

The dual polytope has 20 cells each is made of rhombohedron as shown in Figure 19. The rhombohedron has the symmetry of $D_3 \times C_2$.

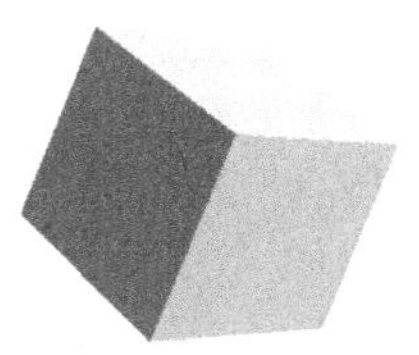

Fig. 19. The rhombohedron, one of the typical cell of the dual polytope of the polytope $W(A_4)(1001)$.

6. Quasi regular polygonal tiling with the Coxeter symmetry D_n

Before we conclude we would also show how the plane can be tiled with the isogonal and isotoxal hexagons with the dihedral symmetry D_3. The rank-2 Coxeter groups are the dihedral groups D_n which can be represented by the Coxeter diagram shown in Figure 20.

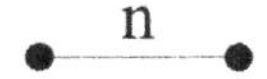

Fig. 20. The Coxeter diagram representing the dihedral symmetry.

The regular and quasi regular polygons possessing the dihedral symmetry D_n has been worked out in the references 32 & 33. Here we give isogonal and isotoxal hexagons having the symmetry D_3 and the tiling of the plane with them. An isogonal hexagon has two types of lengths with 120° interior angles as shown in Figure 21(a). Its dual isotoxal hexagon has all edge lengths the same but two different interior angles alternating at each vertex and complementing each other to 240°. It is shown in Figure 21(b).

Fig. 21. The isogonal hexagon (a) and its dual isotoxal hexagon (b) possessing the dihedral symmetry D_3.

The tilings with isogonal hexagons are shown in Figure 22.
Tiling of the plane with isotoxal hexagons and regular hexagons are shown in Figure 23.

7. Concluding remarks

We have presented a few examples of the 4D polytopes along with their duals. Their cells are made of Platonic and Archimedean polyhedra. A systematic study of the 4D polytopes and their dual polytopes will follow.The 4D polytopes and their duals with $W(A_4)$ symmetry has been already worked out in the reference 30 where we also projected the polytopes into 3D with the symmetry $W(A_3)$. The analyses related with the

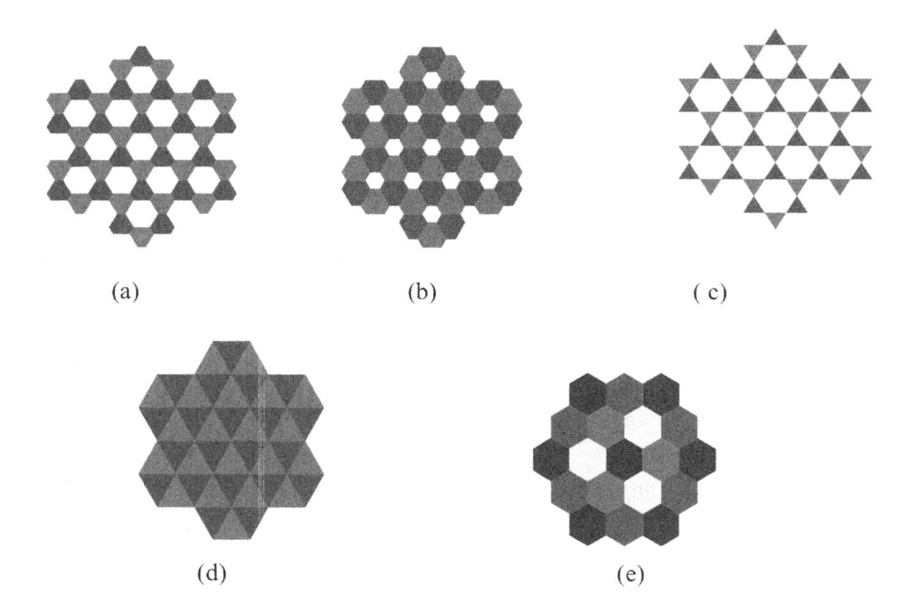

Fig. 22. Tiling of the plane with isogonal hexagons-regular hexagons (a-b), regular hexagons and triangles (c), tiling with triangles (d) and the honeycomb lattice (e).

Fig. 23. Tiling of the plane with isotoxal and regular hexagons.

$W(H_4), W(B_4)$, and $W(F_4)$ symmetries are under preparation[26,34,35]. The tilings of the plane with isogonal and isotoxal hexagons could be interesting for the graphene physics if short and long bonds of carbon atoms are taken into account.

References

1. A.B. Zamolodchikov, *Int.J. Mod. Phys.A* **4**, 4235 (1989).

2. T.J.Hollowood and P. Mansfield, *Phys.Lett. B* **224**, 373 (1989); M. Koca and G. Mussardo, *Int. J. Mod. Phys. A* **6**, 1543 (1991).
3. B. Kostant, *Sel. Math. New Ser.* **16** 416 (2010); arXiv:1003.0046v1.
4. M. Koca, R. Koç and M. Al-Barwani, *J. Phys. A: Math. General* **34**, 11201 (2001); M. Koca and N. O. Koca, 'The role of H4 in E8 in relating the experimental masses to the radii of the Gosset circles' (in preparation).
5. R. Coldea et al., *Science* **327**, 177 (2010).
6. J. A. Kjall, F. Pollmann and J.E. Moore, *Phys. Rev. B* **83**, 020407(R) (2011).
7. M. Atiyah and P. Sutcliffe, *Milan Journal of Math.* **71**, 33 (2003); arXiv:math-ph/0303071 v1 (Mar. 2003).
8. E. Catalan, *J. l'École Polytechnique (Paris)* **41**, 1 (1865).
9. L. Schlaffli, *Journal de Mathematiques* **20**, 359 (1855).
10. T. Gosset, *Messenger of Mathematics*, **29** 43 (1900).
11. W. A. Wythoff, *Koninklijke Akademie van Wetenshappen te Amsterdam, Proc. of the Section of Sciences* **9** 529 (1907).
12. H. S. M. Coxeter, *Regular Polytopes*, (Third Edition, Dover Publications, New York, 1973).
13. J. H. Conway, H. Burgiel and C. Goodman-Strauss, *The Symmetries of Things*, (A K Peters Ltd. Wellesley, Massachusetts, 2008).
14. M. Koca, R. Koç R and M. Al-Ajmi, *J. Math. Phys.* **48**, 113514 (2007).
15. M. Koca, N. O. Koca and R. Koç, *J. Math. Phys.* **51**, 043501 (2010).
16. J.H. Conway and D.A. Smith, *On Quaternions and Octonions* (A K Peters Ltd, Natick, Massachusetts 2003); M. Koca, R. Koç and M. Al-Barwani, *J. Phys. A: Math. Gen.* **34**, 11201 (2001).
17. H.S.M. Coxeter, *Regular Complex Polytopes* (Cambridge Univ. Press, 1974); P. du Val, *Homographies, Quaternions and Rotations* (Oxford University Press, 1964).
18. M. Koca, N.O.Koca and M. Al-Ajmi, *J. Phys. A: Math. Theor*, **42**, 495201 (2009).
19. M. Koca, R. Koç and M. Al-Barwani, *J. Math. Phys.* **44**, 3123 (2003); ibid **47**, 04307 (2006).
20. M. Koca, R. Koç and M. Al-Ajmi, *J. Phys. A: Math. Theor*, **40**, 7633 (2007).
21. M. Koca, R. Koç, M. Al-Barwani and S. Al-Farsi *Linear Algebra Appl.* **412**, 441 (2006).
22. M. Koca, N.O. Koca and M. Al-Suheili, *SQU Journal for Science*, **16** 63 (2011); arXiv: 1006.3149.
23. M. Koca, R. Koç and M. Al-Barwani, *J. Math. Phys.* **47**, 3123 (2003).
24. M. Koca, M. Al-Ajmi and R. Koç, *African Phys. Rev.* **2**, 0010 (2008).
25. M. Koca, M. Al-Ajmi and R. Koç, *African Phys. Rev.* **2**, 0009 (2008).
26. M. Koca, M. Al-Ajmi, N. O. Koca arXiv: 1106.3433 (2011) [to be published in *Tr. J. Phys.* (2012)].
27. V. Elser and N. J. A. Sloane, *J. Phys. A: Math. Gen.* **20**, 6161 (1987).
28. J. H. Conway and M. J. T. Guy, *Proc. of the Colloquium on Convexity at Copenhagen*, p.38, (Kobenhavns University Mathematics Institute, Copenhagen, 1967).
29. M. Koca, M. Al-Ajmi M and N. O. Koca, *Linear Algebra Appl.* **434** 977

(2011).
30. M. Koca, N.O. Koca and M. Al-Ajmi, *IJGMMP* **9** 1250035 (2012); arXiv: 1102.1132. (submitted for publication).
31. M Koca, *J.Phys. Conf. Ser.* **284**, 012040 (2011).
32. M. Koca and N. O. Koca, *J.Phys. Conf. Ser.* **284**, 012039 (2011).
33. M. Koca, N.O. Koca and R. Koç, *The African Review of Phys.* **6**:0006, 41 (2011); arXiv: 1006.2434.
34. M. Koca, M. Al-Ajmi M and N. O. Koca, 'The polytopes with $W(B_4)$ symmetry and their projections in 3D' (under preparation).
35. M. Koca, M. Al-Ajmi M and N. O. Koca, 'Quaternionic Construction of the $W(F_4)$ Polytopes with Their Dual Polytopes and Branching under the Subgroups $B(B_4)$ and $W(B_3)\times W(A_1)$'; arXiv: 1203.4574 (2012) (submitted for publication).

HERMITE-BERNOULLI 2D POLYNOMIALS

BURAK KURT

Department of Mathematics, Faculty of Education,
University of Akdeniz,
TR-07058 Antalya, Turkey
E-mail: burakkurt@akdeniz.edu.tr

VELİ KURT

Department of Mathematics, Faculty of Science,
University of Akdeniz,
TR-07058 Antalya, Turkey
E-mail: vkurt@akdeniz.edu.tr

In this work, we introduce and investigate the Hermite-Apostol-Bernoulli polynomials by means of a suitable generating function. We establish several interesting properties of these general polynomials. Furthermore, we prove Raabe relation and give some relations between 2-dimensional Hermite polynomials and Apostol-Bernoulli polynomials.

Keywords: Apostol-Bernoulli polynomials; Hermite polynomials; Generalized Hermite polynomials; 2D Hermite polynomials; Hermite-Apostol-Bernoulli polynomials.

1. Introduction, Definitions and Notations

The Bernoulli polynomials have been studied since the 18th century. There are many applications of these numbers and polynomials in mathematics and in physics. Many functions are used to obtain generating functions of the Euler polynomials, Bernoulli polynomials and Genocchi polynomials.

Bernoulli polynomials are defined by the following generating function

$$\sum_{n=0}^{\infty} B_n(x)\frac{t^n}{n!} = \frac{te^{xt}}{e^t - 1}, \ |t| < 2\pi. \tag{1}$$

For $x = 0$, $B_n(0) = B_n$; i.e. the Bernoulli numbers are defined as follows

$$\sum_{n=0}^{\infty} B_n\frac{t^n}{n!} = \frac{t}{e^t - 1}, \ |t| < 2\pi. \tag{2}$$

Some of these Bernoulli polynomials and Bernoulli numbers are given respectively as follows. For $n \in \mathbb{Z}^+$ we have

$$B_0(x) = 1,\ B_1(x) = x - \frac{1}{2},\ B_2(x) = x^2 - x + \frac{1}{6},\ B_3(x) = x^3 - \frac{3}{2}x^2 + \frac{1}{2}x, \cdots$$

and

$$B_0 = 1, B_1 = -\frac{1}{2}, B_2 = \frac{1}{6}, B_3 = 0, B_4 = -\frac{1}{30},$$
$$B_5 = 0, B_6 = \frac{1}{42}, \cdots, B_{2k+1} = 0$$

for detail cf. Refs. 1–16 .

The following equalities can be obtained from (1) and (2) easily

$$B_n(x) = \sum_{k=0}^{n} \binom{n}{k} B_k x^{n-k},\ n \geqslant 0, \quad B_n(0) = B_n(1) = B_n,$$

$$B_n(x+1) - B_n(x) = nx^{n-1},\ n \geqslant 1, \quad B_n'(x) = nB_{n-1}(x).$$

On the other hand, Bernoulli polynomials satisfy the following differential equation[12]

$$\frac{B_{n+1}}{n+1}y^{(n+1)} + \frac{B_n}{n}y^{(n)} + \frac{B_{n-1}}{n-1}y^{(n-1)} + \cdots + \frac{B_2}{2}y'' + \left(\frac{1}{2} - x\right)y' + ny = 0.$$

In Refs. 1 & 2, Bretti et al. worked on generalizations of the Bernoulli polynomials. They gave some examples from Appell polynomials. They defined the multidimensional extension of the Bernoulli polynomials. In Refs. 3–6 Dattoli et al., defined Hermite-Bernoulli polynomials. They gave some relations. Dadonov[7] gave integral representation for the two-dimensional Hermite polynomials. He investigated some properties of this representation. Khan et al.[8,9] defined Hermite-based Appell polynomials. Some theorems are proven by them. Luo[10], Srivastava et al.[13], and Wang et al.[15] investigated Apostol-Bernoulli polynomials. At last, Wünsche[17–19] defined the general Hermite-Laguerre two-dimensional polynomials. He proved the orthonormalization and completness relations. He also applied Hermite two-dimensionals polynomial to Jacobi polynomials.

The α-order Apostol-Bernoulli polynomials and α-order Apostol-Bernoulli numbers are defined respectively by[10,13,15]

$$\sum_{n=0}^{\infty} B_n^{\alpha}(x, \lambda)\frac{t^n}{n!} = \left(\frac{t}{\lambda e^t - 1}\right)^{\alpha} e^{xt},\ |t + \ln \lambda| < 2\pi \tag{3}$$

and

$$\sum_{n=0}^{\infty} B_n^{\alpha}(\lambda)\frac{t^n}{n!} = \left(\frac{t}{\lambda e^t - 1}\right)^{\alpha}, \ |t + \ln \lambda| < 2\pi. \tag{4}$$

Some of these Apostol-Bernoulli polynomials and Apostol-Bernoulli numbers are given respectively as follows. For $n \in \mathbb{Z}^+$, $\alpha = 1$ we have

$$B_0(x,\lambda) = 0,\ B_1(x,\lambda) = \frac{1}{\lambda - 1},\ B_2(x,\lambda) = \frac{2}{\lambda - 1}\left(x - \frac{\lambda}{\lambda - 1}\right)$$

$$B_3(x,\lambda) = \frac{3x^2}{\lambda - 1} - \frac{6\lambda}{(\lambda-1)^2}\left(x - \frac{\lambda}{\lambda - 1}\right) - \frac{3\lambda}{(\lambda-1)^2}, \ \cdots$$

and

$$B_0(\lambda) = 0,\ B_1(\lambda) = \frac{1}{\lambda - 1},\ B_2(\lambda) = \frac{-2\lambda}{(\lambda-1)^2},\ B_3(\lambda) = \frac{3\lambda\,(\lambda - 1)}{(\lambda-1)^3},\ \cdots .$$

Hermite polynomials with one variable are defined as follows

$$\sum_{n=0}^{\infty} H_n(x)\frac{t^n}{n!} = e^{2xt+t^2}. \tag{5}$$

Hermite polynomial is a solution of Hermite-differential equation

$$y''(x) - 2xy'(x) + 2\lambda y(x) = 0. \tag{6}$$

From (5), we write

$$H_n(x) = \sum_{m=0}^{[n/2]} \frac{(-1)^m n!}{m!(n-2m)!}. \tag{7}$$

There are different solutions of (6) in references 11 and 12. Also the general solution of the inhomogeneous Hermite differential equation of the form

$$y''(x) - 2xy'(x) + 2\lambda y(x) = \sum_{m=0}^{\infty} a_m x^m$$

is given by Jung in Ref. 11.

Two-variable Hermite polynomials are defined by Refs. 3,5,7,8,11,17–19

$$\sum_{n=0}^{\infty} H_n^{(2)}(x,y)\frac{t^n}{n!} = e^{xt+yt^2}. \tag{8}$$

From (8), we have easily

$$\frac{\partial}{\partial y} H_n^{(2)}(x,y) = \frac{\partial^2}{\partial x^2} H_n^{(2)}(x,y).$$

Also from (8), we write

$$H_n^{(2)}(x,y) = \sum_{m=0}^{[n/2]} \frac{n! y^r x^{n-2r}}{r!(n-2r)!}. \tag{9}$$

On the other hand, Hermite polynomial satisfies the following differential equation which is known as heat equation[5]

$$\begin{cases} \frac{\partial}{\partial y} F(x,y) = \frac{\partial^2}{\partial x^2} F(x,y) \\ F(x,0) = x^n \end{cases}.$$

By using the following definition[14] of usual Hermite polynomials $H_m(x)$

$$\begin{aligned} H_m(x) &= \exp\left(-\frac{1}{4}\frac{\partial^2}{\partial x^2}\right)(2x)^m \\ &= \sum_{k=0}^{[m/2]} \frac{(-1)^k m!}{k!(m-2k)!}(2x)^{m-2k}, \end{aligned}$$

the simplest kind of Hermite 2D polynomial is then the product of two usual Hermite polynomials $H_m(x)$ and $H_n(y)$ of two different variables x and y according to

$$\begin{aligned} H_{m,n}(I;x,y) &= \exp\left(-\frac{1}{4}\left(\frac{\partial^2}{\partial x^2}+\frac{\partial^2}{\partial y^2}\right)\right)(2x)^m(2y)^n \\ &= H_m(x)H_n(y) \end{aligned}$$

where I denotes the 2D identity matrix. The general Hermite 2D polynomials can be introduced using a general 2D matrix U in the following way

$$\begin{aligned} H_{m,n}(U;x,y) &= \exp\left(-\frac{1}{4}\left(\frac{\partial^2}{\partial x^2}+\frac{\partial^2}{\partial y^2}\right)\right)\left(2x'\right)^m\left(2y'\right)^n \\ &= 2^{m+n}\exp\left(-\frac{1}{4}\left(\frac{\partial^2}{\partial x^2}+\frac{\partial^2}{\partial y^2}\right)\right)(U_{11}x+U_{12}y)^m(U_{21}x+U_{22}y)^n \end{aligned}$$

where

$$U = \begin{bmatrix} U_{11} & U_{12} \\ U_{21} & U_{22} \end{bmatrix}.$$

The advantage of using the last equation of the general Hermite 2D polynomials is the linear transformations of the vector components (x,y) of 2D vector given in the representation theory of the two-dimensional general linear group $GL(2,\mathbb{C})$. There are a lot of applications of 2D Hermite polynomials and 2D Laguerre polynomials in physics, especially in harmonic oscillator, quantum optics, classical optics.

2. Hermite-Bernoulli Polynomials, Hermite-Apostol-Bernoulli Polynomials

Definition 2.1. Hermite-Bernoulli polynomials and Hermite-Bernoulli numbers are defined respectively as[4]

$$\sum_{n=0}^{\infty} \left({}_HB_n(x,y)\right) \frac{t^n}{n!} = \frac{t}{e^t-1} e^{xt+yt^2}, \ |t| < 2\pi \tag{10}$$

and

$$\sum_{n=0}^{\infty} \left({}_HB_n\right) \frac{t^n}{n!} = \frac{t}{e^t-1}, \ |t| < 2\pi. \tag{11}$$

From (8), (10) and (11), we have

$$\begin{aligned} {}_HB_n(x,y) &= \sum_{l=0}^{n} \binom{n}{l} \left({}_HB_{n-l}\right) H_l(x,y) \\ &= \sum_{l=0}^{n} \binom{n}{l} \left({}_HB_l\right) H_{n-l}(x,y) \end{aligned} \tag{12}$$

and

$${}_HB_n(mx,y) = \sum_{k=0}^{n} \binom{n}{k} \left({}_HB_k(x,y)\right) \left((m-1)\,x\right)^{n-k}. \tag{13}$$

Also, we can write

$$\frac{\partial}{\partial x} \left({}_HB_n(x,y)\right) = n \left({}_HB_{n-1}(x,y)\right),$$

$$\frac{\partial}{\partial y} \left({}_HB_n(x,y)\right) = n\,(n-1) \left({}_HB_{n-2}(x,y)\right),$$

and

$$\frac{\partial}{\partial y} \left({}_HB_n(x,y)\right) = \frac{\partial^2}{\partial x^2} \left({}_HB_n(x,y)\right).$$

On the other hand, from (1) and (10) we have

$${}_HB_n(x,y) = n! \sum_{h=0}^{n} \frac{B_{n-h}}{(n-h)!} \sum_{s=0}^{[h/2]} \frac{x^{n-2s} y^s}{s!(n-2s)!}.$$

Some Hermite-Bernoulli polynomials are given by

$${}_HB_0(x,y) = 1, \ \ {}_HB_1(x,y) = \frac{1}{2}, \ {}_HB_2(x,y) = -2\left(-\frac{5}{12} + \frac{x^2}{2} + y\right)$$

$${}_HB_3(x,y) = 3!\left(\frac{B_3}{3!} + \frac{1}{12} - \frac{1}{2}\left(\frac{x^2}{2} + y\right) + \left(x^3 + y^2\right)\right), \ \cdots .$$

Theorem 2.1. *Hermite-Bernoulli polynomials satisfy the following differential equation*

$$\left[\frac{B_n}{n!}D_x^n + \cdots + \frac{B_4}{4!}D_x^4 + \frac{B_3}{3!}D_x^3 + \left(\frac{B_2}{2!} - 2y\right)D_x^2 + \left(\frac{1}{2} - x\right)D_x + n\right]B_n^2(x,y) = 0,$$

$$\left[\left(x - \frac{1}{2}\right)D_y + 2D_x^{-1}D_y + 2yD_x^{-1}D_y^2 - \sum_{k=0}^{n-1}\frac{B_{n-k+1}}{(n-k+1)!}D_x^{-(n-k)}D_y^{n-k+1} - (n+1)D_x\right]B_n^2(x,y) = 0.$$

The proof of this theorem is given by Bretti and Ricci[2].

We consider the definition of Hermite-Apostol-Bernoulli polynomial and Hermite-Apostol-Bernoulli number.

Definition 2.2. For arbitrary real or complex parameters α and λ, Hermite-Apostol-Bernoulli polynomials of order α and Hermite-Apostol-Bernoulli numbers of order α are defined respectively as

$$\sum_{n=0}^{\infty}\left({}_HB_n^{(\alpha)}(x,y;\lambda)\right)\frac{t^n}{n!} = \left(\frac{t}{\lambda e^t - 1}\right)^{\alpha} e^{xt+yt^2}, \quad |t + \log\lambda| < 2\pi \tag{14}$$

and

$$\sum_{n=0}^{\infty}\left({}_HB_n^{(\alpha)}(\lambda)\right)\frac{t^n}{n!} = \left(\frac{t}{\lambda e^t - 1}\right)^{\alpha}, \quad |t + \log\lambda| < 2\pi. \tag{15}$$

From (8), (14) and (15) we have

$${}_HB_n^{(\alpha)}(x,y;\lambda)) = \sum_{l=0}^{n}\binom{n}{l}\left({}_HB_{n-l}^{(\alpha)}\right)H_l^2(x,y). \tag{16}$$

Substituting $\alpha = 1$ into Definition 2, we have

$$\begin{aligned}
\sum_{n=0}^{\infty} \left({}_H B_n(x,y;\lambda) \right) \frac{t^n}{n!} &= \frac{t}{\lambda e^t - 1} e^{xt+yt^2} \\
&= \left(\sum_{k=0}^{\infty} B_n(x,\lambda) \frac{t^k}{k!} \right) \left(\sum_{r=0}^{\infty} \frac{y^r t^{2r}}{r!} \right) \\
&= \sum_{k=0}^{\infty} \sum_{r=0}^{\infty} \frac{B_n(x,\lambda) y^r t^{k+2r}}{k! r!} \\
&= \sum_{n=0}^{\infty} \left(\sum_{r=0}^{\left[\frac{n}{2}\right]} \frac{B_{n-2r}(x,\lambda) y^r}{r!(n-2r)!} \right) \frac{t^n}{n!}.
\end{aligned}$$

Comparing the coefficients $\frac{t^n}{n!}$ on both sides of the above equation, we have

$$ {}_H B_n(x,y;\lambda) = \sum_{r=0}^{[n/2]} \frac{B_{n-2r}(x,\lambda) y^r}{r!(n-2r)!}. $$

By using Apostol-Bernoulli polynomials, we can find some of the Hermite-Apostol-Bernoulli polynomials as follows

$$\begin{aligned}
&{}_H B_0(x,y;\lambda) = 1, \ {}_H B_1(x,y;\lambda) = B_1(x,\lambda) = \frac{1}{\lambda - 1}, \\
&{}_H B_2(x,y;\lambda) = 2! \left(B_2(x,\lambda) + B_0(x,\lambda) y \right) = 2! \left(\frac{2x}{\lambda - 1} - \frac{2\lambda}{\lambda - 1} \right), \\
&{}_H B_3(x,y;\lambda) = \frac{1}{\lambda - 1} \left(3x^2 - \frac{6\lambda}{\lambda - 1} \left(x - \frac{\lambda}{\lambda - 1} \right) - \frac{3\lambda}{\lambda - 1} + 6y \right), \ \cdots .
\end{aligned}$$

Proposition 2.1. *Hermite-Apostol-Bernoulli polynomials satisfy the following equation*

$$ {}_H B_n^{(\alpha+\beta)}(x,y;\lambda) = \sum_{l=0}^{n} \binom{n}{l} \left({}_H B_n^{(\alpha)}(\lambda) \right) \left({}_H B_{n-l}^{(\beta)}(x,y;\lambda) \right). \tag{17} $$

Proof. It follows from (14)

$$\begin{aligned}
\sum_{n=0}^{\infty} \left({}_H B_n^{(\alpha+\beta)}(x,y;\lambda) \right) \frac{t^n}{n!} &= \left(\frac{t}{\lambda e^t - 1} \right)^{\alpha+\beta} e^{xt+yt^2} \\
&= \left(\left(\frac{t}{\lambda e^t - 1} \right)^{\alpha} e^{xt+yt^2} \right) \left(\frac{t}{\lambda e^t - 1} \right)^{\beta}
\end{aligned}$$

$$= \left(\sum_{n=0}^{\infty} \left({}_H B_n^{(\alpha)}(x, y; \lambda)\right) \frac{t^n}{n!}\right) \left(\sum_{n=0}^{\infty} \left({}_H B_n^{(\beta)}(\lambda)\right) \frac{t^n}{n!}\right)$$
$$= \sum_{n=0}^{\infty} \left(\sum_{l=0}^{n} \binom{n}{l} \left({}_H B_n^{(\alpha)}(\lambda)\right) \left({}_H B_{n-l}^{(\beta)}(x, y; \lambda)\right)\right) \frac{t^n}{n!}.$$

Comparing the coefficients of $\frac{t^n}{n!}$ on both sides of the above equation, we have (17). □

Corollary 2.1. *There is a following relation between Hermite-Apostol-Bernoulli number and two-variable Hermite polynomials*

$${}_H B_n^{(\alpha+\beta)}(x, y; \lambda) = \sum_{k=0}^{n} \binom{n}{k} \left({}_H B_n^{(\alpha+\beta)}(\lambda)\right) H_{n-k}^{(2)}(x, y).$$

Proposition 2.2. *There is a following relation for the Hermite-Apostol-Bernoulli polynomials*

$$\lambda_H B_n^{(\alpha)}(x+1, y; \lambda) -_H B_n^{(\alpha)}(x, y; \lambda) = n_H B_{n-1}^{(\alpha-1)}(x, y; \lambda).$$

We put $\alpha = 1$, $y = 0$ in last equation. We have the classical result as

$$\lambda_H B_n(x+1; \lambda) -_H B_n(x; \lambda) = nx^{n-1}.$$

Proposition 2.3. *Analogue of the Raabe relation for the Hermite-Apostol-Bernoulli polynomials is*

$${}_H B_n^{(\alpha)}(mx, my; \lambda) = \sum_{l=0}^{n} \binom{n}{l} \left({}_H B_{n-l}^{(\alpha)}(x, y; \lambda)\right) H_l^{(2)}((m-1)x, (m-1)y).$$

Theorem 2.2. *There is a different relation between Hermite-Apostol-Bernoulli polynomials and two-variable Hermite polynomials as*

$${}_H B_{n+l}^{(l)}(x, y; \lambda) = (-1)^l \sum_{m=0}^{\infty} \binom{m+l-1}{m} \sum_{k=0}^{n} \binom{n}{k} H_k^{(2)}(x, y) m^{n-k} \binom{n+l}{l} l!. \tag{18}$$

Proof. It follows from (14)

$$\sum_{n=0}^{\infty}\left({}_{H}B_{n}^{(l)}(x,y;\lambda)\right)\frac{t^{n}}{n!}=\left(\frac{t}{\lambda e^{t}-1}\right)^{l}e^{xt+yt^{2}}$$
$$=(-1)^{l}t^{l}\left(1-\lambda e^{t}\right)^{-l}\sum_{k=0}^{\infty}H_{k}^{(2)}(x,y)\frac{t^{k}}{k!}$$
$$=(-1)^{l}t^{l}\sum_{m=0}^{\infty}\binom{m+l-1}{m}\lambda^{m}e^{tm}\sum_{k=0}^{\infty}H_{k}^{(2)}(x,y)\frac{t^{k}}{k!}$$
$$=(-1)^{l}t^{l}\sum_{m=0}^{\infty}\binom{m+l-1}{m}\lambda^{m}\left(\sum_{k=0}^{\infty}\frac{m^{k}t^{k}}{k!}\right)\left(\sum_{k=0}^{\infty}H_{k}^{(2)}(x,y)\frac{t^{k}}{k!}\right)$$
$$=\sum_{n=0}^{\infty}\left((-1)^{l}\sum_{m=0}^{\infty}\binom{m+l-1}{m}\lambda^{m}\sum_{k=0}^{n}\binom{n}{k}H_{k}^{(2)}(x,y)m^{n-k}\right)\frac{t^{n+l}}{l!}$$
$$=\sum_{n=0}^{\infty}\left((-1)^{l}\sum_{m=0}^{\infty}\binom{m+l-1}{m}\lambda^{m}\sum_{k=0}^{n}\binom{n}{k}H_{k}^{(2)}(x,y)m^{n-k}\binom{n+l}{l}l!\right)$$
$$\times\frac{t^{n+l}}{(n+l)!}.$$

From the relations

$${}_{H}B_{n}^{(l)}(x,y;\lambda)=0,\ {}_{H}B_{n+1}^{(l)}(x,y;\lambda)=0,\ \cdots,\ {}_{H}B_{n+l-1}^{(l)}(x,y;\lambda)=0,$$

comparing the coefficients of $\frac{t^n}{n!}$ on both sides of the above equation, we have (18). □

Acknowledgments

The present investigation was supported by the Scientific Research Project Administration of Akdeniz University.

References

1. G. Bretti, P. Natalini and E. P. Ricci, *Abstr. Appl. Anal.* **2004**, 613 (2004).
2. G. Bretti and E. P. Ricci, *Taiwanese J. Math.* **8**, 415 (2004).
3. G. Dattoli, *J. Math. Anal. Appl.* **284**, 447, (2003).
4. G. Dattoli, S. Lorenzutta and C. Cesarano, *Rend. Mat. Appl. (7)* **19**, 385 (1999).
5. G. Dattoli and S. Khan, *J. Difference Equ. Appl.* **13**, 671 (2007).
6. G. Dattoli, S. Lorenzutta and C. Cesarano, *Rend. Mat. Appl. (7)* **22**, 193 (2002).

7. V. V. Dodonov, *J. Phys. A* **27**, 6191 (1994).
8. S. Khan, N. A. Makboul Hassan and G. Yasmin, *J. Math. Anal. Appl.* **351**, 756 (2009)
9. S. Khan, G. Yasmin and and N. A. Makboul Hassan, *Appl. Math. and Comp.* **217**, 2169 (2010).
10. Q. M. Luo and H. M. Srivastava, *Comput. Math. Appl.* **51**, 631 (2006).
11. S. M. Jung, *Bull. Sci. Math.* **133**, 756 (2009).
12. M. X. He and P. E. Ricci, *J. Comput. Appl. Math.* **139**, 231 (2002).
13. H. M. Srivastava, M. Gang and S. Choudhary, *Russ. J. of. Math. Phys.* **17**, 251 (2010).
14. N. Saran, S. D. Sharma and T. N. Triverdi, *Special Functions*, (Pragati Prakashan Pub., 2006).
15. W. Wang, C. Jia and T. Wang, *Comput. Math. Appl.* **55**, 1322 (2008).
16. D. V. Widder, *The Heat Equation* (Academic Press, New York, 1955).
17. A. Wünsche, *J. Phys. A* **33**, 1603 (2000).
18. A. Wünsche, *J. Comput. Appl. Math.* **133**, 665 (2001).
19. A. Wünsche, *J. Comput. Appl. Math.* **153**, 521 (2003).

SYMMETRY CLASSIFICATION OF COUPLED HEAT-DIFFUSION SYSTEMS VIA LOW DIMENSIONAL LIE ALGEBRAS

F. M. MAHOMED

Centre for Differential Equations, Continuum Mechanics and Applications, School of Computational and Applied Mathematics, University of the Witwatersrand, Wits 2050, South Africa
E-mail: Fazal.Mahomed@wits.ac.za

M. MOLATI

Department of Mathematics and Computer Science, National University of Lesotho, PO Roma 180, Lesotho
E-mail: m.molati@nul.ls

We review second-order system of coupled heat-diffusion equations. The system under consideration contains several arbitrary elements the forms of which are specified via the method of group classification based on the use of low-dimensional Lie algebras. We collect and present several results.

Keywords: Group classification; Lie algebra; Realization; Invariant system.

1. Introduction

There are many important applications of heat-diffusion equations in finance, biology, environmental science and engineering, just to mention a few. The physical phenomena involved are described by differential equations (DEs). Many DEs that arise as mathematical models contain arbitrary elements the forms of which are determined by experiment or simplicity conditions. Often one can require that the DE system has an additional symmetry group. This is the essence of the Lie group classification method.

The results of earlier works on the Lie symmetry approach to both linear and nonlinear equations are compiled in Ref. 1 and different forms of the heat-diffusion equations continue to be studied by various authors. Below we mention various approaches on how one can effect group classification. The direct method only requires the knowledge of the basic Lie algorithm[2,3].

Another approach on group classification utilizes equivalence transformations[3]. These are changes in the independent and dependent variables of a family of DEs that leave the family invariant. Also we have the method known as the preliminary group classification[4]. Lastly we have the approach based upon the Lie algebras of low dimensions. The procedure is extended to the Lie algebras of higher dimensions when possible until the arbitrary elements are specified completely. We revisit the latter pertaining to heat-diffusion systems in this work (consult Ref. 5 for more details). In all these approaches except the direct analysis a prior knowledge of the equivalence group is required.

The earliest work on symmetry group classification dates back to Lie[6,7] himself in 1881 and 1883. Lie[6] classified scalar linear (1+1) partial DEs according to their symmetry groups. In particular, he obtained four canonical forms for scalar linear (1+1) parabolic DEs.

Much later in the late nineteen-fifties Ovsiannikov[3] gave impetus to group classification with his work on nonlinear heat equations.

Lie[7] also obtained a complete point symmetry group classification of scalar second-order ordinary differential equations (ODEs)

$$y'' = F(x, y, y')$$

which admit non-similar complex Lie algebras of dimension 0, 1, 2, 3 or 8. Dimension 0 means that the equation admits no point symmetry at all. Lie noted that a second-order equation cannot admit a maximal 4, 5, 6 or 7 dimensional symmetry Lie algebra.

Over a century since the work of Lie on symmetry classification of scalar second-order ODEs in the complex plane, Mahomed and Leach[8] in 1989; Ibragimov and Mahomed[9] in 1996 investigated point symmetry group classification in the real domain.

2. Symmetry classification

The desire is to perform the complete group classification, but in some problems of interest the direct analysis becomes ineffective especially when the underlying equation involves arbitrary functions which depend upon the dependent variables or even their first order derivatives. In such cases the method of preliminary group classification can be used, although the arbitrary elements may not be specified completely. In order to solve completely the group classification problem one needs to combine both the standard Lie approach and the approach based upon the Lie algebras of low dimension. Basarab–Horwath *et al*[10] employed the procedure which combines these two approaches for heat-conductivity partial differential equations (PDEs) of the form

$$u_t = F(t, x, u, u_x)u_{xx} + G(t, x, u, u_x).$$

However, similar ideas were exploited by Mahomed and Leach[8] earlier for scalar second-order ODEs given above.

We consider the generalized coupled heat-diffusion system[5] of the form

$$\begin{aligned} u_t &= F^1(t, x, u, v, u_x, v_x)u_{xx} \\ &\quad + F^2(t, x, u, v, u_x, v_x)v_{xx} + G^1(t, x, u, v, u_x, v_x), \\ v_t &= F^3(t, x, u, v, u_x, v_x)u_{xx} \\ &\quad + F^4(t, x, u, v, u_x, v_x)v_{xx} + G^2(t, x, u, v, u_x, v_x), \end{aligned} \tag{1}$$

where $u(t, x)$ and $v(t, x)$ represent physical processes. The functions F^1, F^2, F^3 and F^4 are the nonzero arbitrary elements which represent diffusion coefficients or thermal conductivities. The nonzero arbitrary elements G^1 and G^2 are the source terms or heat transfer coefficients.

3. Classification results

In the sequel we present the results of symmetry classification of system (1). The proofs of the lemmas and theorems are to be found in Ref. 5.

Lemma 3.1. *The generator of symmetry group for system (1) is of the form*

$$Y = a(t)\partial_t + b(t, x)\partial_x + c(t, x, u, v)\partial_u + d(t, x, u, v)\partial_v, \tag{2}$$

where $a, \ldots, d$ are smooth functions satisfying the equations

$$\begin{aligned} c_{vv}F^1 + d_{vv}F^2 &= 0, \\ c_{uv}F^1 + d_{uv}F^2 &= 0, \\ c_{uu}F^1 + d_{uu}F^2 &= 0, \\ c_{vv}F^3 + d_{vv}F^4 &= 0, \\ c_{uv}F^3 + d_{uv}F^4 &= 0, \end{aligned}$$

$$
\begin{aligned}
c_{uu}F^3 + d_{uu}F^4 &= 0,\\
(d_v - b_x)\,F^1_{v_x} + c_v F^1_{u_x} &= 0,\\
(b_x - c_u)\,F^1_{u_x} - d_u F^1_{v_x} &= 0\\
(d_v - b_x)\,F^2_{v_x} + c_v F^2_{u_x} &= 0,\\
(b_x - c_u)\,F^2_{u_x} - d_u F^2_{v_x} &= 0,\\
(d_v - b_x)\,F^3_{v_x} + c_v F^3_{u_x} &= 0,\\
(b_x - c_u)\,F^3_{u_x} - d_u F^3_{v_x} &= 0,\\
(d_v - b_x)\,F^4_{v_x} + c_v F^4_{u_x} &= 0,\\
(b_x - c_u)\,F^4_{u_x} - d_u F^4_{v_x} &= 0,\\
2c_{xv}F^1 + (d_v - b_x)\,G^1_{v_x} + c_v G^1_{u_x} - (b_{xx} - 2d_{xv})\,F^2 &= 0,\\
b_t + 2d_{xu}F^2 + d_u G^1_{v_x} + (c_u - b_x)\,G^1_{u_x} - (b_{xx} - 2c_{xu})\,F^1 &= 0,\\
b_t + 2c_{xv}F^3 + (d_v - b_x)\,G^2_{v_x} + c_v G^2_{u_x} - (b_{xx} - 2d_{xv})\,F^4 &= 0,\\
2d_{xu}F^4 + d_u G^2_{v_x} + (c_u - b_x)\,G^2_{u_x} - (b_{xx} - 2c_{xu})\,F^3 &= 0,\\
(\dot{a} - 2b_x)\,F^1 + d_x F^1_{v_x} + c_x F^1_{u_x} &\\
+dF^1_v + cF^1_u + bF^1_x + aF^1_t + d_u F^2 - c_v F^3 &= 0,\\
c_v F^1 + (\dot{a} - 2b_x + d_v - c_u)\,F^2 + d_x F^2_{v_x} + c_x F^2_{u_x} &\\
+dF^2_v + cF^2_u + bF^2_x + aF^2_t - c_v F^4 &= 0,\\
d_u F^4 + (\dot{a} - 2b_x - d_v + c_u)\,F^3 + d_x F^3_{v_x} + c_x F^3_{u_x} &\\
+dF^3_v + cF^3_u + bF^3_x + aF^3_t - d_u F^1 &= 0,\\
(\dot{a} - 2b_x)\,F^4 + d_x F^4_{v_x} + c_x F^4_{u_x} &\\
+dF^4_v + cF^4_u + bF^4_x + aF^4_t + c_v F^3 - d_u F^2 &= 0,\\
c_{xx}F^1 + d_{xx}F^2 + d_x G^1_{v_x} + c_x G^1_{u_x} &\\
+dG^1_v + cG^1_u + bG^1_x + aG^1_t - (c_u - \dot{a})\,G^1 - c_v G^2 - c_t &= 0,\\
c_{xx}F^3 + d_{xx}F^4 + (\dot{a} - d_v)\,G^2 + d_x G^2_{v_x} + c_x G^2_{u_x} &\\
+dG^2_v + cG^2_u + bG^2_x + aG^2_t - d_u G^1 - d_t &= 0.
\end{aligned}
\tag{3}
$$

In the above classifying equations the overdot represents derivative with respect to t and the subscripts denote partial differentiation.

Lemma 3.2. *The equivalence group for system (1) is given by the transformations*

$$\bar{t} = T(t), \quad \bar{x} = X(t,x), \quad \bar{u} = U(t,x,u,v), \quad \bar{v} = V(t,x,u,v), \tag{4}$$

where

$$\dot{T} \neq 0, \quad \frac{D(X)}{D(t,x)} \neq 0, \quad \frac{D(U,V)}{D(t,x,u,v)} \neq 0.$$

Theorem 3.1. *There are four inequivalent systems which admit the one-dimensional Lie algebras. The corresponding realizations of one-dimensional Lie algebras are denoted*[a]*:* $A_1^1 = \langle \partial_t \rangle$, $A_1^2 = \langle \partial_x \rangle$, $A_1^3 = \langle \partial_u \rangle$ *and* $A_1^4 = \langle \partial_v \rangle$*. The functional forms of the arbitrary functions contained in these systems are such that they are independent of the variables appearing in each realization. For instance all the arbitrary functions corresponding to* A_1^1 *are independent of* t.

Theorem 3.2. *There are ten invariant systems admitting the two-dimensional solvable Lie algebras. The realizations of the two-dimensional Lie algebra* $A_{2,1}$ *are* $A_{2,1}^1 = \langle \partial_t, \partial_x \rangle$, $A_{2,1}^2 = \langle \partial_t, \partial_u \rangle$, $A_{2,1}^3 = \langle \partial_t, \partial_v \rangle$, $A_{2,1}^4 = \langle \partial_x, \partial_u \rangle$, $A_{2,1}^5 = \langle \partial_x, \partial_v \rangle$, $A_{2,1}^6 = \langle \partial_u, \partial_v \rangle$*. The corresponding forms of the arbitrary elements are such that they are independent of the variables appearing in each realization.*

Moreover, the two-dimensional Lie algebra $A_{2,2}$ *has the realizations* $A_{2,2}^1 = \langle -t\partial_t - x\partial_x, \partial_t \rangle$, $A_{2,2}^2 = \langle -t\partial_t - x\partial_x, \partial_x \rangle$, $A_{2,2}^3 = \langle -x\partial_x - u\partial_u, \partial_x \rangle$, $A_{2,2}^4 = \langle -x\partial_x - v\partial_v, \partial_x \rangle$*. The corresponding invariant systems*[b] *are given as follows.*

$$\begin{aligned} A_{2,2}^1 :\ u_t &= x\tilde{F}^1(u,v)u_{xx} + x\tilde{F}^2(u,v)v_{xx} + x^{-1}\tilde{G}^1(u,v), \\ v_t &= x\tilde{F}^3(u,v)u_{xx} + x\tilde{F}^4(u,v)v_{xx} + x^{-1}\tilde{G}^2(u,v). \end{aligned}$$

$$\begin{aligned} A_{2,2}^2 :\ u_t &= t\tilde{F}^1(u,v)u_{xx} + t\tilde{F}^2(u,v)v_{xx} + t^{-1}\tilde{G}^1(u,v), \\ v_t &= t\tilde{F}^3(u,v)u_{xx} + t\tilde{F}^4(u,v)v_{xx} + t^{-1}\tilde{G}^2(u,v). \end{aligned}$$

$$\begin{aligned} A_{2,2}^3 :\ u_t &= u^2\tilde{F}^1(t,v,u_x)u_{xx} + u^3\tilde{F}^2(t,v,u_x)v_{xx} + u\tilde{G}^1(t,v,u_x), \\ v_t &= u\tilde{F}^3(t,v,u_x)u_{xx} + u^2\tilde{F}^4(t,v,u_x)v_{xx} + \tilde{G}^2(t,v,u_x). \end{aligned}$$

$$\begin{aligned} A_{2,2}^4 :\ u_t &= v^2\tilde{F}^1(t,u,v_x)u_{xx} + v^3\tilde{F}^2(t,u,v_x)v_{xx} + \tilde{G}^1(t,u,v_x), \\ v_t &= v\tilde{F}^3(t,u,v_x)u_{xx} + v^2\tilde{F}^4(t,u,v_x)v_{xx} + v\tilde{G}^2(t,u,v_x). \end{aligned}$$

[a] The superscript distinguishes one realization from the other while the subscript denotes the dimension of the Lie algebra.

[b] The invariant systems for $A_{2,2}^3$ and $A_{2,2}^4$ can be regarded as one system provided a suitable choice of equivalence transformations exists for transforming one into another. This note also applies to other similar cases throughout this work.

The above realizations are the maximal symmetry Lie algebra for system (1) provided $\tilde{F}^1$, $\tilde{F}^2$, $\tilde{F}^3$, $\tilde{F}^4$, $\tilde{G}^1$ and $\tilde{G}^2$ are different arbitrary functions of their respective arguments.

Theorem 3.3. *There are twelve invariant systems admitting the decomposable three-dimensional solvable Lie algebras and seventy one which admit the nondecomposable three-dimensional solvable Lie algebras.*

The realizations and their corresponding invariant systems for the decomposable three-dimensional solvable Lie algebras are given below.

$$
\begin{aligned}
A^1_{3,1} &= \langle \partial_t, \partial_x, \partial_u \rangle : \\
&\quad u_t = F^1(v, u_x, v_x)u_{xx} + F^2(v, u_x, v_x)v_{xx} + G^1(v, u_x, v_x), \\
&\quad v_t = F^3(v, u_x, v_x)u_{xx} + F^4(v, u_x, v_x)v_{xx} + G^2(v, u_x, v_x). \\
\tilde{A}^1_{3,1} &= \langle \partial_t, \partial_x, \partial_v \rangle : \\
&\quad u_t = F^1(x, u_x, v_x)u_{xx} + F^2(x, u_x, v_x)v_{xx} + G^1(x, u_x, v_x), \\
&\quad v_t = F^3(x, u_x, v_x)u_{xx} + F^4(x, u_x, v_x)v_{xx} + G^2(x, u_x, v_x). \\
A^2_{3,1} &= \langle \partial_t, \partial_u, \partial_v \rangle : \\
&\quad u_t = F^1(u, u_x, v_x)u_{xx} + F^2(u, u_x, v_x)v_{xx} + G^1(u, u_x, v_x), \\
&\quad v_t = F^3(u, u_x, v_x)u_{xx} + F^4(u, u_x, v_x)v_{xx} + G^2(u, u_x, v_x). \\
\tilde{A}^2_{3,1} &= \langle \partial_x, \partial_u, \partial_v \rangle : \\
&\quad u_t = F^1(t, u_x, v_x)u_{xx} + F^2(t, u_x, v_x)v_{xx} + G^1(t, u_x, v_x), \\
&\quad v_t = F^3(t, u_x, v_x)u_{xx} + F^4(t, u_x, v_x)v_{xx} + G^2(t, u_x, v_x). \\
A^1_{3,2} &= \langle -t\partial_t - x\partial_x, \partial_t, x\partial_x + u\partial_u \rangle : \\
&\quad u_t = xu\hat{F}^1(v)u_{xx} + xu^2\hat{F}^2(v)v_{xx} + x^{-1}u^2\hat{G}^1(v), \\
&\quad v_t = x\hat{F}^3(v)u_{xx} + xu\hat{F}^4(v)v_{xx} + x^{-1}u\hat{G}^2(v). \\
\tilde{A}^1_{3,2} &= \langle -t\partial_t - x\partial_x, \partial_t, x\partial_x + v\partial_v \rangle : \\
&\quad u_t = xv\hat{F}^1(u)u_{xx} + x\hat{F}^2(u)v_{xx} + x^{-1}v\hat{G}^1(u)x^{-1}v\hat{G}^1(u), \\
&\quad v_t = xv^2\hat{F}^3(u)u_{xx} + xv\hat{F}^4(u)v_{xx} + x^{-1}v^2\hat{G}^2(u). \\
A^2_{3,2} &= \langle -t\partial_t - x\partial_x, \partial_x, t\partial_t + u\partial_u \rangle : \\
&\quad u_t = tu^{-2}\hat{F}^1(v)u_{xx} + tu^{-1}\hat{F}^2(v)v_{xx} + t^{-1}u\hat{G}^1(v), \\
&\quad v_t = tu^{-3}\hat{F}^3(v)u_{xx} + u^{-2}\hat{F}^4(v)v_{xx} + t^{-1}\hat{G}^2(v). \\
\tilde{A}^2_{3,2} &= \langle -t\partial_t - x\partial_x, \partial_x, t\partial_t + v\partial_v \rangle : \\
&\quad u_t = tv^{-2}\hat{F}^1(u)u_{xx} + tv^{-3}\hat{F}^2(u)v_{xx} + t^{-1}\hat{G}^1(u), \\
&\quad v_t = tv^{-1}\hat{F}^3(u)u_{xx} + tv^{-2}\hat{F}^4(u)v_{xx} + t^{-1}v\hat{G}^2(u).
\end{aligned}
$$

$$
\begin{aligned}
A^3_{3,2} &= \langle -x\partial_x - u\partial_u, \partial_x, t\partial_t + u\partial_u\rangle : \\
& u_t = t^{-3}u^2\hat{F}^1(v)u_{xx} + t^{-3}u^3\hat{F}^2(v)v_{xx} + t^{-1}u\hat{G}^1(v), \\
& v_t = t^{-3}u\hat{F}^3(v)u_{xx} + t^{-3}u^2\hat{F}^4(v)v_{xx} + t^{-1}\hat{G}^2(v). \\
\tilde{A}^3_{3,2} &= \langle -x\partial_x - u\partial_u, \partial_x, u\partial_u + v\partial_v\rangle : \\
& u_t = u^2v^{-2}\hat{F}^1(t)u_{xx} + u^3v^{-3}\hat{F}^2(t)v_{xx} + u\hat{G}^1(t), \\
& v_t = uv^{-1}\hat{F}^3(t)u_{xx} + u^2v^{-2}\hat{F}^4(t)v_{xx} + v\hat{G}^2(t). \\
A^4_{3,2} &= \langle -x\partial_x - v\partial_v, \partial_x, t\partial_t + v\partial_v\rangle : \\
& u_t = t^{-3}v^2\hat{F}^1(u)u_{xx} + t^{-3}v\hat{F}^2(u)v_{xx} + t^{-1}\hat{G}^1(u), \\
& v_t = t^{-3}v^3\hat{F}^3(u)u_{xx} + t^{-3}v^2\hat{F}^4(u)v_{xx} + t^{-1}v\hat{G}^2(u). \\
\tilde{A}^4_{3,2} &= \langle -x\partial_x - v\partial_v, \partial_x, u\partial_u + v\partial_v\rangle : \\
& u_t = u^{-2}v^2\hat{F}^1(t)u_{xx} + uv\hat{F}^2(t)v_{xx} + u\hat{G}^1(t), \\
& v_t = u^{-3}v^3\hat{F}^3(t)u_{xx} + u^{-2}v^2\hat{F}^4(t)v_{xx} + v\hat{G}^2(t).
\end{aligned}
$$

The unknowns $\hat{F}^1$, $\hat{F}^2$, $\hat{F}^3$, $\hat{F}^4$, $\hat{G}^1$ and $\hat{G}^2$ are in general different arbitrary functions of their arguments.

The realizations of the nondecomposable three-dimensional solvable Lie algebras and the corresponding forms of the arbitrary elements can be found in Ref. 5.

Theorem 3.4. *There are twelve invariant systems which admit the decomposable four-dimensional solvable Lie algebras and fifty three admitting the nondecomposable four-dimensional solvable Lie algebras.*

The realizations of the decomposable and nondecomposable four-dimensional solvable Lie algebras and their corresponding invariant systems are given in Tables 1 and 2 respectively.

Generally the arbitrary functions, $\mathcal{F}^1, \ldots, \mathcal{F}^4$, $\mathcal{G}^1$ and $\mathcal{G}^2$, of their respective arguments are different. Also the arbitrary constants K_1, $K_2, \ldots, K_6$ are in general distinct.

4. Conclusion

We have reviewed group classification of a general second-order system of diffusion equations with respect to solvable Lie algebras of low dimensions up to dimension four. In order to achieve the complete group classification of the system, a further study employing Lie algebras of dimensions five and more is required. Moreover, other types of Lie algebras such as semisimple

Table 1. Decomposable 4D solvable Lie algebras.

Algebra	Realization	Invariant System
$A_{3,2} \oplus A_1^1$	$\langle -t\partial_t - x\partial_x, \partial_t, x\partial_x + u\partial_u, x\partial_x + u\partial_u + v\partial_v \rangle$	$u_t = \bar{K}_1 xuu_{xx} + \bar{K}_2 xuv^{-1}v_{xx} + \bar{K}_5 x^{-1}u,$ $v_t = \bar{K}_3 xu_{xx} + \bar{K}_4 xuv_{xx} + \bar{K}_6 x^{-1}uv.$
$4A_1^1$	$\langle \partial_t, \partial_x, \partial_u, v\partial_v \rangle$	$u_t = \bar{\mathcal{F}}^1(u_x)u_{xx} + v^{-1}\bar{\mathcal{F}}^2(u_x)v_{xx} + \bar{\mathcal{G}}^1(u_x),$ $v_t = v\bar{\mathcal{F}}^3(u_x)u_{xx} + \bar{\mathcal{F}}^4(u_x)v_{xx} + \bar{\mathcal{G}}^2(u_x).$
$4A_1^2$	$\langle \partial_t, \partial_x, u\partial_u, \partial_v \rangle$	$u_t = \tilde{\mathcal{F}}^1(v_x)u_{xx} + u\tilde{\mathcal{F}}^2(v_x)v_{xx} + u\tilde{\mathcal{G}}^1(v_x),$ $v_t = u^{-1}\tilde{\mathcal{F}}^3(v_x)u_{xx} + \tilde{\mathcal{F}}^4(v_x)v_{xx} + \tilde{\mathcal{G}}^2(v_x).$
$4A_1^3$	$\langle \partial_t, \partial_x, \partial_u, \partial_v \rangle$	$u_t = F^1(u_x, v_x)u_{xx} + F^2(u_x, v_x)v_{xx} + G^1(u_x, v_x),$ $v_t = F^3(u_x, v_x)u_{xx} + F^4(u_x, v_x)v_{xx} + G^2(u_x, v_x).$
$\tilde{A}_{3,2} \oplus \tilde{A}_1^1$	$\langle -t\partial_t - x\partial_x, \partial_t, x\partial_x + v\partial_v, x\partial_x + u\partial_u + v\partial_v \rangle$	$u_t = \tilde{K}_1 xvu_{xx} + \tilde{K}_2 xuv_{xx} + \tilde{K}_5 x^{-1}uv,$ $v_t = \tilde{K}_3 xv^2u_{xx} + \tilde{K}_4 xvv_{xx} + \tilde{K}_6 x^{-1}v^2.$
$A_{3,3} \oplus A_1^3$	$\langle \partial_t + x\partial_x + \ln x\partial_u, \partial_x, \partial_u, \partial_v \rangle$	$u_t = \tilde{K}_1 e^{2t}u_{xx} + \tilde{K}_2 e^{2t}v_{xx} + \tilde{K}_5 + \tilde{K}_1 \frac{e^{2t}}{2x^2},$ $v_t = \tilde{K}_3 e^{2t}u_{xx} + \tilde{K}_4 e^{2t}v_{xx} + \tilde{K}_6 + \tilde{K}_3 \frac{e^{2t}}{2x^2}.$
$\tilde{A}_{3,3} \oplus \tilde{A}_1^3$	$\langle \partial_t + x\partial_x + \ln x\partial_v, \partial_x, \partial_u, \partial_v \rangle$	$u_t = \tilde{K}_1 e^{2t}u_{xx} + \tilde{K}_2 e^{2t}v_{xx} + \tilde{K}_5 + \tilde{K}_2 \frac{e^{2t}}{2x^2},$ $v_t = \tilde{K}_3 e^{2t}u_{xx} + \tilde{K}_4 e^{2t}v_{xx} + \tilde{K}_6 + \tilde{K}_4 \frac{e^{2t}}{2x^2}.$
$A_{3,4} \oplus A_1^3$	$\langle \partial_t + x\partial_x + x\partial_u + x\partial_v, \partial_x, \partial_u, \partial_v \rangle$	$u_t = \tilde{K}_1 e^{2t}u_{xx} + \tilde{K}_2 e^{2t}v_{xx} + \tilde{K}_5,$ $v_t = \tilde{K}_3 e^{2t}u_{xx} + \tilde{K}_4 e^{2t}v_{xx} + \tilde{K}_6.$
$A_{3,6} \oplus A_1^3$	$\langle \partial_x + \partial_u + \ln x\partial_v, \partial_x, \partial_u, \partial_v \rangle$	$u_t = \tilde{\mathcal{F}}^1(t, u_x)u_{xx} + \tilde{\mathcal{F}}^2(t, u_x)v_{xx} + x^{-1}v_x\tilde{\mathcal{F}}^2(t, u_x) + \tilde{\mathcal{G}}^1(t, u_x),$ $v_t = \tilde{\mathcal{F}}^3(t, u_x)u_{xx} + \tilde{\mathcal{F}}^4(t, u_x)v_{xx} + x^{-1}v_x\tilde{\mathcal{F}}^4(t, u_x) + \tilde{\mathcal{G}}^2(t, u_x).$
$\tilde{A}_{3,6} \oplus \tilde{A}_1^3$	$\langle \partial_x + \ln x\partial_u + \partial_v, \partial_x, \partial_u, \partial_v \rangle$	$u_t = \tilde{\mathcal{F}}^1(t, v_x)u_{xx} + \tilde{\mathcal{F}}^2(t, v_x)v_{xx} + x^{-1}u_x\tilde{\mathcal{F}}^1(t, v_x) + \tilde{\mathcal{G}}^1(t, v_x),$ $v_t = \tilde{\mathcal{F}}^3(t, v_x)u_{xx} + \tilde{\mathcal{F}}^4(t, v_x)v_{xx} + x^{-1}u_x\tilde{\mathcal{F}}^3(t, v_x) + \tilde{\mathcal{G}}^2(t, v_x).$
$A_{3,7} \oplus A_1^3$	$\langle \partial_x + \partial_u + qx^{q-1}\partial_v, \partial_x, \partial_u, \partial_v \rangle$	$u_t = \tilde{\mathcal{F}}^1(t, u_x)u_{xx} + \tilde{\mathcal{F}}^2(t, u_x)v_{xx} - (q-2)x^{-1}v_x\tilde{\mathcal{F}}^2(t, u_x) + \tilde{\mathcal{G}}^1(t, u_x),$ $v_t = \tilde{\mathcal{F}}^3(t, u_x)u_{xx} + \tilde{\mathcal{F}}^4(t, u_x)v_{xx} - (q-2)x^{-1}v_x\tilde{\mathcal{F}}^4(t, u_x) + \tilde{\mathcal{G}}^2(t, u_x).$
$\tilde{A}_{3,7} \oplus \tilde{A}_1^3$	$\langle \partial_x + qx^{q-1}\partial_u + \partial_v, \partial_x, \partial_u, \partial_v \rangle$	$u_t = \tilde{\mathcal{F}}^1(t, v_x)u_{xx} + \tilde{\mathcal{F}}^2(t, v_x)v_{xx} - (q-2)x^{-1}u_x\tilde{\mathcal{F}}^1(t, v_x) + \tilde{\mathcal{G}}^1(t, v_x),$ $v_t = \tilde{\mathcal{F}}^3(t, v_x)u_{xx} + \tilde{\mathcal{F}}^4(t, v_x)v_{xx} - (q-2)x^{-1}u_x\tilde{\mathcal{F}}^3(t, v_x) + \tilde{\mathcal{G}}^2(t, v_x).$

The unknowns, $\bar{\mathcal{F}}^1, \ldots, \bar{\mathcal{F}}^4$, $\tilde{\mathcal{F}}^1, \ldots, \tilde{\mathcal{F}}^4$, $\bar{\mathcal{G}}^1$, $\tilde{\mathcal{G}}^1$, $\bar{\mathcal{G}}^2$ and $\tilde{\mathcal{G}}^2$, are arbitrary functions of their respective arguments and $\bar{K}_1, \bar{K}_2, \ldots, \bar{K}_6$, $\tilde{K}_1, \ldots, \tilde{K}_6$ are arbitrary constants.

Table 2. Nondecomposable 4D solvable Lie algebras.

Algebra	Realization	Invariant System
$A^6_{4,1}$	$\langle \partial_t, \partial_u, \partial_v, x\partial_x + v\partial_u + t\partial_v \rangle$	$u_t = x^2(K_1 + K_3 \ln x)u_{xx} + x^2[K_2 + \ln x(K_4 - K_3 \ln x - K_1)]v_{xx} + K_5 + \ln x\left(K_6 + \frac{\ln x}{2}\right)$ $v_t = K_3 x^2 u_{xx} + x^2(K_4 - K_3 \ln x)v_{xx} + K_6 + \ln x.$
$\tilde{A}^6_{4,1}$	$\langle \partial_t, \partial_u, \partial_v, x\partial_x + t\partial_u + u\partial_v \rangle$	$u_t = x^2(K_1 - K_2 \ln x)u_{xx} + K_2 x^2 v_{xx} + K_5 + \ln x,$ $v_t = x^2[K_3 + \ln x(K_1 - K_2 \ln x - K_4)]u_{xx} + x^2(K_4 + K_2 \ln x)v_{xx} + K_6 + \ln x\left(K_5 + \frac{\ln x}{2}\right)$
$\hat{A}^6_{4,1}$	$\langle \partial_x, \partial_u, \partial_v, v\partial_u + x\partial_v \rangle$	$u_t = [\mathcal{F}^1(t) + v_x \mathcal{F}^3(t)]u_{xx} + [\mathcal{F}^2(t) + v_x\left(\mathcal{F}^4(t) - v_x\mathcal{F}^3(t) - \mathcal{F}^1(t)\right)]v_{xx} + \mathcal{G}^1(t) + v_x\mathcal{G}^2(t)$ $v_t = F^3(t)u_{xx} + [\mathcal{F}^4(t) - v_x\mathcal{F}^3(t)]v_{xx} + \mathcal{G}^2(t).$
$\bar{A}^6_{4,1}$	$\langle \partial_x, \partial_u, \partial_v, x\partial_u + u\partial_v \rangle$	$u_t = [\mathcal{F}^1(t) - u_x\mathcal{F}^2(t)]u_{xx} + \mathcal{F}^2(t)v_{xx} + \mathcal{G}^1(t),$ $v_t = [\mathcal{F}^3(t) + u_x\left(\mathcal{F}^1(t) - \mathcal{F}^4(t) - u_x\mathcal{F}^t(t)\right)]u_{xx} + [\mathcal{F}^4(t) + u_x\mathcal{F}^2(t)]v_{xx} + \mathcal{G}^2(t) + u_x\mathcal{G}^1(t)$
$A^1_{4,2}$	$\langle \partial_t, \partial_u, \partial_x, qt\partial_t + x\partial_x + (x+u)\partial_u + v\partial_v \rangle$	$u_t = v^{2-q}\mathcal{F}^1(\gamma)u_{xx} + v^{2-q}\mathcal{F}^2(\gamma)v_{xx} + v^{1-q}\mathcal{G}^1(\gamma),$ $v_t = v^{2-q}\mathcal{F}^3(\gamma)u_{xx} + v^{2-q}\mathcal{F}^4(\gamma)v_{xx} + v^{1-q}\mathcal{G}^2(\gamma).$
$\tilde{A}^1_{4,2}$	$\langle \partial_t, \partial_u, \partial_x, qt\partial_t + x\partial_x + (x+u)\partial_u \rangle$	$u_t = e^{(2-q)u_x}\mathcal{F}^1(v)u_{xx} + e^{(3-q)u_x}\mathcal{F}^2(v)v_{xx} + e^{(1-q)u_x}\mathcal{G}^1(v),$ $v_t = e^{(1-q)u_x}\mathcal{F}^3(v)u_{xx} + e^{(2-q)u_x}\mathcal{F}^4(v)v_{xx} + e^{-qu_x}\mathcal{G}^2(v).$
$\hat{A}^1_{4,2}$	$\langle \partial_x, \partial_u, \partial_t, t\partial_t + qx\partial_x + (t+u)\partial_u + v\partial_v \rangle$	$u_t = K_1 v^{2q-1}u_{xx} + K_2 v^{2q-1}v_{xx} + \ln v + K_5,$ $v_t = K_3 v^{2q-1}u_{xx} + K_4 v^{2q-1}v_{xx} + K_6.$
$A^2_{4,2}$	$\langle \partial_t, \partial_v, \partial_x, qt\partial_t + x\partial_x + u\partial_u + (x+v)\partial_v \rangle$	$u_t = u^{2-q}\mathcal{F}^1(\lambda)u_{xx} + u^{2-q}\mathcal{F}^2(\lambda)v_{xx} + u^{1-q}\mathcal{G}^1(\lambda),$ $v_t = u^{2-q}\mathcal{F}^3(\lambda)u_{xx} + u^{2-q}\mathcal{F}^4(\lambda)v_{xx} + u^{1-q}\mathcal{G}^2(\lambda).$
$\tilde{A}^2_{4,2}$	$\langle \partial_t, \partial_v, \partial_x, qt\partial_t + x\partial_x + (x+v)\partial_v \rangle$	$u_t = e^{(2-q)v_x}\mathcal{F}^1(u)u_{xx} + e^{(3-q)v_x}\mathcal{F}^2(u)v_{xx} + e^{(1-q)v_x}\mathcal{G}^1(u),$ $v_t = e^{(1-q)v_x}\mathcal{F}^3(u)u_{xx} + e^{(2-q)v_x}\mathcal{F}^4(u)v_{xx} + e^{(1-q)v_x}\mathcal{G}^2(u).$
$\hat{A}^2_{4,2}$	$\langle \partial_x, \partial_v, \partial_t, t\partial_t + qx\partial_x + u\partial_u + (t+v)\partial_v \rangle$	$u_t = K_1 u^{2q-1}u_{xx} + K_2 u^{2q-1}v_{xx} + K_5,$ $v_t = K_3 u^{2q-1}u_{xx} + K_4 u^{2q-1}v_{xx} + \ln u + K_6.$

The arguments γ and λ are respectively defined by $\gamma = u_x - \ln v$, $\lambda = v_x - \ln u$.

Table 2 (Continued).

Algebra	Realization	Invariant System
$A^6_{4,2}$	$\langle\partial_t,\partial_u,\partial_v,t\partial_x+qu\partial_u+(t+v)\partial_v\rangle$	$u_t=K_1u_{xx}+K_2e^{(\frac{q-1}{t})x}v_{xx}+K_5e^{\frac{qx}{t}}+\dfrac{u_x}{q}$, $v_t=K_3e^{(\frac{q-1}{t})x}u_{xx}+K_4v_{xx}+K_6e^{\frac{x}{t}}+v_x-1.$
$\tilde{A}^6_{4,2}$	$\langle\partial_t,\partial_u,\partial_v,t\partial_x+(t+u)\partial_u+qv\partial_v\rangle$	$u_t=K_1u_{xx}+K_2e^{(\frac{1-q}{t})x}v_{xx}+K_5e^{\frac{x}{t}}+u_x-1$, $v_t=K_3e^{(\frac{1-q}{t})x}u_{xx}+K_4v_{xx}+K_6e^{\frac{qx}{t}}+\dfrac{v_x}{q}.$
$\hat{A}^6_{4,2}$	$\langle\partial_t,\partial_u,\partial_v,qt\partial_x+(u+v)\partial_u+v\partial_v\rangle$	$u_t=\left(K_1+\frac{K_3x}{qt}\right)u_{xx}+\left[K_2+\frac{x}{qt}\left(K_4-\frac{K_3x}{qt}-K_1\right)\right]v_{xx}+e^{\frac{x}{qt}}\left[K_5+\frac{1}{qt}\left(K_6x-2q^2tv_xe^{-\frac{x}{qt}}\right)\right]$, $v_t=K_3u_{xx}+\left(K_4-\frac{K_3x}{qt}\right)v_{xx}+K_6e^{\frac{x}{qt}}+qv_x.$
$\bar{A}^6_{4,2}$	$\langle\partial_t,\partial_u,\partial_v,qt\partial_x+u\partial_u+(u+v)\partial_v\rangle$	$u_t=\left(K_1-\frac{K_2x}{qt}\right)u_{xx}+K_2v_{xx}+K_5e^{\frac{x}{qt}}+qu_x$, $v_t=\left[K_3+\frac{x}{qt}\left(K_1-\frac{K_2x}{qt}-K_4\right)\right]u_{xx}+\left(K_4+\frac{K_2x}{qt}\right)v_{xx}+e^{\frac{x}{qt}}\left[K_6+\frac{1}{qt}\left(K_5x-2q^2tu_xe^{-\frac{x}{qt}}\right)\right].$
$\breve{A}^6_{4,2}$	$\langle\partial_x,\partial_u,\partial_v,x\partial_x+qu\partial_u+(x+v)\partial_v\rangle$	$u_t=\mathcal{F}^1(t)e^{2v_x}u_{xx}+\mathcal{F}^2(t)e^{(q+1)v_x}v_{xx}+\mathcal{G}^1(t)e^{qv_x}$, $v_t=\mathcal{F}^3(t)e^{(3-q)v_x}u_{xx}+\mathcal{F}^4(t)e^{2v_x}v_{xx}+\mathcal{G}^2(t)e^{v_x}.$
$\grave{A}^6_{4,2}$	$\langle\partial_x,\partial_u,\partial_v,x\partial_x+(x+u)\partial_u+qv\partial_v\rangle$	$u_t=\mathcal{F}^1(t)e^{2u_x}u_{xx}+\mathcal{F}^2(t)e^{(3-q)u_x}v_{xx}+\mathcal{G}^1(t)e^{u_x}$, $v_t=\mathcal{F}^3(t)e^{(q+1)u_x}u_{xx}+\mathcal{F}^4(t)e^{2u_x}v_{xx}+\mathcal{G}^2(t)e^{qu_x}.$
$A^1_{4,3}$	$\langle\partial_t,\partial_u,\partial_x,t\partial_t+x\partial_u+v\partial_v\rangle$	$u_t=v^{-1}\mathcal{F}^1(\gamma)u_{xx}+v^{-2}\mathcal{F}^2(\gamma)v_{xx}+v^{-1}\mathcal{G}^1(\gamma)$, $v_t=\mathcal{F}^3(\gamma)u_{xx}+v^{-1}\mathcal{F}^4(\gamma)v_{xx}+\mathcal{G}^2(\gamma).$
$\tilde{A}^1_{4,3}$	$\langle\partial_t,\partial_u,\partial_x,t\partial_t+x\partial_u\rangle$	$u_t=e^{-u_x}\mathcal{F}^1(v,v_x)u_{xx}+e^{-u_x}\mathcal{F}^2(v,v_x)v_{xx}+e^{-u_x}\mathcal{G}^1(v,v_x)$, $v_t=e^{-u_x}\mathcal{F}^3(v,v_x)u_{xx}+e^{-u_x}\mathcal{F}^4(v,v_x)v_{xx}+e^{-u_x}\mathcal{G}^2(v,v_x).$
$\hat{A}^1_{4,3}$	$\langle\partial_x,\partial_u,\partial_t,x\partial_x+t\partial_u+v\partial_v\rangle$	$u_t=v^2\mathcal{F}^1(v_x)u_{xx}+v\mathcal{F}^2(v_x)v_{xx}+\mathcal{G}^1(v_x)+\ln v$, $v_t=v^3\mathcal{F}^3(v_x)u_{xx}+v^2\mathcal{F}^4(v_x)v_{xx}+v\mathcal{G}^2(v_x).$
$A^2_{4,3}$	$\langle\partial_t,\partial_v,\partial_x,t\partial_t+u\partial_u+x\partial_v\rangle$	$u_t=u^{-1}\mathcal{F}^1(\lambda)u_{xx}+\mathcal{F}^2(\lambda)v_{xx}+\mathcal{G}^1(\lambda)$, $v_t=u^{-2}\mathcal{F}^3(\lambda)u_{xx}+u^{-1}\mathcal{F}^4(\lambda)v_{xx}+u^{-1}\mathcal{G}^2(\lambda).$

Table 2 (Continued).

Algebra	Realization	Invariant System
$\tilde{A}^2_{4,3}$	$\langle \partial_t, \partial_v, \partial_x, t\partial_t + x\partial_v \rangle$	$u_t = e^{-v_x}\mathcal{F}^1(u,u_x)u_{xx} + e^{-v_x}\mathcal{F}^2(u,u_x)v_{xx} + e^{-v_x}\mathcal{G}^1(u,u_x),$ $v_t = e^{-v_x}\mathcal{F}^3(u,u_x)u_{xx} + e^{-v_x}\mathcal{F}^4(u,u_x)v_{xx} + e^{-v_x}\mathcal{G}^2(u,u_x).$
$\hat{A}^2_{4,3}$	$\langle \partial_x, \partial_v, \partial_t, x\partial_x + u\partial_u + t\partial_v \rangle$	$u_t = u^2\mathcal{F}^1(u_x)u_{xx} + u^3\mathcal{F}^2(u_x)v_{xx} + u\mathcal{G}^1(u_x),$ $v_t = u\mathcal{F}^3(u_x)u_{xx} + u^2\mathcal{F}^4(u_x)v_{xx} + \mathcal{G}^2(u_x) + \ln u.$
$A^6_{4,3}$	$\langle \partial_t, \partial_u, \partial_v, x\partial_x + u\partial_u + t\partial_v \rangle$	$u_t = x^2\mathcal{F}^1(u_x)u_{xx} + x^3\mathcal{F}^2(u_x)v_{xx} + x\mathcal{G}^1(u_x),$ $v_t = x\mathcal{F}^3(u_x)u_{xx} + x^2\mathcal{F}^4(u_x)v_{xx} + \mathcal{G}^2(u_x) + \ln x.$
$\tilde{A}^6_{4,3}$	$\langle \partial_t, \partial_u, \partial_v, x\partial_x + t\partial_u + v\partial_v \rangle$	$u_t = x^2\mathcal{F}^1(v_x)u_{xx} + x\mathcal{F}^2(v_x)v_{xx} + \mathcal{G}^1(v_x) + \ln x,$ $v_t = x^3\mathcal{F}^3(v_x)u_{xx} + x^2\mathcal{F}^4(v_x)v_{xx} + x\mathcal{G}^2(v_x).$
$\hat{A}^6_{4,3}$	$\langle \partial_x, \partial_u, \partial_v, u\partial_u + x\partial_v \rangle$	$u_t = \mathcal{F}^1(t)u_{xx} + \mathcal{F}^2(t)e^{-v_x}v_{xx} + \mathcal{G}^1(t)e^{v_x},$ $v_t = \mathcal{F}^3(t)e^{-v_x}u_{xx} + \mathcal{F}^4(t)v_{xx} + \mathcal{G}^2(t).$
$\bar{A}^6_{4,3}$	$\langle \partial_x, \partial_u, \partial_v, x\partial_u + v\partial_v \rangle$	$u_t = \mathcal{F}^1(t)u_{xx} + \mathcal{F}^2(t)e^{-u_x}v_{xx} + \mathcal{G}^1(t),$ $v_t = \mathcal{F}^3(t)e^{u_x}u_{xx} + \mathcal{F}^4(t)v_{xx} + \mathcal{G}^2(t)e^{u_x}.$
$A^6_{4,4}$	$\langle \partial_t, \partial_u, \partial_v, t\partial_x + (u+v)\partial_u + (t+v)\partial_v \rangle$	$u_t = \left(K_1 + \frac{K_3 x}{t}\right)u_{xx} + \left[K_2 + \frac{x}{t}\left(K_4 - \frac{K_3 x}{t} - K_1\right)\right]v_{xx} + e^{\frac{x}{t}}\left[K_5 + \frac{1}{t}\left(K_6 x - (2v_x - 1)t\,e^{-\frac{x}{t}}\right)\right],$ $v_t = K_3 u_{xx} + \left(K_1 + \frac{K_3 x}{t}\right)v_{xx} + K_6 e^{\frac{x}{t}} + v_x - 1.$
$\tilde{A}^6_{4,4}$	$\langle \partial_t, \partial_u, \partial_v, t\partial_x + (t+u)\partial_u + (u+v)\partial_v \rangle$	$u_t = \left(K_1 - \frac{K_2 x}{t}\right)u_{xx} + K_2 v_{xx} + K_5 e^x + u_x - 1,$ $v_t = \left[K_3 + \frac{x}{t}\left(K_1 - \frac{K_2 x}{t} - K_4\right)\right]u_{xx} + \left(K_4 + \frac{K_2 x}{t}\right)v_{xx} + \left(\frac{K_5 e^x}{t-1}\right) + K_6 e^{\frac{x}{t}} - 2u_x + 1.$
$\hat{A}^6_{4,4}$	$\langle \partial_x, \partial_u, \partial_v, x\partial_x + (u+v)\partial_u + (x+v)\partial_v \rangle$	$u_t = e^{2v_x}\left\{[\mathcal{F}^1(t) + v_x\mathcal{F}^3(t)]u_{xx} + \left[\mathcal{F}^2(t) + v_x\left(\mathcal{F}^4(t) - v_x\mathcal{F}^3(t) - \mathcal{F}^1(t)\right)\right]v_{xx} + \mathcal{G}^1(t) + v_x\mathcal{G}^2(t)\right\},$ $v_t = \mathcal{F}^3(t)e^{2v_x}u_{xx} + e^{v_x}[\mathcal{F}^4(t) - v_x\mathcal{F}^3(t)]v_{xx} + \mathcal{G}^2(t)e^{v_x}.$
$\bar{A}^6_{4,4}$	$\langle \partial_x, \partial_u, \partial_v, t\partial_x + (x+u)\partial_u + (u+v)\partial_v \rangle$	$u_t = e^{2u_x}[\mathcal{F}^1(t) - u_x\mathcal{F}^2(t)]u_{xx} + \mathcal{F}^2(t)e^{2u_x}v_{xx} + \mathcal{G}^1(t)e^{u_x},$ $v_t = e^{2u_x}\left\{\left[\mathcal{F}^3(t) + u_x(\mathcal{F}^1(t) - u_x\mathcal{F}^2(t) - \mathcal{F}^4(t))\right]u_{xx} + [\mathcal{F}^4(t) + u_x\mathcal{F}^2(t)]v_{xx} + \mathcal{G}^2(t) + u_x\mathcal{G}^1(t)\right\}.$

Table 2 (Continued).

Algebra	Realization	Invariant System
$A^1_{4,5}$	$\langle\partial_t, \partial_x, \partial_u, t\partial_t + qx\partial_x + pu\partial_u + v\partial_v\rangle$	$u_t = K_1 v^{2q-1} u_{xx} + K_2 v^{p+2q-2} v_{xx} + K_5 v^{p-1},$
		$v_t = K_3 v^{2q-p} u_{xx} + K_4 v^{2q-1} v_{xx} + K_6.$
$\tilde{A}^1_{4,5}$	$\langle\partial_t, \partial_u, \partial_x, t\partial_t + px\partial_x + qu\partial_u + v\partial_v\rangle$	$u_t = K_1 v^{2p-1} u_{xx} + K_2 v^{2p+q-2} v_{xx} + K_5 v^{q-1},$
		$v_t = K_3 v^{2p-q} u_{xx} + K_4 v^{2p-1} v_{xx} + K_6.$
$\hat{A}^1_{4,5}$	$\langle\partial_x, \partial_t, \partial_u, qt\partial_t + x\partial_x + pu\partial_u + v\partial_v\rangle$	$u_t = v^{2-q}\mathcal{F}^1(v_x)u_{xx} + v^{p-q+1}\mathcal{F}^2(v_x)v_{xx} + v^{p-q}\mathcal{G}^1(v_x),$
		$v_t = v^{3-p-q}\mathcal{F}^3(v_x)u_{xx} + v^{2-q}\mathcal{F}^4(v_x)v_{xx} + v^{1-q}\mathcal{G}^2(v_x).$
$\bar{A}^1_{4,5}$	$\langle\partial_x, \partial_u, \partial_t, pt\partial_t + x\partial_x + qu\partial_u + v\partial_v\rangle$	$u_t = v^{2-p}\mathcal{F}^1(v_x)u_{xx} + v^{1-p+q}\mathcal{F}^2(v_x)v_{xx} + v^{q-p}\mathcal{G}^1(v_x),$
		$v_t = v^{3-p-q}\mathcal{F}^3(v_x)u_{xx} + v^{2-p}\mathcal{F}^4(v_x)v_{xx} + v^{1-p}\mathcal{G}^2(v_x).$
$\check{A}^1_{4,5}$	$\langle\partial_u, \partial_t, \partial_x, qt\partial_t + px\partial_x + u\partial_u + v\partial_v\rangle$	$u_t = K_1 v^{2p-q} u_{xx} + K_2 v^{2p-q} v_{xx} + K_5 v^{1-q},$
		$v_t = K_3 v^{2p-q} u_{xx} + K_4 v^{2p-q} v_{xx} + K_6 v^{1-q}.$
$\grave{A}^1_{4,5}$	$\langle\partial_u, \partial_x, \partial_t, pt\partial_t + qx\partial_x + u\partial_u + v\partial_v\rangle$	$u_t = K_1 v^{2q-p} u_{xx} + K_2 v^{2q-p} v_{xx} + K_5 v^{1-p},$
		$v_t = K_3 v^{2q-p} u_{xx} + K_4 v^{2q-p} v_{xx} + K_6 v^{1-p}.$
$A^2_{4,5}$	$\langle\partial_t, \partial_x, \partial_v, t\partial_t + qx\partial_x + u\partial_u + pv\partial_v\rangle$	$u_t = K_1 u^{2q-1} u_{xx} + K_2 u^{2q-p} v_{xx} + K_5,$
		$v_t = K_3 u^{p+2q-2} u_{xx} + K_4 u^{2q-1} v_{xx} + K_6 u^{p-1}.$
$\tilde{A}^2_{4,5}$	$\langle\partial_t, \partial_v, \partial_x, t\partial_t + px\partial_x + u\partial_u + qv\partial_v\rangle$	$u_t = K_1 u^{2p-1} u_{xx} + K_2 u^{2p-q} v_{xx} + K_5,$
		$v_t = K_3 u^{2p+q-2} u_{xx} + K_4 u^{2p-1} v_{xx} + K_6 u^{q-1}.$
$\hat{A}^2_{4,5}$	$\langle\partial_x, \partial_t, \partial_v, qt\partial_t + x\partial_x + u\partial_u + pv\partial_v\rangle$	$u_t = u^{2-q}\mathcal{F}^1(u_x)u_{xx} + u^{3-p-q}\mathcal{F}^2(u_x)v_{xx} + u^{1-q}\mathcal{G}^1(u_x),$
		$v_t = u^{p-q+1}\mathcal{F}^3(u_x)u_{xx} + u^{2-q}\mathcal{F}^4(u_x)v_{xx} + u^{p-q}\mathcal{G}^2(u_x).$
$\bar{A}^2_{4,5}$	$\langle\partial_x, \partial_v, \partial_t, pt\partial_t + x\partial_x + u\partial_u + qv\partial_v\rangle$	$u_t = u^{2-p}\mathcal{F}^1(u_x)u_{xx} + u^{3-p-q}\mathcal{F}^2(u_x)v_{xx} + u^{1-p}\mathcal{G}^1(u_x),$
		$v_t = u^{1-p+q}\mathcal{F}^3(u_x)u_{xx} + u^{2-p}\mathcal{F}^4(u_x)v_{xx} + u^{q-p}\mathcal{G}^2(u_x).$
$\check{A}^2_{4,5}$	$\langle\partial_v, \partial_t, \partial_x, qt\partial_t + px\partial_x + u\partial_u + v\partial_v\rangle$	$u_t = K_1 u^{2p-q} u_{xx} + K_2 u^{2p-q} v_{xx} + K_5 u^{1-q},$
		$v_t = K_3 u^{2p-q} u_{xx} + K_4 u^{2p-q} v_{xx} + K_6 u^{1-q}.$
$\grave{A}^2_{4,5}$	$\langle\partial_v, \partial_x, \partial_t, pt\partial_t + qx\partial_x + u\partial_u + v\partial_v\rangle$	$u_t = K_1 u^{2q-p} u_{xx} + K_2 u^{2q-p} v_{xx} + K_5 u^{1-p},$
		$v_t = K_3 u^{2q-p} u_{xx} + K_4 u^{2q-p} v_{xx} + K_6 u^{1-p}.$
$A^6_{4,5}$	$\langle\partial_t, \partial_u, \partial_v, pt\partial_x + u\partial_u + qv\partial_v\rangle$	$u_t = K_1 u_{xx} + K_2 e^{(\frac{1-q}{pt})x} v_{xx} + K_5 e^{\frac{x}{pt}} + pu_x,$
		$v_t = K_3 e^{(\frac{q-1}{pt})x} u_{xx} + K_4 v_{xx} + K_6 e^{\frac{qx}{pt}} + \dfrac{pv_x}{q}.$

Table 2 (Continued).

Algebra	Realization	Invariant System
$\tilde{A}^6_{4,5}$	$\langle \partial_t, \partial_u, \partial_v, pt\partial_x + qu\partial_u + v\partial_v \rangle$	$u_t = K_1 u_{xx} + K_2 e^{(\frac{q-1}{pt})x} v_{xx} + K_5 e^{\frac{qx}{pt}} + pu_x/q,$ $v_t = K_3 e^{(\frac{1-q}{pt})x} u_{xx} + K_4 v_{xx} + K_6 e^{\frac{x}{pt}} + pv_x.$
$\hat{A}^6_{4,5}$	$\langle \partial_t, \partial_u, \partial_v, t\partial_x + qu\partial_u + pv\partial_v \rangle$	$u_t = K_1 u_{xx} + K_2 e^{(\frac{q-p}{t})x} v_{xx} + K_5 e^{\frac{qx}{t}} + \frac{u_x}{q},$ $v_t = K_3 e^{(\frac{p-q}{t})x} u_{xx} + K_4 v_{xx} + K_6 e^{\frac{px}{t}} + v_x/p.$
$\bar{A}^6_{4,5}$	$\langle \partial_t, \partial_u, \partial_v, t\partial_x + pu\partial_u + qv\partial_v \rangle$	$u_t = K_1 u_{xx} + K_2 e^{(\frac{p-q}{t})x} v_{xx} + K_5 e^{\frac{px}{t}} + \frac{u_x}{p},$ $v_t = K_3 e^{(\frac{q-p}{t})x} u_{xx} + K_4 v_{xx} + K_6 e^{\frac{qx}{t}} + \frac{v_x}{q}.$
$\breve{A}^6_{4,5}$	$\langle \partial_t, \partial_u, \partial_v, qt\partial_x + u\partial_u + pv\partial_v \rangle$	$u_t = K_1 u_{xx} + K_2 e^{(\frac{1-p}{qt})x} v_{xx} + K_5 e^{\frac{x}{qt}} + qu_x,$ $v_t = K_3 e^{(\frac{p-1}{qt})x} u_{xx} + K_4 v_{xx} + K_6 e^{\frac{px}{qt}} + \frac{qv_x}{p}.$
$\grave{A}^6_{4,5}$	$\langle \partial_t, \partial_u, \partial_v, qt\partial_x + pu\partial_u + v\partial_v \rangle$	$u_t = K_1 u_{xx} + K_2 e^{(\frac{p-1}{qt})x} v_{xx} + K_5 e^{\frac{px}{qt}} + \frac{qu_x}{p},$ $v_t = K_3 e^{(\frac{1-p}{qt})x} u_{xx} + K_4 v_{xx} + K_6 e^{\frac{x}{qt}} + qv_x.$
$A^6_{4,6}$	$\langle \partial_t, \partial_u, \partial_v, qt\partial_x + (pu+v)\partial_u + (pv-u)\partial_v \rangle$	$u_t = \mathcal{F}^1(x)u_{xx} + \mathcal{F}^2(x)v_{xx} + \mathcal{G}^1(x, v_x) - \frac{pqu_x}{p^2+1},$ $v_t = \mathcal{F}^3(x)u_{xx} + \mathcal{F}^4(x)v_{xx} + \mathcal{G}^2(x, v_x) - \frac{qu_x}{p^2+1}.$
$\tilde{A}^6_{4,6}$	$\langle \partial_t, \partial_u, \partial_v, qt\partial_x + (pu-v)\partial_u + (u+pv)\partial_v \rangle$	$u_t = \mathcal{F}^1(x)u_{xx} + \mathcal{F}^2(x)v_{xx} + \mathcal{G}^1(x, u_x) - \frac{pv_x}{p^2+1},$ $v_t = \mathcal{F}^3(x)u_{xx} + \mathcal{F}^4(x)v_{xx} + \mathcal{G}^2(x, u_x) - \frac{pqv_x}{p^2+1}.$
$A^2_{4,8}$	$\langle \partial_u, \partial_t, t\partial_u + v\partial_v, t\partial_t + x\partial_x + (1+q)u\partial_u + qv\partial_v \rangle$	$u_t = K_1 x u_{xx} + K_2 x^{q+2} v^{-1} v_{xx} + K_5 x^q,$ $v_t = K_3 x^{-q} v u_{xx} + K_4 x v_{xx} + K_6 x^{-1} v.$
$A^4_{4,8}$	$\langle \partial_u, \partial_x, t\partial_t + x\partial_u, qt\ln t\partial_t + x\partial_x + (1+q)u\partial_u + v\partial_v \rangle$	$u_t = K_1 t^{-1} v^{2-q} u_{xx} + K_2 t^{-1} v^2 v_{xx} + K_5 t^{-1} v,$ $v_t = t^{-1}\left\{K_3 v^{2(1-q)} u_{xx} + K_4 v^{2-q} v_{xx} + K_6 v^{1-q}\right\}.$
$A^5_{4,8}$	$\langle \partial_v, \partial_x, t\partial_t + x\partial_v, qt\ln t\partial_t + x\partial_x + u\partial_u + (1+q)v\partial_v \rangle$	$u_t = t^{-1}\left\{K_1 u^{2-q} u_{xx} + K_2 u^{2(1-q)} v_{xx} + K_5 u^{1-q}\right\},$ $v_t = K_3 t^{-1} u^2 u_{xx} + K_4 t^{-1} u^{2-q} v_{xx} + K_6 t^{-1} u.$

Lie algebras and the algebras which are semidirect sums of semisimple algebras and solvable Lie algebras need to be considered.

Acknowledgements

We thank Prof. Dr. Ugur Camci for his kind invitation to the 13th Regional Conference on Mathematical Physics (REGCONF13). We are also grateful to The Scientific and Technological Research Council of Turkey (TUBITAK) for support in the scientific visit of Turkey and Akdeniz University, Antalya. We finally thank the Organizing Committee for its organization and great hospitality.

References

1. N. H. Ibragimov (Ed.), *CRC Handbook of Lie Group Analysis of Differential Equations* (CRC Press, Boca Raton, Vol. 1: 1994; Vol. 2: 1995; Vol. 3: 1996).
2. P. J. Olver, *Applications of Lie Groups to Differential Equations* (Springer, New York, 1986).
3. L. V. Ovsiannikov, *Group Analysis of Differential Equations* (Academic, New York, 1982).
4. N. H. Ibragimov, M. Torrisi and A. Valenti, *J. Math. Phys.* **32**, 2988 (1991).
5. M. Molati and F. M. Mahomed *J. Phys. A: Math. and Theor.* **43**, 415203 (2010).
6. S. Lie, *Arch für Math.* **VI** Heft 3, 328 (1881).
7. S. Lie, *Arch für Math.* **VIII**, 187 (1883).
8. F. M. Mahomed and P. G. L. Leach, *J. Math. Phys.* **30**, 2770 (1989).
9. N. H. Ibragimov and F. M. Mahomed, Ordinary Differential Equations in *CRC Handbook of Lie Group Analysis of Differential Equations* (CRC Press, Boca Raton, Vol. 3: 1996).
10. P. Basarab–Horwath, V. Lahno and R. Zhdanov, *Acta Applicandae Mathematicae* **69** 43 (2001).

GREEN'S FUNCTIONS AND TRANSITION AMPLITUDES FOR TIME-DEPENDENT LINEAR HARMONIC OSCILLATOR WITH LINEAR TIME-DEPENDENT TERMS ADDED TO THE HAMILTONIAN

M. A. RASHID* and M. U. FAROOQ†

Center for Advanced Mathematics and Physics, National University of Sciences and Technology, H-12, Islamabad, Pakistan
**E-mail: muneerrshd@yahoo.com*
†E-mail: m_ufarooq@yahoo.com

We calculate the Green's function for the time-dependent linear harmonic oscillator with linear time-dependent terms added to the Hamiltonian and use it to calculate the transition amplitudes between states of the time-independent linear harmonic oscillator. In the absence of the linear terms, transitions are only allowed from even (odd) to even (odd) parity states. However, in the case when the linear terms are present, transitions from even (odd) to odd (even) states are also allowed.

1. Introduction

Linear harmonic oscillator is very important from the point of view of both classical and quantum mechanics with applications in every walk of life. In quantum mechanics, it is one of the models for which the corresponding Schrödinger equation can be solved analytically to give the bound state energies and the corresponding wave-functions. This model thus acts as a laboratory for those working in quantum mechanics. In every book on quantum mechanics, time-independent linear harmonic oscillator is dealt with quite extensively. The time-dependent linear harmonic oscillator is not dealt with in the same detail as generally analytic solutions do not exist.

Time-dependent harmonic oscillators have, however, been considered by many authors[1–4]. From the point of physical applications, Parker[5] applied the *alpha* and *beta* coefficients to the cosmological creation of particles in an expanding universe. Earlier Kanai[2] had considered a simple form of the time-dependent linear oscillator. Londonwitz et al.[6] proceeded to calculate the Green's function for the general form of Kanai's model and used it to calculate the corresponding transition amplitudes[7].

Rashid and Mahmood bypassed the Green's function formalism and used standard operator techniques to obtain the transition amplitudes[8]. In Ref. 9, Rashid calculated, using the same operator techniques, the transition amplitudes for the time-dependent linear harmonic oscillator with the addition of linear terms.

In this paper, we complete the calculations by calculating the Green's function for this case, i.e. when the time-dependent oscillator contains linear terms. This technique, as a byproduct, gives the transition amplitudes exactly as calculated earlier.

In the absence of linear terms, the Hamiltonian is invariant under parity transformation. Thus transitions are only allowed from even (odd) to even (odd) parity states. In the presence of linear terms, the transitions from even (odd) to odd (even) states are also allowed as the Hamiltonian is not invariant under parity transformation.

This paper is organized as follows: In Section 2, we introduce the model and express the effect of the time-dependence in terms of a time-dependent unitary operator. In terms of this operator, we define $x_\pm(t)$, $p_\pm(t)$ operators and calculate the coefficients which appear in their expressions as inhomogeneous linear combination of x and p. In Section 3, we define the Green's function and calculate it explicitly in terms of the coefficients mentioned before. Section 4 is devoted to the calculation of the transition amplitudes.

2. The Model

We consider the Hamiltonian

$$H(t) = f(t)\frac{p^2}{2m} + \frac{1}{2}g(t)m\omega^2x^2 + u(t)p + v(t)x. \tag{1}$$

The usual time-independent harmonic oscillator has $f(t) = g(t) = 1$, $u(t) = v(t) = 0$. The time-dependent linear harmonic oscillator considered above has t-dependent continuous real functions $f(t)$, $g(t)$, $u(t)$, $v(t)$ where $u(t)$, $v(t)$ are the coefficients of the linear terms which when absent make the Hamiltonian parity conserving (as it is then quadratic in x and p). The reality of the time-dependent functions makes the Hamiltonian hermitian.

The effect of time-dependence can be expressed in terms of a unitary (time-dependent) operator $U(t)$ which transform the wave function $\psi(x,0)$ to $\psi(x,t)$. In other words

$$\psi(x,t) = U(t)\psi(x,0), \tag{2}$$

where

$$U(t)U^\dagger(t) = U^\dagger(t)U(t) = 1. \tag{3}$$

The time-dependent Schrödinger equation

$$i\hbar\frac{\partial}{\partial t}\psi(x,t) = H(t)\psi(x,t), \tag{4}$$

results in

$$i\hbar\frac{d}{dt}U(t) = H(t)U(t), \tag{5}$$

as an equation for $U(t)$. In the case, when H is time-independent, we can formally solve it to obtain

$$U(t) = e^{-i\frac{H}{\hbar}t}. \tag{6}$$

However if H depends upon time, no simple expression for $U(t)$ exists. The time-dependence of x and p can be expressed by defining the operator $x_\pm(t)$ and $p_\pm(t)$ in terms of $U(t)$ as

$$x_+(t) = U^\dagger(t)xU(t), \tag{7}$$

$$x_-(t) = U(t)xU^\dagger(t), \tag{8}$$

$$p_+(t) = U^\dagger(t)pU(t), \tag{9}$$

$$p_-(t) = U(t)pU^\dagger(t). \tag{10}$$

We may invert these equations to obtain, for example,

$$x = U(t)x_+(t)U^\dagger(t), \tag{11}$$

and

$$x = U^\dagger(t)x_-(t)U(t). \tag{12}$$

In the presence of linear terms, we have

$$x_+(t) = a(t)x + b(t)p + y_1(t), \tag{13}$$

$$p_+(t) = c(t)x + d(t)p + y_2(t), \tag{14}$$

where $a(0) = 1 = d(0)$, $b(0) = c(0) = 0$, $y_1(0) = y_2(0) = 0$. As $[x_+(t), p_+(t)] = [x,p]$, we have

$$a(t)d(t) - b(t)c(t) = 1, \tag{15}$$

or the determinant $\begin{vmatrix} a(t) & b(t) \\ c(t) & d(t) \end{vmatrix}$ is time-independent. Using the equation

$$\frac{\partial x_+}{\partial t} = \frac{1}{i\hbar}[x_+(t), H_+(t)], \tag{16}$$

where

$$H_+(t) = U^\dagger(t)H(t)U(t), \tag{17}$$

in terms of $H(t)$ defined in (1), we have

$$\frac{\partial x_+}{\partial t} = f(t)\frac{p_+(t)}{m} + u(t). \tag{18}$$

In this equation, we substitute from Eqs. (13), (14) for $x_+(t)$, $p_+(t)$ to arrive at

$$\dot{a}(t) = \frac{f(t)}{m}c(t), \tag{19}$$

$$\dot{b}(t) = \frac{f(t)}{m}d(t), \tag{20}$$

$$\dot{y}_1(t) = \frac{f(t)}{m}y_2(t) + u(t). \tag{21}$$

Similarly from

$$\frac{\partial p_+}{\partial t} = \frac{1}{i\hbar}[p_+(t), H_+(t)] = -g(t)m\omega^2 x_+(t) - v(t), \tag{22}$$

we obtain

$$\dot{c}(t) = -m\omega^2 g(t)a(t), \tag{23}$$

$$\dot{d}(t) = -m\omega^2 g(t)b(t), \tag{24}$$

$$\dot{y}_2(t) = -m\omega^2 g(t)y_1(t) - v(t). \tag{25}$$

The first-order coupled differential equations in Eqs. (19)-(21), (23)-(25) may be decoupled to obtain second-order equations in the form

$$\ddot{Z}(t) - \frac{f'(t)}{f(t)}\dot{Z}(t) + \omega^2 f(t)g(t)Z(t) = 0, \tag{26}$$

for the functions $a(t)$ and $b(t)$ and

$$\ddot{Z}(t) - \frac{g'(t)}{g(t)}\dot{Z}(t) + \omega^2 f(t)g(t)Z(t) = 0, \tag{27}$$

for $c(t)$ and $d(t)$. For $y_1(t)$, $y_2(t)$, we obtain second-order inhomogeneous linear differential equations with the homogeneous parts the same as in Eqs. (26) and (27) respectively.

Using the initial conditions $a(0) = d(0) = 1$, $b(0) = c(0) = 1$, these have unique solutions for $a(t)$, $b(t)$, $c(t)$, $d(t)$, $y_1(t)$, $y_2(t)$ in terms of the known $f(t)$, $g(t)$, $u(t)$ and $v(t)$ in the model.

Similar techniques result in

$$x_-(t) = d(t)x - b(t)p - d(t)y_1(t) + b(t)y_2(t), \tag{28}$$

$$p_-(t) = -c(t)x + a(t)p + c(t)y_1(t) - a(t)y_2(t). \tag{29}$$

In particular, for the time-independent case and without the linear terms($f(t) = g(t) = 1$, $u(t) = v(t) = 0$), these can be immediately solved to give

$$a(t) = d(t) = \cos\omega t,\ b(t) = \frac{1}{m\omega}\sin\omega t,\ c(t) = -m\omega\sin\omega t, \tag{30}$$

which satisfy the initial conditions $a(0) = d(0) = 1$, $b(0) = c(0) = 0$, and the determinantal condition $\begin{vmatrix} a(t) & b(t) \\ c(t) & d(t) \end{vmatrix} = 1$.

3. The Green's Function

The Green's function is defined by

$$< x|U(t)|x' > = G(x, x'; t), \tag{31}$$

and $\psi(x,t)$ is obtained from $\psi(x,0)$ by

$$\psi(x,t) = \int G(x, x'; t)\psi(x', 0)dx', \tag{32}$$

subject to

$$\int \psi^\dagger(x,t)\psi(x,t)dx = \int \psi^\dagger(x,0)\psi(x,0)dx, \tag{33}$$

and

$$Lt_{t\to 0}\ \psi(x,t) = \psi(x,0). \tag{34}$$

These equations translated in terms of the Green's function become

$$\int G^\dagger(x, x''; t)G(x, x'; t)dx = \delta(x'' - x'), \tag{35}$$

and

$$Lt_{t\to 0}\ G(x, x'; t) = \delta(x' - x). \tag{36}$$

These non-linear conditions normalize the Green function.

Now Eq. (7) implies $U(t)x_+(t) = xU(t)$ and using (13) and $p = -i\hbar\frac{\partial}{\partial x}$ in the x-representation results in

$$\frac{\partial}{\partial x'}G(x, x'; t) = \frac{1}{i\hbar b(t)}(a(t)x' - x + y_1(t))G(x, x'; t), \tag{37}$$

which can be immediately integrated to yield

$$G(x, x'; t) = \exp\left[\frac{1}{i\hbar b(t)}\left(a(t)\frac{x'^2}{2} - xx' + y_1(t)x'\right)\right] g(x, t), \tag{38}$$

where $g(x, t)$ is, so far, an arbitrary function of its variables x and t. In Eq. (38), we have the x' dependence of $G(x, x'; t)$ completely determined.

Next we utilize (8), (28) to arrive at

$$\frac{\partial}{\partial x} G(x, x'; t) = \frac{1}{i\hbar b(t)} \left[d(t)x - x' - d(t)y_1(t) + b(t)y_2(t)\right]. \tag{39}$$

In this equation, we substitute the expression for $G(x, x'; t)$ determined earlier Eq. (38) in terms of a single unknown function $g(x, t)$ to find

$$\frac{\partial g(x,t)}{\partial x} - \frac{g(x,t)}{i\hbar b(t)} x' = \frac{1}{i\hbar b(t)} \left[d(t)x - x' - d(t)y_1(t) + b(t)y_2(t)\right] g(x, t). \tag{40}$$

Note that the only terms in x' on both sides of this equation cancel to leave an equation for $g(x, t)$. This is a consistency requirement as $g(x, t)$ does not depend upon x'. We can integrate the equation for $g(x, t)$ to find

$$g(x, t) = F(t) \exp\left[\frac{1}{i\hbar b(t)}\left(\frac{1}{2}d(t)x^2 - d(t)y_1(t)x + b(t)y_2(t)x\right)\right], \tag{41}$$

which when substituted in Eq. (38) yields

$$G(x, x'; t) = F(t) \exp\Big[\frac{1}{i\hbar b(t)}\Big(\frac{1}{2}d(t)x^2 + \frac{1}{2}a(t)x'^2 - xx' + y_1(t)x' - d(t)y_1(t)x + b(t)y_2(t)x\Big)\Big]. \tag{42}$$

Finally, the boundary conditions on the Green's function result in

$$F(t) = \frac{1}{\sqrt{2\pi i\hbar b(t)}}. \tag{43}$$

Thus we arrive at the Green's function (sometimes called the propagator) for the time-dependent linear harmonic oscillator with the addition of the linear term as

$$G(x, x'; t) = \frac{1}{\sqrt{2\pi i\hbar b(t)}} \exp\Big[\frac{1}{i\hbar b(t)}\Big(\frac{1}{2}d(t)x^2 + \frac{1}{2}a(t)x'^2 - xx' - d(t)y_1(t)x + b(t)y_2(t)x + y_1(t)x'\Big)\Big]. \tag{44}$$

The Green's function for the time-dependent linear harmonic oscillator without the linear terms was obtained by Londonwitz et al in 1979[7]. In Eq. (44), we have given the Green's function for the case which includes the time-dependent linear terms.

4. Transition Amplitudes

The Green's function contains, in a concise form, all the information about the time-dependent linear harmonic oscillator. The ususal time-independent harmonic oscillator with no linear terms can now act as our laboratory and we attempt to calculate the transition amplitudes between the stationary states. The time-independent linear harmonic oscillator has the energy spectrum given by

$$E_n = \hbar\omega(n + \frac{1}{2}), \tag{45}$$

and the corresponding wave-function

$$\psi_n(x) = \Big(\frac{\alpha}{\pi 2^n n!}\Big)^{\frac{1}{2}} H_n(\alpha x) e^{-\frac{1}{2}\alpha^2 x^2}, \tag{46}$$

where

$$\alpha = \sqrt{\frac{m\omega}{\hbar}}, \tag{47}$$

in terms of the angular frequency ω appearing in the Hamiltonian. The transition amplitudes we need to obtain are given by

$$a_{nm}(t) = \int_{-\infty}^{\infty} dx \int_{-\infty}^{\infty} dx' \psi_n(x) G(x, x'; t) \psi_m(x'). \tag{48}$$

The calculation of these integrals is not straightforward. The usual method is to use a generating function

$$S(\alpha x, s) = e^{-s^2 + 2\alpha x s} = \sum \frac{H_n(\alpha x)}{n!} s^n, \tag{49}$$

and calculate the integral by replacing the Hermite polynomials in the wave-function by $S(\alpha x, s)$ and $S(\alpha x', s')$. Then $a_{nm}(t) = \frac{\alpha}{\pi\sqrt{2^{n+m} n! m!}}$ multiplied by the coefficient of $\frac{s^n s'^m}{n! m!}$ in the integral

$$\int\int dx dx' \exp\left[-\frac{1}{2}\alpha^2(x^2 + x'^2) e^{2\alpha(sx + s'x') - s^2 - s'^2}\right] G(x, x'; t), \tag{50}$$

or in the integral

$$\begin{aligned} \frac{1}{\sqrt{2\pi i\hbar b(t)}} e^{-(s^2 + s'^2)} \int_{-\infty}^{\infty}\int_{-\infty}^{\infty} dx dx' \exp\Big[&-\frac{1}{2}(\alpha^2 + i\beta d) x^2 \\ &- \frac{1}{2}(\alpha^2 + i\beta a) x'^2) + i\beta x x' + i\beta(dy_1 - by_2) x \\ &- i\beta y_1 x' + 2\alpha(sx + s'x')\Big], \end{aligned} \tag{51}$$

where $\beta(t) = \frac{1}{\hbar b(t)}$. Note that in the above integral, the imaginary quantities are explicitely mentioned.

Evaluation of the integral

There is a standard procedure for evaluating the integral in Eq. (51). We have to rotate the x, x' axes to y, y' axes given by

$$\begin{aligned} x &= y\sin\theta + y'\cos\theta, \\ x' &= -y\cos\theta + y'\sin\theta, \end{aligned} \tag{52}$$

where we choose the angle which leaves no term in yy' in the exponential. If we write the quadratic terms as $ax^2 + bx'^2 + cxx'$, then transformation to y, y' gives

$$\begin{aligned} & a(y\sin\theta + y'\cos\theta)^2 + b(-y\cos\theta + y'\sin\theta)^2 \\ & + c(y\sin\theta + y'\cos\theta)(-y\cos\theta + y'\sin\theta) \\ = & (a\sin^2\theta + b\cos^2\theta - c\cos\theta\sin\theta)y^2 + (a\cos^2\theta + b\sin^2\theta + c\cos\theta\sin\theta)y'^2 \\ & + \left((2a\sin\theta\cos\theta - 2b\cos\theta\sin\theta + c(\sin^2\theta - \cos^2\theta)\right) yy'. \end{aligned} \tag{53}$$

The coefficient of yy' can be simplified to $(a-b)\sin 2\theta - c\cos 2\theta$ and will vanish if we choose the angle θ such that

$$\tan\left(2\theta(t)\right) = \frac{c(t)}{a(t) - b(t)}. \tag{54}$$

In our case $a(t) = -\frac{1}{2}\left(\alpha^2 + i\beta d\right)$, $b(t) = -\frac{1}{2}\left(\alpha^2 + i\beta a\right)$, $c(t) = i\beta$. Thus

$$\tan\left(2\theta(t)\right) = \frac{2}{a(t) - d(t)}. \tag{55}$$

Then since $a(t)$, $d(t)$ are both real, this is a real angle of rotation. The coefficients of y^2 and y'^2 can then be obtained using

$$\begin{aligned} \cos 2\theta &= \frac{a(t) - d(t)}{\sqrt{4 + \left(a(t) - d(t)\right)^2}} \quad \text{and} \\ \sin 2\theta &= \frac{2}{\sqrt{4 + \left(a(t) - d(t)\right)^2}}. \end{aligned} \tag{56}$$

Noting that the transformation is an orthogonal transformation $dxdx' \to dydy'$, we finally have to absorb the linear terms in y and y' to make complete squares and use the integral

$$\int_{-\infty}^{\infty} e^{-\alpha(y+k)^2} dy = \frac{\pi}{\sqrt{\alpha}}, \tag{57}$$

to complete the evaluation. The final answer will be an exponential involving a quadratic expression in s and s'. For the case when linear terms are absent, this takes the form

$$\exp\left[\frac{1}{\lambda}(\mu^* s^2 - 4i\sigma ss' + \mu s'^2)\right], \tag{58}$$

where $\sigma(t) = \frac{1}{m\omega b(t)}$ and

$$\begin{aligned} \mu &= 1 + i\sigma(t)\,(a(t) - d(t)) - \sigma^2(t)\,(1 - a(t)d(t)) \quad \text{and} \\ \lambda &= 1 - i\sigma(t)\,(a(t) + d(t)) + \sigma^2(t)\,(1 - a(t)d(t)) \,. \end{aligned} \tag{59}$$

Examining the integral in Eq. (51), we note that the presence of the linear term merely modifies s, s' to s_1 and s_1' where

$$\begin{aligned} s_1 &= s + \frac{i\beta}{2\alpha}\,(dy_1 - by_2) \\ s_1' &= s_1 - \frac{i\beta y_1}{2\alpha}, \end{aligned}$$

in the part without the external factor $e^{-(s^2+s'^2)}$.

The remaining calculation is now obvious. After evaluating the integral, we have to evaluate $a_{nm}(t)$ given in Eq. (49). We have illustrated the procedure. The final answer will be the same as presented earlier in Eqs. (39, 42) in Ref 9. As remarked earlier, the Hamiltonian without the linear terms is quadratic in x and p and is therefore invariant under parity transformation. This allows for transitions from even (odd) parity states to even (odd) parity states only. Thus the only coefficients $a_{nm}(t)$ which will be non-zero are those with $n - m =$ an even integer. In the presence of linear terms, we will have transitions from even (odd) parity to odd (even) parity states also. Thus $a_{nm}(t)$ are non-zero for all non-negative integral values of n and m.

References

1. P. Camiz et al., A *J. Math. Phys.* **12**, 2040 (1971).
2. E. Kanai, *Prog. Theo. Phys. Kyoto* **3**, 440 (1948).
3. Z. H. Kerner, *Can. J. Phys.* **36**, 3719 (1958).
4. W. K. H. Stevens, *Proc. Phys. Soc. London* **72**, 1027 (1958).
5. L. Parker, *Phys. Rev.* **183**, 1057 (1969) (see in particular Eq. (13)).
6. Londonvitz, A. M. Levine and W. M. Schreiber, *Phys. Rev. A* **20**, 1162 (1979).
7. Londonvitz, A. M. Levine and W. M. Schreiber, *J. Math. Phys.* **21**, 2159 (1988).
8. M. A. Rashid and A. Mahmood, *J. Phys. A: Math. Gen.* **34**, 8185 (2001).
9. M. A. Rashid, in the *Proceeding of the 12th Regional Conference on Mathematical Physics* (World Scientific, 2006) p. 39.

TIME DEPENDENT HARMONIC OSCILLATOR AND QUASI-COHERENT STATES

NURİ ÜNAL
Akdeniz University, Department of Physics,
PK 07058, Antalya, Turkey
E-mail: nuriunal@akdeniz.edu.tr

We reduce the solution of the Schrödinger equation to the Riccati equation for the harmonic oscillators with time dependent parameters, the mass and the frequency. We also discuss the linear coupling of the harmonic oscillator with the external driving force. We solve the Riccati equation for three special cases: (i) A time dependent harmonic oscillator with $M(t)\omega(t) = \text{const.}$, (ii) a driven harmonic oscillator, (iii) a harmonic oscillator with exponentially increasing mass parameter. In the first and second cases, we obtain the coherent states. In third case, we find a quasi-coherent state.

Keywords: Harmonic oscillator; Quasi-coherent states.

1. Introduction

In classical and quantum mechanics, time dependent Hamiltonians, such as the Caldirola–Kanai Hamiltonian, are used to represent the non-conservative systems[1,2]. In order to discuss the time dependent harmonic oscillator, the usual approach is to use the invariants of the system[3] and many authors pursued this approach[4−7]. In quantum mechanics, the action of the driving force on the harmonic oscillator and the time dependent harmonic oscillator were discussed by Feynman[8,9] and Abdalla, respectively[10]. There has been much interest in time-dependent harmonic oscillator in different areas of physics[11−35].

Dodonov and Manko tried to construct the coherent states for the harmonic oscillators with time dependent parameters by using the two invariants of the problem[5]. They derived two constant hermitian conjugate operators A and $A^\dagger$ from the invariants and found the constant eigenvalues and eigenstates of the lowering operator, A.

In this study, we reduce the Schrödinger equation for time-dependent harmonic oscillator into Riccati equation by introducing a conformal time and new time dependent coordinate and momentum. We solve the classical equations of motion and the Schrödinger equation for three special cases; A time dependent harmonic oscillator with $M(t)\omega(t)$ =const., a driven harmonic oscillator and a harmonic oscillator with exponentially increasing mass parameter. We find the coherent or quasi-coherent states of the time dependent harmonic oscillators and discuss the uncertainties for these examples.

2. Time dependent harmonic oscillators

2.1. *Action*

For the time dependent harmonic oscillator with the linear coupling external time dependent force, $F(t)$, the Lagrangian is given as

$$L = p\dot{x} - \frac{\omega(t)}{2}\left(\frac{p^2}{M(t)\,\omega(t)} + M(t)\,\omega(t)\,x^2\right) + M(t)F(t)x.$$

Here, $M(t)$ and $\omega(t)$ represents time dependent parameters, mass and frequency. We introduce the new coordinate, X as

$$X = \sqrt{M(t)\,\omega(t)}x. \tag{1}$$

Then, P is given as $P = (M(t)\omega(t))^{-1/2}\,p$ and we write the Lagrangian as

$$L = P\dot{X} - H. \tag{2}$$

Here, H is the Hamiltonian given as

$$H = \frac{\omega(t)}{2}\left(P^2 + X^2\right) + \frac{1}{4}\frac{d}{dt}\ln\left[M(t)\,\omega(t)\right](PX + XP) - \sqrt{\frac{M(t)}{\omega(t)}}F(t)X. \tag{3}$$

We introduce a conformal time, τ, as

$$d\tau = \omega(t)\,dt.$$

Then, the action becomes

$$A = \int d\tau\left[P\frac{dX}{d\tau} - \frac{1}{2}\left(P^2 + X^2\right) - \jmath(\tau)(PX + XP) - k(\tau)X\right],$$

where $\jmath(\tau)$ is

$$\jmath(\tau) = \frac{1}{4}\frac{d}{d\tau}\ln\left[M(t)\,\omega(t)\right] \quad \text{and} \quad k(\tau) = \sqrt{\frac{M(t)}{\omega^3(t)}}F(t).$$

2.2. *Quantum equations*

The Schrödinger equation is given as

$$i\frac{\partial\psi(X,\tau)}{\partial\tau} = \left[-\frac{\partial^2}{\partial X^2} + X^2 - ij(\tau)(2X\frac{\partial}{\partial X}+1) + k(\tau)X\right]\psi(X,\tau). \tag{4}$$

We propose a non-stationary solution of the Schrödinger equation in the following form:

$$\psi(a^*,\tau) = N(\tau)\exp\left[-\frac{a_1(\tau)X^2}{2} + a_2(\tau)X\right], \tag{5}$$

and substitute it into Eq. (4) $a_i(\tau)$ for $i = 1, 2$:

$$\begin{aligned} -i\dot{a}_1 &= -a_1^2 + 1 + 2ij(\tau)a_1, \\ i\dot{a}_2 &= a_1a_2 - ij(\tau)a_2 + k(\tau). \end{aligned} \tag{6}$$

Eq. (6) shows that $a_2(t)$ satisfies a Riccati equation. Thus, the Schrödinger equation is reduced to Riccati equation.

2.3. *The expectation values and uncertainties*

In order to evaluate the expectation values, we need the following scalar product:

$$(\psi,\psi) = |N(\tau)|^2\sqrt{\frac{2\pi}{M\omega(a_1+a_1^*)}}\exp\left[\frac{(a_2+a_2^*)^2}{2(a_1+a_1^*)}\right],$$

where

$$\sqrt{M\omega} = 2\int^{\tau} j(\tau')\,d\tau'.$$

Then, we evaluate the expectation values for X and P :

$$<X> = \frac{(a_2+a_2^*)}{(a_1+a_1^*)} \quad \text{and} \quad <P> = \frac{(a_2a_1^*-a_1a_2^*)}{i(a_1+a_1^*)}.$$

The uncertainties are given as

$$(\Delta X)^2 = \frac{1}{(a_1+a_1^*)} \quad \text{and} \quad (\Delta P)^2 = \frac{a_1a_1^*}{(a_1+a_1^*)}. \tag{7}$$

Since a_1 depends on $j(\tau)$ Eq. (7) shows that the uncertainties, ΔX and ΔP, are depend on $j(\tau)$. Therefore, the linear dipole coupling in Hamiltonian does not change ΔX and ΔP.

3. Special cases

Case 1:

If $j(t) = k(t) = 0$, then, $\int^t \omega(t')\,dt' = \tau$. In this case, $a_1(\tau) = 1$ and it corresponds to harmonic oscillator and the nonstationary solution of Schrödinger equation gives the minimum uncertainty, coherent state solution,

$$\psi_0(X,\tau) = N(\tau)\exp\left[-\frac{X^2}{2} + a_2(0)\,e^{-i\tau}X\right]. \tag{8}$$

Case 2:

If $j(t) = 0$, then this case corresponds to driven harmonic oscillator and for the nonstationary solution of Schrödinger equation, we choose the solution $a_1(\tau) = 1$. Then, the solution of Schrödinger equation is given in the following form:

$$\psi(X,\tau) = N\exp\left[-\frac{X^2}{2} + a_2(\tau)\,X\right], \tag{9}$$

where $a_2(\tau)$ is given as

$$a_2(\tau) = a_2(0)\,e^{-i\tau} - ie^{-i\tau}\int d\tau' e^{i\tau'}k(\tau'). \tag{10}$$

Eq. (9) shows that in this case also if the initial state is a coherent state it will always stay as the coherent state. The eigenvalues of the coherent state, $a_2(\tau)$, will change according to Eq. (10). Thus, the linear dipole coupling changes the eigenvalue, $a_2(\tau)$, like in the classical harmonic oscillator.

If we expand $\psi(X,\tau)$ into the power series of $\sqrt{2}a_2(\tau)$, we find the expression of coherent states in terms of the energy eigenstates as

$$\psi(X,\tau) = N\exp\left(-\frac{X^2}{2}\right)\sum_{n,m}\frac{H_n(X)}{\sqrt{2^n n!}}\frac{\left(\sqrt{2}a_2(\tau)\right)^n}{\sqrt{n!}}. \tag{11}$$

Case 3:

If $M(t) = M_0 e^{\mu t}$, and $\omega(t) = \omega_0$=constant and $F(t) = 0$. Then,

$$j(t) = \frac{d}{4dt}\ln\left(M(t)\,\omega(t)\right) = \frac{\mu}{4} = \text{constant}. \tag{12}$$

The momentum is

$$p = M(t)\frac{dx}{dt}.$$

For the position, $x(t)$, the classical equations of motion are the same with the damped harmonic oscillator:

$$\frac{d^2x}{dt^2} = -\mu\frac{dx}{dt} - \omega_0^2 x. \tag{13}$$

The solution for position, $x(t)$, is given as

$$x(t) = Re(x_0 e^{-\frac{\mu}{2}t + i\omega_1 t}), \quad \text{with } \omega_1^2 = \omega_0^2 - \left(\frac{\mu}{2}\right)^2.$$

Then,

$$p(t) = Re\left[\left(i\omega_1 - \frac{\mu}{2}\right) M_0 x_0 e^{\frac{\mu t}{2} + i\omega_1 t}\right].$$

Solutions of the quantum equations:

$$-i\frac{da_1}{dt} = (-ia_1)^2 + 1 - \mu a_1 \tag{14}$$

gives

$$a_1 = \omega_1 + \frac{i\mu}{2}.$$

We substitute a_1 into Eq. (6). Then, the solution of the equation for a_2 is given as

$$a_2(t) = a_2(0)\, e^{-i\omega_1 t}. \tag{15}$$

In the similiar way, we evaluate the uncertainties

$$(\Delta x)^2 = \frac{e^{-\mu t}\omega_0}{2\omega_1}, \qquad (\Delta p)^2 = \frac{e^{\mu t}\omega_0}{2\omega_1}. \tag{16}$$

This shows that (Δx) decreases and (Δp) increases exponentially. However, the product of uncertainties is constant and it is given as

$$\Delta x \Delta p = \frac{1}{2}\frac{\omega_0}{\omega_1}. \tag{17}$$

This value is smaller than the value given in Ref. 5 and the product is larger than the minimum value, $\frac{1}{2}$.

4. Conclusion

In this study, we discuss the properties of harmonic oscillators with time dependent parameters, the mass and the frequency. First, we introduce a conformal time and define a new time dependent coordinate and momenta. Then, we write the Schrödinger equation in terms of $X$and τ and reduce the solution of the Schrödinger equation into the Riccati equation.

As the special examples we discussed three cases: In the first example, there is no external force and $M(t)\,\omega(t)$ is constant. Then, if the system is in a coherent state in the beginning it evolves as a coherent state in terms of a conformal time (generalized phase function), $\int^t \omega(t')\,dt' = \tau$. In the second example, we discussed the driven harmonic oscillator and show that the presence of the external force with linear couplings changes only the eigenvalues of the coherent states according to Eq. (13), but it does not change the coherency property of the harmonic oscillator.

As a third example, we discussed the harmonic oscillator with exponentially changing mass parameter, the Caldirola–Kanai Hamiltonian, and find the non-stationary solutions of the Schrödinger equation. These solutions correspond to the quasi-coherent states. For the quasi-coherent states proposed in here, (Δx) decreases and (Δp) increases exponentially and the product is larger than minimum value. Thus, for the system with the linear coupling with the external field, a coherent state evolves as a coherent state with a new eigenvalue. But, time dependent parameters means quadratic coupling with the external field and and the system always have quasi-coherent states.

5. Acknowledgements

This work has been supported by Akdeniz University, Scientific Research Projects Unit.

References

1. P. Caldirola, *Nuovo Cimento* **18**, 393 (1941).
2. E. Kanai, *Prog. Theor. Phys.* **3**, 440 (1948).
3. H.R. Lewis and W.B. Riesenfeld, *J. Math. Phys.* **40**, 1458 (1969).
4. R. W. Hasse, *J. Math. Phys.* **16**, 2005 (1975).
5. V. V. Dodonov and V. I. Manko, *Phys. Rev.* **A 20**, 550 (1979).
6. B. Remaud, *J. Phys. A: Math. Gen.* **13**, 2013 (1980).
7. M. S. Abdalla, *Phys. Rev.* **A 33**, 2870 (1986).
8. R. P. Feynman, Ph. D. thesis, Princeton University, 1942 and *Rev. Mod. Phys.* 20, 367 (1948); reprinted in: "*Feynman's Thesis – A New Approach to Quantum Theory*", (L. M. Brown, Editor), (World Scientific Publishers, Singapore, 2005), pp. 1–69 and 71–112.
9. R. P. Feynman and A. R. Hibbs, *Quantum Mechanics and Path Integrals*, (McGraw–Hill, New York, 1965).
10. M. S. Abdalla, *Phys. Rev.* **A 34**, 4598 (1986).
11. C. M. A. Dantas, I. A . Pedrosa and B. Baseia, *Phys. Rev.* **A 45**, 3 (1992).
12. R. K. Colegrave and M. S. Abdalla, *Opt. Acta* **28**, 495 (1981).

13. N. A. Lemos and C. P. Natividade, *Nuovo Cimento* **399**, 211 (1989).
14. Y. Ben-Aryeh and A. Mann, *Pyhs. Rev.* **A 32**, 552 (1985).
15. T. Rentzsch, W. Schmidt, J. A. Maruhn, H. Stoecker, W. Greiner, in: *Proceedings of Gross Properties of Nuclei and Nuclear Excitations*, Vol. 49–66, (Hischegg, 1988), p. 149.
16. G. Peilert, A. Rosenhauer, T. Rentzsch, H. Stocker, J. Aichelin, W. Greiner, The Nuclear Equation of State, Viscosity and Fragment Flow in *High-Energy Heavy Ion Collisions*, Print-88-0121, Nov. 1987, published in Berkeley Heavy Ion, Vol. 43, 1987, QCD 199:H5:1987, p. 43.
17. M. Bleicher, M. Reiter, A. Dumitru, J. Brachmann, C. Spieles, S.A. Bass, H. Stocker, W. Greiner, *Phys. Rev.* **C 59**, 1844 (1999).
18. D.H. Rischke, Y. Pursun, J.A. Maruhn, H. Stocker, W. Greiner, The phase transition to the quark–gluon plasma and its effects on hydrochynamic Iow, published in *Heavy Ion Phys.***1**, 309 (1995).
19. R.S. Kohlman, J. Joo, A.J. Epstein, in: J.E. Mark (Ed.), *Physical Properties of Polymers Handbook*, (AIP Press, Woodbury, New York, 1996).
20. H.N. Nagashima, R.N. Onody, R.M. Faria, *Phys. Rev.* **B 59**, 905 (1999).
21. H.J. Carmichael, *An Open Systems Approach to Quantum Optics*, (Springer, Berlin, 1993).
22. C.W. Gardner, *Quantum Noise*, (Springer, New York, 1991).
23. H.J. Carmichael, *Statistical Methods in Quantum Optics I*, (Springer, Berlin, 1999).
24. S.L. Sondhi, S.M. Girvin, J.P. Carini, D. Shahar, *Rev. Mod. Phys.* **69**, 315 (1997).
25. H.P. Wei, D.C. Tsui, M.A. Paalanen, A.M.M. Pruisker, *Phys. Rev. Lett.* **61**, 1294 (1998).
26. A. Wolter, P. Rannou, J.P. Travers, *Phys. Rev.* **B 58**, 7637 (1998).
27. Fulin Zuo, M. Angelopoulos, A.G. MacDiarmid and A.J. Epstein, *Phys. Rev.* **B 36**, 3475 (1987).
28. A. Gold, *Z. Phys.* **B 83**, 499 (1991) .
29. A. Gold, *Z. Phys.* **B 81**,155 (1990).
30. A.P. Betenev and V.V. Kurin, *Phys. Rev.* **B 56**, 7855 (1997).
31. A.A. Golubov, B.A. Malomed and A.V. Ustinov, *Phys. Rev.* **B 54**, 3047 (1996).
32. H.G. Schuster, *Deterministic Chaos*, 2nd Edition, (VCH, Weinheim, 1989).
33. E. Ott, *Chaos in Dynamical Systems*, (Cambridge University Press, Cambridge, 1993).
34. J.R. Ackerhalt, P.W. Milonni, M.-L. Shih, *Phys. Rep.* **128**, 205 (1985).
35. A.K. Dhara and S.V. Lawande, *Phys. Rev.* **A 30**, 560 (1984).

PART B
General Relativity and Cosmology

INTEGRABILITY CONDITIONS FOR CONFORMAL RICCI COLLINEATION EQUATIONS

M. AFZAL*, U. CAMCI** and K. SAIFULLAH*,†

*Department of Mathematics,
Quaid-i-Azam University,
Islamabad, Pakistan
†E-mail: saifullah@qau.edu.pk

**Akdeniz University, Department of Physics & TUG,
PK 07058 Antalya, Turkey
E-mail: ucamci@tug.tubitak.gov.tr

We derive integrability conditions in the form of differential constraints for solving the conformal Ricci collineation equations for cylindrically symmetric static spacetimes. Once these constraint equations are solved one gets the Lie algebras for conformal collineation vectors for these spacetimes.

Keywords: Conformal Ricci collineations; Cylindrically symmetric spacetimes.

1. Introduction

In Einstein's theory of general relativity the relation between spacetime geometry and distribution and motion of matter embodied in the Einstein field equations (EFEs)

$$G_{ab} = R_{ab} - \frac{1}{2}Rg_{ab} = \kappa T_{ab}. \tag{1}$$

Here G_{ab} is the Einstein tensor, R is the trace of the Ricci tensor, R_{ab}, which itself is a trace of the Riemann curvature tensor, R^a_{bcd}, and T_{ab} is the stress-energy tensor which gives the matter field in the spacetime. Also $\kappa = 8\pi G/c^2$, where G is the gravitational constant and c is the speed of light.

On taking variation of indices $a, b = 0, 1, 2, 3$, Eq. (1) is a system of ten coupled highly non-linear second order partial differential equations along with twenty unknown functions that is, ten components of the stress-energy tensor T_{ab} and ten components of the metric tensor g_{ab} of four spacetime coordinates x^a. To solve these equations we need to determine the ten components of the stress-energy tensor as well as ten components of g_{ab} of four spacetime variables. Also R_{ab} and its first and second derivatives are nonlinear functions of g_{ab}, so analytically solutions are very difficult. One way to tackle this problem is by making certain assumptions on the matter contents of the space. For example by considering different types of T_{ab}, the Schwarzchild, Reissner-Nordstrom and Robertson-Walker[1] solutions were obtained. These solutions have played an important role in the discussion of many physical problems. On the other hand, if no particular matter distribution is assumed then solutions can be obtained by imposing symmetry conditions on the algebraic structure of the metric, the Ricci tensor or the Riemann tensor[1,2].

Spacetimes were classified[3–5] according to the minimal symmetry of their isometries or Killing vectors (KVs) and a list of metrics admitting these isometries or collineations was provided. Because of the physical relevance of homotheties, a complete classification by homotheties was also obtained[6]. Conformal Killing vectors (CKVs) generate constraints of motion along null geodesic for massless particles and provide conservation laws. The general solutions of the conformal Killing equation in static spherically symmetric and plane symmetric spacetimes were also provided[7,8].

It is of mathematical interest to look into the classification by collineations and conformal collineations of other tensors. Since the Ricci tensor arises naturally from the Riemann curvature tensor (with components R^a_{bcd} and $R^c_{acb} = R_{ab}$) and hence from the connection, the study of symmetries of Ricci tensor R_{ab} has a natural geometrical significance[9–15]. A complete classification of static spherically symmetric spacetimes according to their conformal Ricci collineations (CRCs)[16] showed that the set of all CRCs is a vector space but it may be infinite dimensional when the Ricci tensor is degenerate. But if R_{ab} is non-degenerate then Lie algebra of CRCs is finite dimensional and the maximum dimension of the group of CRCs is 15 and this occurs if and only if the Ricci tensor regarded as metric is conformally flat[2]. Our interest is to investigate the CRCs of static cylindrically symmetric spacetimes. The stationary cylindrically symmetric fields admit three Killing vectors, ∂_t, ∂_θ, ∂_z, as the minimal symmetry fields. In this paper after setting up the conformal Ricci collineation equations in static cylindrically symmetric spacetimes, we provide the constraint equations that need to be solved in order to obtain a complete solution of the CRC equations.

2. Spacetime symmetries

Here we formally define some of the spacetime symmetries[2,17–19].

- *Killing vectors or isometries (KVs)*:

The geometry is given by the metric tensor. A vector field ξ is called a Killing vector (KV) or isometry if Lie derivative of the metric tensor along the vector field remains invariant, that is

$$\pounds_{\xi} g_{ab} = 0. \tag{2}$$

In component form this can be written as

$$g_{ab,c}\xi^{c} + g_{ac}\xi^{c}_{,b} + g_{cb}\xi^{c}_{,a} = 0. \tag{3}$$

Its solutions are KVs or isometries. The set of all solutions form a Lie algebra and generate a Lie group of transformations. Most explicit solutions of EFEs have been found using Killing symmetries. KVs have been used to derive the most general axially symmetric stationary metric. These symmetries help in describing the kinematics and dynamic properties of spaces.

- *Ricci collineations (RCs)*:

A vector ξ^a is a Ricci collineation (RC) if Lie derivative of R_{ab} with respect to ξ^a vanishes

$$\pounds_{\xi} R_{ab} = 0. \tag{4}$$

Since the Ricci tensor is built from the metric tensor, it must inherit its symmetries. Thus if the Lie derivative of g_{ab} vanishes, it must vanish for R_{ab} also. Hence every KV is an RC but the converse may not be true. The RCs which are not KVs are called proper RCs[12]. For the Einstein spaces, $R_{ab} \propto g_{ab}$, the RCs and isometries coincide. For static spherically symmetric spacetime it has been shown that the RCs and isometries, for higher than minimal, are identical if the determinant of the Ricci tensor is non-zero (and finite)[4].

- *Matter collinations (MCs)*:

If R_{ab} in Eq. (4) is replaced by the stress-energy tensor T_{ab}, then the vector ξ is called matter collineation (MC). In component form this becomes

$$T_{ab,c}\xi^{c} + T_{ac}\xi^{c}_{,b} + T_{cb}\xi^{c}_{,a} = 0.$$

By virtue of Einstein's field equations the stress-energy tensor and Ricci tensor have same algebraic structure. Obviously every KV is a MC, but converse may not be true, in general. Matter collineations for different spacetimes have been discussed in literature[14,19–21].

- *Homothetic motions (HMs)*:

A vector ξ is said to be homothetic motion (HM) if the right hand side of Eq. (2) is replaced by a constant times the metric tensor

$$\pounds_{\xi} g_{ab} = 2\sigma g_{ab}, \tag{5}$$

where σ is a non-zero constant. These symmetries are also called homotheties. The spherically symmetric[6], plane symmetric[22] and cylindrically symmetric[23] static Lorentzian manifolds have been classified according to their homotheties and metrics.

- *Conformal Killing vectors (CKVs)*:

If σ defined in Eq. (5) is not a constant but an arbitrary function of spacetime coordinates, the symmetries are called conformal isometries or conformal Killing vectors (CKVs).

- *Conformal Ricci collineations (CRCs)*:

When the metric tensor of Eq. (5) is replaced by a relevant tensor then we get conformal collineations of that tensor. We define conformal Ricci collineations[16,24] if the Lie derivative of Ricci tensor along the vector field ξ remains invariant up to a conformal factor ψ i.e.

$$\pounds_{\xi} R_{ab} = 2\psi R_{ab}, \tag{6}$$

or in component form

$$R_{ab,c}\xi^c + R_{ac}\xi^c_{,b} + R_{cb}\xi^c_{,a} = 2\psi R_{ab}, \tag{7}$$

where the conformal factor ψ is a function of all spacetime coordinates. If ψ is constant then ξ gives homothetic Ricci collineations (HRCs)[25] and if ψ is zero then it gives Ricci collineations (RCs).

3. Conformal Ricci collineation equations

The most general static cylindrically symmetric metric can be written as[1]

$$ds^2 = e^{\nu(\rho)}dt^2 - d\rho^2 - a^2 e^{\lambda(\rho)}d\theta^2 - e^{\mu(\rho)}dz^2, \tag{8}$$

where a has the dimensions of length. As the above metric is diagonal and the metric coefficients depend on ρ only, the only non-zero components of the Ricci tensor are

$$R_{00} = \frac{e^{\nu(\rho)}}{2}\left(2\nu^{''} + \nu^{'2} + \nu^{'}\lambda^{'} + \nu^{'}\mu^{'}\right), \tag{9}$$

$$R_{11} = -\left(\frac{\nu^{''}}{2} + \frac{\lambda^{''}}{2} + \frac{\mu^{''}}{2} + \frac{\nu^{'2}}{4} + \frac{\lambda^{'2}}{4} + \frac{\mu^{'2}}{4}\right), \tag{10}$$

$$R_{22} = -a^2\frac{e^{\lambda(\rho)}}{2}\left(2\lambda^{''} + \lambda^{'2} + \nu^{'}\lambda^{'} + \lambda^{'}\mu^{'}\right), \tag{11}$$

$$R_{33} = \frac{e^{-\mu(\rho)}}{2}\left(2\mu^{''} + \mu^{'2} + \nu^{'}\mu^{'} + \lambda^{'}\mu^{'}\right). \tag{12}$$

Here prime denotes differentiation with respect to the radial coordinate ρ. The Ricci scalar is given by

$$R = \nu^{''} + \lambda^{''} + \mu^{''} + \frac{1}{2}\left(\nu^{'2} + \lambda^{'2} + \mu^{'2} + \nu^{'}\lambda^{'} + \nu^{'}\mu^{'} + \lambda^{'}\mu^{'}\right). \tag{13}$$

Using the EFE defined by Eq. (1), the general form of the stress-energy tensor, T^a_b, in our case becomes

$$T^0_0 = -\frac{1}{4}\left(2\lambda^{''} + 2\mu^{''} + \lambda^{'2} + \mu^{'2} + \lambda^{'}\mu^{'}\right), \tag{14}$$

$$T^1_1 = -\frac{1}{4}\left(\nu^{'}\lambda^{'} + \nu^{'}\mu^{'} + \lambda^{'}\mu^{'}\right), \tag{15}$$

$$T^2_2 = -\frac{1}{4}\left(2\nu^{''} + 2\mu^{''} + \nu^{'2} + \mu^{'2} + \nu^{'}\mu^{'}\right), \tag{16}$$

$$T^3_3 = -\frac{1}{4}\left(2\nu^{''} + 2\lambda^{''} + \nu^{'2} + \lambda^{'2} + \nu^{'}\lambda^{'}\right). \tag{17}$$

For the sake of brevity R_{aa} will be written as R_a $\forall\ a = 0, 1, 2, 3$ and for the static cylindrically symmetric metric (8), Eq. (7) takes the following form

$$R_0^{'}\xi^1 + 2R_0\xi^0_{,0} = 2R_0\psi, \tag{18}$$

$$R_1^{'}\xi^1 + 2R_1\xi^1_{,1} = 2R_1\psi, \tag{19}$$

$$R_2^{'}\xi^1 + 2R_2\xi^2_{,2} = 2R_2\psi, \tag{20}$$

$$R_3^{'}\xi^1 + 2R_3\xi^3_{,3} = 2R_3\psi, \tag{21}$$

$$R_0\xi^0_{,1} + R_1\xi^1_{,0} = 0, \tag{22}$$

$$R_0\xi^0_{,2} + R_2\xi^2_{,0} = 0, \tag{23}$$

$$R_0\xi^0_{,3} + R_3\xi^3_{,0} = 0, \tag{24}$$

$$R_1\xi^1_{,2} + R_2\xi^2_{,1} = 0, \tag{25}$$

$$R_1\xi^1_{,3} + R_3\xi^3_{,1} = 0, \tag{26}$$

$$R_2\xi^2_{,3} + R_3\xi^3_{,2} = 0. \tag{27}$$

Eqs. (18) to (27) are a system of first order, non-linear coupled partial differential equations involving four components of the arbitrary conformal Ricci collineation (CRC) vector $\xi = \left(\xi^0, \xi^1, \xi^2, \xi^3\right)$, conformal factor ψ, four components of the Ricci tensor R_0, R_1, R_2, R_3 and their partial derivatives. The ξ^a's and ψ depend on all spacetime coordinates $x^a = \left(x^0, x^1, x^2, x^3\right) = (t, \rho, \theta, z)$; and R_a $(a = 0, 1, 2, 3)$ on ρ only. Also "," in lower indices will show partial derivative with respect to spacetime variables and 0, 1, 2 and 3 after "," in lower indices will represent the partial derivatives with respect to t, ρ, θ and z respectively.

4. Integrability conditions

The general solution of the CRC equations for the static cylindrically symmetric metric defined in Eq. (8), when Ricci tensor is non-degenerate, that is, when $\det(R_{ab}) \neq 0$, can be obtained by solving the system given in Eqs. (18)-(27) simultaneously. To get a general solution of CRC equations, we proceed as follows; From Eqs. (22)-(24), we can write ξ^1, ξ^2 and ξ^3 in term of ξ^0 and arbitrary functions $A_i(\rho, \theta, z)$, $i = 1, 2, 3$ as

$$\xi^1 = -\frac{R_0}{R_1} \int \xi^0_{,1} dt + A_1(\rho, \theta, z), \tag{28}$$

$$\xi^2 = -\frac{R_0}{R_2} \int \xi^0_{,2} dt + A_2(\rho, \theta, z), \tag{29}$$

$$\xi^3 = -\frac{R_0}{R_3} \int \xi^0_{,3} dt + A_3(\rho, \theta, z). \tag{30}$$

Now by using the values of ξ^1 and ξ^2 in Eq. (25) we get

$$R_1 \left\{ -\frac{R_0}{R_1} \int \xi^0_{,12} dt + A_{1,2}(\rho, \theta, z) \right\}$$
$$+ R_2 \left\{ -\frac{R_0}{R_2} \int \xi^0_{,21} dt - \left[\frac{R_0}{R_2}\right]' \int \xi^0_{,2} dt + A_{2,1}(\rho, \theta, z) \right\} = 0,$$

which implies

$$-2R_0 \int \xi^0_{,12} dt - R_2 \left[\frac{R_0}{R_2}\right]' \int \xi^0_{,2} dt + R_1 A_{1,2}(\rho, \theta, z) + R_2 A_{2,1}(\rho, \theta, z) = 0. \tag{31}$$

On differentiating with respect to t, we can write ξ^0 in the form of arbitrary functions $B_0(t, \theta, z)$ and $f(t, \rho, z)$ as

$$\xi^0 = \sqrt{\frac{R_2}{R_0}} \int B_0(t, \theta, z)\, dy + f(t, \rho, z). \tag{32}$$

Now putting Eq. (32) back into Eq. (31) we obtain

$$R_1 A_{1,2}(\rho,\theta,z) + R_2 A_{2,1}(\rho,\theta,z) = 0. \tag{33}$$

In the same way using the values of ξ^1, ξ^2 and ξ^3 from Eqs. (28), (29) and (30) respectively, in Eqs. (26) and (27), yields

$$\xi^0_{,3} = \sqrt{\frac{R_3}{R_0}} B_1(t,\theta,z)\,dz, \tag{34}$$

$$R_1 A_{1,3}(\rho,\theta,z) + R_3 A_{3,1}(\rho,\theta,z) = 0, \tag{35}$$

$$\xi^0_{,23} = 0, \tag{36}$$

$$R_2 A_{2,3}(\rho,\theta,z) + R_3 A_{3,2}(\rho,\theta,z) = 0. \tag{37}$$

Now invoking Eq. (32) into Eqs. (34) and (36) gives

$$B_0 = B_0(t,\theta),\ B_1 = B_1(t,z),$$
$$f(t,\rho,z) = \sqrt{\frac{R_3}{R_0}} \int B_1(t,z)\,dz + B_2(t,\rho). \tag{38}$$

Putting Eq. (38) in Eq. (32) we get the value of ξ^0 as

$$\xi^0 = \sqrt{\frac{R_2}{R_0}} \int B_0(t,\theta)\,d\theta + \sqrt{\frac{R_3}{R_0}} \int B_1(t,z)\,dz + B_2(t,\rho). \tag{39}$$

On substituting this value into Eqs. (28), (29) and (30) respectively we find

$$\xi^1 = -\frac{R_0}{R_1}\left(\sqrt{\frac{R_2}{R_0}}\right)' \int B_0(t,\theta)\,d\theta dt - \frac{R_0}{R_1}\left(\sqrt{\frac{R_3}{R_0}}\right)' \int B_1(t,z)\,dzdt$$
$$-\frac{R_0}{R_1}\int B_{2,1}(t,\rho)\,dt + A_1(\rho,\theta,z), \tag{40}$$

$$\xi^2 = -\sqrt{\frac{R_0}{R_2}} \int B_0(t,\theta)\,dt + A_2(\rho,\theta,z), \tag{41}$$

$$\xi^3 = -\sqrt{\frac{R_0}{R_3}} \int B_1(t,z)\,dt + A_3(\rho,\theta,z), \tag{42}$$

with conformal factor given by

$$\psi = -\frac{R_0'}{2R_1}\left(\sqrt{\frac{R_2}{R_0}}\right)' \int B_0(t,\theta)\,d\theta dt - \frac{R_0'}{2R_1}\left(\sqrt{\frac{R_3}{R_0}}\right)' \int B_1(t,z)\,dzdt\xi^1$$
$$-\frac{R_0'}{2R_1}\int B_{2,1}(t,\rho)\,dt + \frac{R_0'}{2R_0}A_1(\rho,\theta,z) + \sqrt{\frac{R_2}{R_0}}\int B_{0,0}(t,\theta)\,d\theta$$
$$+\sqrt{\frac{R_3}{R_0}}\int B_{1,0}(t,z)\,dz + B_{2,0}(t,\rho).$$

With these values of ξ^0, ξ^1, ξ^2 and ξ^3 Eqs. (22)–(27) are identically satisfied while Eqs. (18) – (21) yield

$$\xi^0_{,0} - \xi^1_{,1} + \frac{1}{2}\left(\frac{R_0'}{R_0} - \frac{R_1'}{R_1}\right)\xi^1 = 0, \tag{43}$$

$$\xi^0_{,0} - \xi^2_{,2} + \frac{1}{2}\left(\frac{R_0'}{R_0} - \frac{R_2'}{R_2}\right)\xi^1 = 0, \tag{44}$$

$$\xi^0_{,0} - \xi^3_{,3} + \frac{1}{2}\left(\frac{R_0'}{R_0} - \frac{R_3'}{R_3}\right)\xi^1 = 0, \tag{45}$$

$$\xi^1_{,1} - \xi^2_{,2} + \frac{1}{2}\left(\frac{R_1'}{R_1} - \frac{R_2'}{R_2}\right)\xi^1 = 0, \tag{46}$$

$$\xi^1_{,1} - \xi^3_{,3} + \frac{1}{2}\left(\frac{R_1'}{R_1} - \frac{R_3'}{R_3}\right)\xi^1 = 0, \tag{47}$$

$$\xi^2_{,2} - \xi^3_{,3} + \frac{1}{2}\left(\frac{R_2'}{R_2} - \frac{R_3'}{R_3}\right)\xi^1 = 0. \tag{48}$$

Now we will substitute the values of ξ^0, ξ^1, ξ^2 and ξ^3 from Eqs. (39) – (42) in the above system and perform back and forth substitution. We first put the values of ξ^0 and ξ^1 in Eq. (24) to get

$$\begin{aligned}
&B_{2,0}(t,\rho) + \sqrt{\frac{R_2}{R_0}}\int B_{0,0}(t,\theta)\,d\theta + \left[\frac{R_0}{R_1}\left(\sqrt{\frac{R_2}{R_0}}\right)'\right]'\int B_0(t,\theta)\,d\theta dt \\
&\quad + \sqrt{\frac{R_3}{R_0}}\int B_{1,0}(t,z)\,dz + \left[\frac{R_0}{R_1}\left(\sqrt{\frac{R_3}{R_0}}\right)'\right]'\int B_1(t,z)\,dzdt \\
&\quad - A_{1,1}(\rho,\theta,z) + \frac{R_0}{R_1}\int B_{2,11}(t,\rho)\,dt + \left[\frac{R_0}{R_1}\right]'\int B_{2,1}(t,\rho)\,dt \\
&\quad - \frac{1}{2}\left(\frac{R_0'}{R_0} - \frac{R_1'}{R_1}\right)\left\{\frac{R_0}{R_1}\left(\sqrt{\frac{R_2}{R_0}}\right)'\int B_0(t,\theta)\,d\theta dt + A_1(\rho,\theta,z)\right. \\
&\quad \left. - \frac{R_0}{R_1}\left(\sqrt{\frac{R_3}{R_0}}\right)'\int B_1(t,z)\,dzdt - \frac{R_0}{R_1}\int B_{2,1}(t,\rho)\,dt\right\} = 0.
\end{aligned}$$

Differentiating the above equation with respect to t, θ and z, one by one and checking consistency at each step with the values by using back and forth substitution we obtain

$$B_{0,00}\left(t,\theta\right)-\alpha_0^2 B_0\left(t,\theta\right)=0,$$
$$B_{1,00}\left(t,\theta\right)-\alpha_1^2 B_1\left(t,\theta\right)=0,$$

where

$$\alpha_0^2=\sqrt{\frac{R_0}{R_2}}\left\{\frac{R_0}{2R_1}\left(\frac{R_0'}{R_0}-\frac{R_1'}{R_1}\right)\left(\sqrt{\frac{R_2}{R_0}}\right)'-\left[\frac{R_0}{R_1}\left(\sqrt{\frac{R_2}{R_0}}\right)'\right]'\right\},$$

$$\alpha_1^2=\sqrt{\frac{R_0}{R_3}}\left\{\frac{R_0}{2R_1}\left(\frac{R_0'}{R_0}-\frac{R_1'}{R_1}\right)\left(\sqrt{\frac{R_3}{R_0}}\right)'-\left[\frac{R_0}{R_1}\left(\sqrt{\frac{R_3}{R_0}}\right)'\right]'\right\},$$

and

$$B_{2,00}+\frac{R_0}{R_1}B_{2,11}+\left[\left(\frac{R_0}{R_1}\right)'-\frac{R_0}{2R_1}\left(\frac{R_0'}{R_0}-\frac{R_1'}{R_1}\right)\right]B_{2,1}=0,$$
$$A_{1,1}-\frac{1}{2}(\frac{R_0'}{R_0}-\frac{R_1'}{R_1})A_1=0.$$

Now solving Eqs. (44)-(48) on the same pattern we will obtain a list of differential equations in terms of arbitrary functions $B_0\left(t,\theta\right)$, $B_1\left(t,z\right)$, $B_2\left(t,\rho\right)$, $A_1\left(\rho,\theta,z\right)$, $A_2\left(\rho,\theta,z\right)$ and $A_3\left(\rho,\theta,z\right)$ and a set of differential constraints corresponding to these equations. The differential equations are

$$B_{0,00}\left(t,\theta\right)-\alpha_0^2 B_0\left(t,\theta\right)=0, \quad (49)$$
$$B_{0,22}\left(t,\theta\right)-\alpha_2^2 B_0\left(t,\theta\right)=0, \quad (50)$$
$$B_{1,00}\left(t,\theta\right)-\alpha_1^2 B_1\left(t,\theta\right)=0, \quad (51)$$
$$B_{1,33}\left(t,\theta\right)-\alpha_3^2 B_1\left(t,\theta\right)=0, \quad (52)$$

$$B_{2,00}+\frac{R_0}{R_1}B_{2,11}+\left[\left(\frac{R_0}{R_1}\right)'-\frac{R_0}{2R_1}\left(\frac{R_0'}{R_0}-\frac{R_1'}{R_1}\right)\right]B_{2,1}=0, \quad (53)$$

$$B_{2,11}+\left[\frac{R_1}{R_0}\left(\frac{R_0}{R_1}\right)'-\frac{1}{2}\left(\frac{R_1'}{R_1}-\frac{R_3'}{R_3}\right)\right]B_{2,1}=0, \quad (54)$$

$$B_{2,11}+\left[\frac{R_1}{R_0}\left(\frac{R_0}{R_1}\right)'-\frac{1}{2}\left(\frac{R_1'}{R_1}-\frac{R_2'}{R_2}\right)\right]B_{2,1}=0, \quad (55)$$

$$\left(\frac{R_2'}{R_2}-\frac{R_3'}{R_3}\right)B_{2,1}=0, \tag{56}$$

$$B_{2,00}-\frac{R_0}{2R_1}\left(\frac{R_0'}{R_0}-\frac{R_2'}{R_2}\right)B_{2,1}=0, \tag{57}$$

and

$$\begin{aligned}
R_1A_{1,2}+R_2A_{2,1}&=0,\\
R_1A_{1,3}+R_3A_{3,1}&=0,\\
R_2A_{2,3}+R_3A_{3,2}&=0,\\
A_{1,1}-\frac{1}{2}(\frac{R_0'}{R_0}-\frac{R_1'}{R_1})A_1&=0,\\
A_{2,2}-\frac{1}{2}(\frac{R_0'}{R_0}-\frac{R_2'}{R_2})A_1&=0,\\
A_{3,3}-\frac{1}{2}(\frac{R_0'}{R_0}-\frac{R_3'}{R_3})A_1&=0,\\
A_{1,1}-A_{2,2}+\frac{1}{2}(\frac{R_1'}{R_1}-\frac{R_2'}{R_2})A_1&=0,\\
A_{1,1}-A_{3,3}+\frac{1}{2}(\frac{R_1'}{R_1}-\frac{R_3'}{R_3})A_1&=0,\\
A_{2,2}-A_{3,3}+\frac{1}{2}(\frac{R_2'}{R_2}-\frac{R_3'}{R_3})A_1&=0.
\end{aligned}$$

The constraint equations are listed as

$$\sqrt{\frac{R_0}{R_2}}\left\{\frac{R_0}{2R_1}\left(\frac{R_0'}{R_0}-\frac{R_1'}{R_1}\right)\left(\sqrt{\frac{R_2}{R_0}}\right)'-\left[\frac{R_0}{R_1}\left(\sqrt{\frac{R_2}{R_0}}\right)'\right]'\right\}=\alpha_0^2,$$

$$\sqrt{\frac{R_0}{R_3}}\left\{\frac{R_0}{2R_1}\left(\frac{R_0'}{R_0}-\frac{R_1'}{R_1}\right)\left(\sqrt{\frac{R_3}{R_0}}\right)'-\left[\frac{R_0}{R_1}\left(\sqrt{\frac{R_3}{R_0}}\right)'\right]'\right\}=\alpha_1^2,$$

$$\frac{R_0}{2R_1}\left(\frac{R_0'}{R_0}-\frac{R_2'}{R_2}\right)\left(\sqrt{\frac{R_2}{R_0}}\right)'\frac{R_2}{R_0}-\alpha_0^2\frac{R_2}{R_0}=\alpha_2^2,$$

$$\frac{R_0}{2R_1}\left(\frac{R_0'}{R_0}-\frac{R_3'}{R_3}\right)\left(\sqrt{\frac{R_3}{R_0}}\right)\left(\sqrt{\frac{R_3}{R_0}}\right)'-\alpha_1^2\frac{R_3}{R_0}=\alpha_3^2,$$

$$\frac{R_0}{2R_1}\left(\frac{R_0'}{R_0}-\frac{R_3'}{R_3}\right)\left(\sqrt{\frac{R_2}{R_0}}\right)'-\alpha_0^2\sqrt{\frac{R_2}{R_0}}=0,$$

$$\frac{R_0}{2R_1}\left(\frac{R_0'}{R_0}-\frac{R_1'}{R_1}\right)\left(\sqrt{\frac{R_3}{R_0}}\right)'-\alpha_1^2\sqrt{\frac{R_3}{R_0}}=0,$$

$$\frac{R_0}{2R_1}\left(\frac{R_3'}{R_3}-\frac{R_2'}{R_2}\right)\left(\sqrt{\frac{R_2}{R_0}}\right)'-\alpha_2^2\sqrt{\frac{R_0}{R_1}}=0,$$

$$\frac{R_0}{2R_1}\left(\frac{R_3'}{R_3}-\frac{R_2'}{R_2}\right)\left(\sqrt{\frac{R_3}{R_0}}\right)'+\alpha_3^2\sqrt{\frac{R_0}{R_3}}=0,$$

$$\frac{R_0}{2R_1}\left(\frac{R_0'}{R_0}-\frac{R_2'}{R_2}\right)\left(\sqrt{\frac{R_3}{R_0}}\right)'+\left[\frac{R_0}{R_1}\left(\sqrt{\frac{R_3}{R_0}}\right)'\right]'=0,$$

$$\frac{R_0}{2R_1}\left(\frac{R_1'}{R_1}-\frac{R_3'}{R_3}\right)\left(\sqrt{\frac{R_2}{R_0}}\right)+\left[\frac{R_0}{R_1}\left(\sqrt{\frac{R_2}{R_0}}\right)'\right]'=0,$$

$$\frac{R_0}{2R_1}\left(\frac{R_1'}{R_1}-\frac{R_2'}{R_2}\right)\left(\sqrt{\frac{R_3}{R_0}}\right)+\left[\frac{R_0}{R_1}\left(\sqrt{\frac{R_3}{R_0}}\right)'\right]'=0,$$

$$\frac{R_0}{2R_1}\left(\frac{R_1'}{R_1}-\frac{R_3'}{R_3}\right)\left(\sqrt{\frac{R_2}{R_0}}\right)+\left[\frac{R_0}{R_1}\left(\sqrt{\frac{R_2}{R_0}}\right)'\right]'=0,$$

$$\sqrt{\frac{R_0}{R_2}}\alpha_2^2-\frac{R_0}{2R_1}\left(\frac{R_1'}{R_1}-\frac{R_2'}{R_2}\right)\left(\sqrt{\frac{R_2}{R_0}}\right)-\left[\frac{R_0}{R_1}\left(\sqrt{\frac{R_2}{R_0}}\right)'\right]'=0,$$

$$\alpha_3^2\sqrt{\frac{R_0}{R_2}}-\frac{R_0}{2R_1}\left(\frac{R_1'}{R_1}-\frac{R_2'}{R_2}\right)\left(\sqrt{\frac{R_3}{R_0}}\right)-\left[\frac{R_0}{R_1}\left(\sqrt{\frac{R_3}{R_0}}\right)'\right]'=0,$$

$$\alpha_2^2\sqrt{\frac{R_0}{R_2}}-\frac{R_0}{2R_1}\left(\frac{R_1'}{R_1}-\frac{R_2'}{R_2}\right)\left(\sqrt{\frac{R_2}{R_0}}\right)-\left[\frac{R_0}{R_1}\left(\sqrt{\frac{R_2}{R_0}}\right)'\right]'=0,$$

$$\sqrt{\frac{R_0}{R_3}}\alpha_2^2-\frac{R_0}{2R_1}\left(\frac{R_1'}{R_1}-\frac{R_3'}{R_3}\right)\left(\sqrt{\frac{R_3}{R_0}}\right)'-\left[\frac{R_0}{R_1}\left(\sqrt{\frac{R_3}{R_0}}\right)'\right]'=0.$$

The above listed system of differential equations can be solved by making certain assumptions about the components of the Ricci tensor. Usually the degenerate and non-degenerate cases of Ricci tensor are discussed separately, in order to solve these equations[26]. Then the components of the stress-energy tensor are used to get information about the matter contents of the spacetime.

5. Conclusion

The study of CRCs for static spherically symmetric spacetimes[13] showed that the number of independent CRCs for the non-degenerate Ricci tensor are 15, the maximum number for four dimensional manifolds while in the case of degenerate Ricci tensor the static spherically symmetric spacetime admits infinite number of CRCs. In this paper we constructed CRCs equations for static cylindrically symmetric spacetimes. In order to obtain a complete solution of these equations we have constructed a set of constraint equations in the form of integrability conditions. Thus the problem of solving the CRC equations has now been reduced to the solution of these constraints. A complete solution of these equations[26] will be reported elsewhere.

Acknowledgments

MA is thankful to the organizers for providing financial support for participating in the 13th Regional Conference on Mathematical Physics, October 27-31, 2010, Antalya, Turkey. This work has been supported by Akdeniz University, Scientific Research Projects Unit.

References

1. H. Stephani, D. Kramer, M. A. H. MacCallum, C. Hoenselaers and E. Herlt, *Exact Solutions of Einstein Field Equations* (Cambridge University Press, 2003).
2. G. S. Hall, *Symmetries and Curvature Structure in General Relativity* (World Scientific, 2004).
3. A. Qadir and M. Ziad, *Nuovo Cimento* **B 110** , 317 (1995).
4. A. Bokhari and A. Qadir, *J. Math. Phys.* **28**, 1019 (1987).
5. A. Bokhari and A. Qadir, *J. Math. Phys.* **29**, 525 (1988).
6. D. Ahmad and M. Ziad, *J. Math. Phys.* **38**, 2574 (1997).
7. R. Maartens, S. D. Maharaj, and B. O. J. Tupper, *Class. Quantum Grav.* **12**, 225 (1993).
8. K. Saifullah and Shair-e-Yazdan, *Int. J. Mod. Phys. D* **18**, 71 (2009).
9. A. H. Bokhari and A. Qadir, *J. Math. Phys.* **34**, 3543 (1993).

10. T. Bin Farid, A. Qadir and M. Ziad, *J. Math. Phys.* **36**, 5812 (1995).
11. M. Ziad, *Gen. Rel. Grav.* **35**, 915 (2003).
12. A. Qadir, K. Saifullah and M. Ziad, *Gen. Rel. Grav.* **35**, 1927 (2003).
13. I. Yavuz and U. Camci, *Gen. Rel. Grav.* **28**, 691 (1996).
14. G. S. Hall, I. Roy, and E. G. L. R. Vaz, *Gen. Rel. Grav.* **28**, 299 (1996).
15. U. Camci and M. Sharif, *Gen. Rel. Grav.* **35**, 97 (2003).
16. U. Camci, A. Qadir and K. Saifullah, *Commun. Theor. Phys.* **49**, 1527 (2008).
17. K. L. Duggel and R. Sharma, *Symmetries of Spacetimes and Riemannian Manifolds* (Kluwer Academic Publishers, 1999).
18. A. Qadir, in *Applications of Symmetry Methods*, eds. A. Qadir and K. Saifullah (National Centre for Physics, Islamabad, 2006).
19. K. Saifullah, *Nuovo Cimento* **B 122**, 447 (2007).
20. K.Saifullah, *Nuovo Cimento* **B 118**, 1927 (2003).
21. K. Saifullah, *Int. J. Mod. Phys.* **14**, 797 (2005).
22. M. Ziad, *Nuovo Cimento B* **B 114**, 683 (1999).
23. A. Qadir, M. Sharif and M. Ziad, *Class. Quantum Grav.* **17**, 345 (2000).
24. M. Tsamparlis and P. S. Apostolopoulos, *Gen. Rel. Grav.* **36**, 47 (2004).
25. A. H. Bokhari, A. R. Kashif, and A. H. Kara, *Nuovo Cimento* **B 118**, 803 (2003).
26. M. Afzal, *M. Phil. Thesis*, Quid-i-Azam University, Islamabad (2011).

NOETHER SYMMETRIES OF GEODESIC EQUATIONS IN SPACETIMES

IBRAR HUSSAIN

Department of Basic Sciences,
School of Electrical Engineering and Computer Science,
National University of Sciences and Technology,
H-12 Campus, Islamabad, Pakistan
E-mail: ibrar.hussain@seecs.nust.edu.pk
www.nust.edu.pk

For flat spacetimes conformal Killing vectors are symmetries of Lagrangians for Geodesic equations. Here it is shown that in general conformal Killing vectors are not symmetries of Lagrangians for geodesic equations.

Keywords: Noether symmetries; Conformal Killing vectors; Conformally flat spacetimes.

1. Introduction

Symmetries of Lagrangians and symmetries of differential equations (DEs) are useful in finding solutions of DEs[1]. For ordinary differential equations (ODEs) they reduce the order of the equation and for partial differential equations (PDEs) they reduce the number of independent variables[2]. They can also be used to linearize non-linear DEs[3–6]. The Noether symmetries are more useful (for variational problems only) in the sense as they give double reduction of DEs. Noether symmetries are also important as they yield conservation laws via the celebrated Noether theorem[7].

In general relativity (GR) the spacetime symmetries have been extensively studied[8]. Among these, *isometries* or *Killing vectors KVs*, *conformal Killing vectors CKVs*, *homothetic vectors* (*HVs*), *curvature collineations* (*CCs*) and *Ricci collineations RCs* etc have been used in classification of spacetimes[9–16]. These type of symmetries have also used for finding exact solutions of Einstein Field Equations[17]. The set of KVs is always contained in the set of all other types of spacetime symmetries e.g. CKVs, HVs, CCs etc (for detail see Ref. 18). The algebra of KVs is contained in the symmetry algebra of the geodesic equations (Euler-Lagrange equations) of the underlying spaces[19]. Also the set of Noether symmetries always is contained in the set of symmetries of the corresponding Euler-Lagrange equations[20]. A question comes to mind is how the Noether symmetries are related with the spacetime symmetries e.g. CKVs, HVs etc.

Minkowski spacetime which is flat and hence conformally flat, admits 15 CKVs[21]. For this spacetime there are 17 Noether symmetries which properly contains the 15 CKVs[22]. Since KVs and HVs are contained in CKVs, therefore for the Minkowski spacetime all these spacetime symmetries are contained in the set of 17 Noether symmetries. From this result one may expect, that the CKVs form a subalgebra of the symmetries of the Lagrangian that minimizes arc length for any spacetime[22]. An example of a curved spacetime was constructed for which the set of symmetries of the Lagrangian for the geodesic equations only contain the KVs and not the HVs and CKVs[23]. The geodesic Lagrangian depends on the metric tensor and not on its conformal structure, therefore it can be expected that the Lagrangian may only admit the symmetries of the metric tensor i.e. KVs and not the CKVs. Since in conformally flat spacetimes the metric is transformed conformally, the geodesic Lagrangian may admit the CKVs for conformally flat spacetimes. Here a question arises "is it true that the CKVs are always contained in the Noether symmetries for the geodesic equations in conformally flat spacetimes?". Here an answer is found for the this question.

The plan of the paper is as follows. In the next section mathematical formalism to be used is given. In Sec. 3 the symmetries of the Lagrangian for geodesic equations in a conformally flat spacetimes are investigated. In Sec. 4 a summary and discussion are given.

2. Preliminaries

A vector field $\mathbf{Y}$ is called a CKV if the following condition holds[18]

$$\pounds_{\mathbf{Y}} g_{\mu\nu} = \psi g_{\mu\nu}, \tag{1}$$

where $\psi = \psi(x^\sigma)$ is a conformal factor and $\pounds$ denotes the Lie derivative operator. $\mathbf{Y}$ is known as HV if $\psi_{,\sigma} = 0$, and KV if $\psi = 0$, where the comma denotes the partial derivative. If one replaces in (1) the metric tensor $g_{\mu\nu}$ with the Riemann curvature tensor $R^{\mu}{}_{\nu\lambda\sigma}$ and puts $\psi = 0$, then the vector field $\mathbf{Y}$ is known as a CC. Here all the indices μ, ν, λ and σ run from 0 to 3.

Consider a vector field[1]

$$\mathbf{Y}=\xi(s,x^{\mu})\frac{\partial}{\partial s}+\eta^{\nu}(s,x^{\mu})\frac{\partial}{\partial x^{\nu}}. \tag{2}$$

The first prolongation of this vector field is

$$\mathbf{Y}^{[1]}=\mathbf{X}+(\eta^{\nu}_{,s}+\eta^{\nu}_{,\mu}\dot{x}^{\mu}-\xi_{,s}\dot{x}^{\nu}-\xi_{,\mu}\dot{x}^{\mu}\dot{x}^{\nu})\frac{\partial}{\partial \dot{x}^{\nu}}. \tag{3}$$

Then $\mathbf{Y}$ is a Noether symmetry of the Lagrangian

$$L(s,x^{\mu},\dot{x}^{\mu})=g_{\mu\nu}(x^{\sigma})\dot{x}^{\mu}\dot{x}^{\nu}, \tag{4}$$

if there exists some gauge function, $A(s,x^{\mu})$, such that

$$\mathbf{Y}^{[1]}L+(D_s\xi)L=D_sA, \tag{5}$$

where

$$D_s=\frac{\partial}{\partial s}+\dot{x}^{\mu}\frac{\partial}{\partial x^{\mu}}, \tag{6}$$

and " $\cdot$ " denotes differentiation with respect to s.

The importance of Noether symmetries can be seen from the following theorem[1].

Theorem 2.1. *If Y is a Noether point symmetry corresponding to a Lagrangian $L(s,x^{\mu},\dot{x}^{\mu})$ of a second-order ODE $\ddot{x}^{\mu}=g(s,x,\dot{x}^{\mu})$, then*

$$I=\xi L+(\eta^{\mu}-\dot{x}^{\mu}\xi)\frac{\partial L}{\partial \dot{x}^{\mu}}-A, \tag{7}$$

is a first integral of the ODE associated with $\mathbf{Y}$.

The Euler-Lagrange equations for a given Lagrangian are the geodesic equations

$$\ddot{x}^{\mu}+\Gamma^{\mu}_{\nu\lambda}\dot{x}^{\nu}\dot{x}^{\lambda}=0, \tag{8}$$

where

$$\Gamma^{\mu}_{\lambda\nu}=\frac{1}{2}g^{\mu\sigma}(g_{\lambda\sigma,\nu}+g_{\nu\sigma,\lambda}-g_{\lambda\nu,\sigma}). \tag{9}$$

A spacetime is called conformally flat if all the components of the Weyl tensor

$$C^{\mu\nu}_{\ \ \lambda\sigma}=R^{\mu\nu}_{\ \ \lambda\sigma}-2\delta^{[\mu}_{[\lambda}R^{\nu]}_{\sigma]}+\frac{1}{3}\delta^{\mu}_{[\lambda}\delta^{\nu}_{\sigma]}R, \tag{10}$$

are equal to zero, where $R_{\mu\nu}$ is the Ricci tensor and R is its trace known as the Ricci scalar. From (10) it is evident that every flat spacetime is conformally flat but the converse is not true in general.

3. Noether Symmetries of Geodesic Lagrangians in Conformally Flat Spacetimes

The general line element of plane symmetric static spacetime is

$$ds^2 = e^{\nu(x)}dt^2 - dx^2 - e^{\mu(x)}(dy^2 + dz^2), \tag{11}$$

where ν, and μ are arbitrary functions of x. In general, this spacetime admits four KVs with the Lie group $SO(2) \otimes_s R^2 \otimes R$, (where $\otimes_s$ and $\otimes$ denote semi direct and direct product respectively) and the generators are[17]

$$\mathbf{Y}_0 = \frac{\partial}{\partial t}, \quad \mathbf{Y}_1 = \frac{\partial}{\partial y}, \quad \mathbf{Y}_2 = \frac{\partial}{\partial z}, \quad \mathbf{Y}_3 = y\frac{\partial}{\partial z} - z\frac{\partial}{\partial y}. \tag{12}$$

Here we choose the metric coefficients in (11) as[24]

$$e^{\nu} = 1, \quad e^{\mu} = e^{x/a}, \tag{13}$$

where a is a constant having dimensions of length. For this choice of metric coefficients the non-zero components of the Riemann curvature tensor are

$$R^1{}_{212} = -\frac{e^{(x/a)}}{4a^2} = R^1{}_{313} = R^2{}_{323}. \tag{14}$$

While all the components of the Weyl tensor are zero. It is a conformally flat spacetime and therefore admits 15 CKVs. This spacetime admit 7 KVs which form the Lie group $SO(1,3) \otimes R$, with symmetry generators other then given in (12)

$$\mathbf{Y}_4 = y\frac{\partial}{\partial x} - [\frac{1}{2a}(y^2 - z^2) - ae^{(x/a)}]\frac{\partial}{\partial y} - \frac{1}{a}yz\frac{\partial}{\partial z}, \tag{15}$$

$$\mathbf{Y}_5 = z\frac{\partial}{\partial x} - [\frac{1}{2a}(z^2 - y^2) - ae^{(x/a)}]\frac{\partial}{\partial z} - \frac{1}{a}yz\frac{\partial}{\partial y}, \tag{16}$$

$$\mathbf{Y}_6 = \frac{\partial}{\partial x} - \frac{1}{a}[y\frac{\partial}{\partial y} + z\frac{\partial}{\partial z}]. \tag{17}$$

For this spacetime there is no proper HV and infinite number of CCs[12].

The geodesic Lagrangian for the above spacetime is

$$L = \dot{t}^2 - \dot{x}^2 - e^{(x/a)}(\dot{y}^2 + \dot{z}^2). \tag{18}$$

This Lagrangian admits 9 Noether symmetry generators. Seven of these are the KVs given by (12) and (15) - (17) and the other two generators are

$$\mathbf{Z}_0 = \frac{\partial}{\partial s}, \quad \mathbf{Z}_1 = s\frac{\partial}{\partial t}. \tag{19}$$

Here the gauge function is

$$A = c_0 + tc_1, \tag{20}$$

where c_0 and c_1 are constants of integration. The generator $\mathbf{Z}_0$ always exists for the geodesic Lagrangian.[6] The generator $\mathbf{Z}_1$ gives the mixing of the geodetic parameter s with the KV, $\mathbf{Y}_0$, given in (12). Hence here only the KVs are the symmetries of the geodesic Lagrangian and not the CKVs.

4. Summary and Discussion

Here symmetries of the geodesic Lagrangian for a conformally flat spacetimes are investigated. The spacetime discussed in section 3 admits 7 KVs and 15 CKVs. The Lagrangian for the geodesic equation in this conformally flat spacetime admits 9 symmetry generators which includes the 7 KVs given in (12) and (15) - (17) and the other 2 symmetry generators are given by (19). Here we see that the symmetry algebra of the Lagrangian for the geodesic equation in this conformally flat spacetime only contains the KVs and not the CKVs. Hence it is not true in general that the CKVs are symmetries of the geodesic Lagrangians in spacetimes.

Here we see that the Lagrangian (18) admits a symmetry generator Y_1 given in (19) which gives mixing of the geodetic parameter s with the KV $\partial/\partial t$. For a diagonal and static (plane, spherically and cylindrically symmetric) spacetime there are 6 independent non-zero components of the Riemann curvature tensor[25] i.e. $R^{i}{}_{0i0}$, $R^{1}{}_{j1j}$ and $R^{2}{}_{323}$, where $i = 1, 2, 3$ and $j = 2, 3$. For the above discussed spacetime the temporal part of the curvature tensor is zero i.e. $R^{i}{}_{0i0} = 0$, that is, there is a flat section. Therefore we see the mixing of the time-like KV with the geodetic parameter s. This fact is generally proved in[26].

Acknowledgments

Financial support of ICTP and NUST is gratefully acknowledged. I am also grateful to Prof. A. Qadir for his useful discussions.

References

1. N. H. Ibragimov, *Elementary Lie Group Analysis and Ordinary Differential Equations*, (Wiely, Chichester, 1999).
2. G. Bluman, and S. Kumei, *Symmetries and Differential Equations*, (Springer-Verlag, New york, 1989).
3. C. Wafo Soh and F.M. Mahomed, *Int. J. Nonlinear Mech.*, **36**, 671 (2001).
4. N. H. Ibragimov and S. V. Maleshko, *J. Math. Anal. Applic.* **308**, 266 (2005).
5. F. M. Mahomed and A. Qadir, *Nonlinear Dyn.* **48**, 417 (2007).
6. A. Qadir, *SIGMA*. **3**, 103 (2007); arXiv.0711.0814.

7. E. Noether, *Nachr. Konig. Gesell. Wissen., Gottingen, Math.-Phys.Kl.* **2**, 235 (1918); (English translation in *Transport Theory Statist. Phys.* **1** 186 (1971)).
8. G. S. Hall, *Symmetries and Curvature Structure in General Relativity,* (World Scientific, Singapore, 2004).
9. A. Qadir and M. Ziad, *Nuovo Cimeto,* **B110**, 277 (1995).
10. A. H. Bokhari, A. Qadir, M. S. Ahmed and M. Asghar, *J. Math. Phys.* **38**, 3639 (1997).
11. M. Ziad, *Nuovo Cimeto* **B114**, 683 (1999).
12. A. H. Bokhari, A. R. Kashif and A. Qadir, *J. Math. Phys.* **41**, 2167 (2000).
13. A. H. Bokhari, A. R. Kashif and A. Qadir, *Gen. Rel. Grav.***35**, 1059 (2003).
14. I. Hussain, A. Qadir and K. Saifullah K, *Int. J. Mod. Phys.***D14**, 1431 (2005).
15. U. Camci and A. Barnes, *Class. Quant. Grav.***19**, 393 (2002).
16. G. S. Hall and G. Shabbir, *Class. Quant. Grav.* **18**, 907 (2001).
17. D. Kramer, H. Stephani, M. A. H. MacCallum and E. Herlt, *Exact Solutions of Einstein Field Equations,* (Cambridge University Press, Cambridge, 1980).
18. G. H. Katzin, J. Levine and W. R. Davis, *J. Math. Phys.* **10**, 617 (1969).
19. T. Feroze, F. M. Mahomed and A. Qadir, *Nonlinear Dyn.* **45**, 65 (2006).
20. P. J. Olver, *Applications of Lie Groups to Differential Equations,* (Springer-Verlag, New York, 1993).
21. R. Maartens and S. D. Maharaj, *Class. Quant. Grav.* **3**, 1005 (1986).
22. I. Hussain, F. M. Mahomed and A. Qadir A, *Gen. Rel. Grav.***41**, 2399 (2009).
23. I. Hussain, *Gen. Rel. Grav.***42**, 1791 (2010).
24. T. Feroze, A. Qadir and M. Ziad, *J. Math. Phys.***42**, 4947 (2001).
25. A. R. Kashif, (*Curvature Collineations of Some Spacetimes and Their Physical Interpretation* PhD Thesis, Quaid-i-Azam University Islamabad, 2003).
26. T. Feroze and I. Hussain *J. Geom. and Phys.* **61**, 658 (2011).

UNIVERSAL FEATURES OF GRAVITY AND HIGHER DIMENSIONS

NARESH DADHICH

Inter-University Centre for Astronomy & Astrophysics,
Post Bag 4 Pune 411 007, India
E-mail: nkd@iucaa.ernet.in

We study some universal features of gravity in higher dimensions and by universal we mean a feature that remains true in all dimensions ≥ 4. They include: (a) the gravitational dynamics always follow from the Bianchi derivative of a homogeneous polynomial in Riemann curvature and it thereby characterizes the Lovelock polynomial action, (b) all the Λ-vacuum solutions of the Einstein-Lovelock as well as pure Lovelock equation have the same asymptotic limit agreeing with the d dimensional Einstein solution and (c) gravity inside a uniform density sphere is independent of the spacetime dimension and it is always given by the Schwarzschild interior solution.

Keywords: Universality; Bianchi derivative; Vacuum solutions; Uniform density sphere; Gauss-Bonnet; Lovelock; Higher dimensions; Exact solutions.

1. Introduction

In this paper we shall discuss three universal features of gravity in higher dimensions. By universal we mean a feature that is carried over to all dimensions ≥ 4. Gravity is the universal force and hence it could only be described by the curvature of spacetime[1]. That is the gravitational equation of motion should follow from the Riemann curvature. It does indeed do so because the trace of the vanishing Bianchi derivative of the Riemann curvature (Bianchi identity) yields a second order differential operator in terms of the divergence free second rank symmetric tensor, the Einstein tensor. This should always happen even when we include higher order terms in Riemann curvature. There would always exist an analogue of the Einstein tensor for the analogue of Riemann defined as a homogeneous polynomial in Riemann. This is the first universal feature we shall establish by obtaining the gravitational equation from the vanishing of the trace of the Bianchi derivative of the homogeneous polynomial tensor which will characterize the Lovelock polynomial action[2].

Higher order terms in the Lovelock action should make significant contribution only on the high energy end, $r \to 0$ in the vacuum solution while the low energy asymptotic limit, $r \to \infty$ should always tend to the Einstein solution in the corresponding dimension. We shall show that all pure Lovelock as well as Einstein-Lovelock Λ vacuum solutions asymptotically tend to the corresponding Einstein solution[3]. This is the second universal feature. It is well known that gravitational field inside a uniform density sphere is proportional to r in Newtonian gravity as well as in Einstein gravity despite its non-linearity. This is a general feature that should remain true in all dimensions as well as in higher order Einstein-Lovelock gravity[4]. This means it would be always described by the Schwarzschild interior solution. This is the third and final universal feature we would establish.

The paper is organized as follows: Before taking up the main theme of the paper we would also like to allude on some of the classical motivations for higher dimensions. That is what would be done in the next section. Then we have three sections devoted to the three universal features which would be followed to conclude in the discussion.

2. Why higher dimensions?

We shall consider two very general classical motivations for higher dimensions. Since gravity is described by the Riemann curvature of spacetime and hence gravitational equation of motion would remain valid in all dimensions where Riemann curvature is defined. For this the minimum dimension required is 2 and hence $d \geq 2$. However it turns out that 2 and 3 dimensions are not big enough to accommodate free propagation and hence we land in the usual 4 dimension. Now the question arises, are 4 dimensions sufficient for description of gravity? Is there, like the free propagation, any other property that still remains untapped? How about the high energy effects - how do we extend a theory to include high energy modification? The usual method for that is to include the non-linear terms in the basic variable. For gravity, the Riemann curvature is the basic variable and so we should include its square and higher orders in the action. However the resulting equation should still remain second order quasi-linear (second derivative must occur linearly) so that the equation describes a unique evolution for a given initial value formulation. It turns out that this can only happen for a particular combination, known as the Lovelock polynomial[3]. However the higher order Lovelock terms make non-zero contribution in the equation of motion only for $d \geq 4$. Thus we have to go to higher dimensions to realize high energy effects[5].

Next we appeal to the general principle that the total charge for a classical field must always be zero globally[5]. This is true for the electromagnetic field, how about gravity? The charge for gravity is the energy momentum that is always positive, how could that be neutralized? The only way it could be neutralized is that the gravitational field the matter creates itself should have charge of opposite polarity. That means the gravitational interaction energy must be negative and that is why the field has always to be attractive. Note that gravity is attractive to make the total charge zero. The negative charge is however not localizable as it is spread all over the space. When it is integrated over the whole space for a mass point, it would perfectly balance the positive mass. This is what was rigorously established in the famous ADM calculation[6]. Consider a neighborhood of radius R around a mass point. In this region, the total charge is not zero and there is overdominance of positive charge as the negative charge lying in the field outside R has been cut off. Whenever the charge is non zero on any surface (like for an electrically charged sphere, the field propagates off the sphere) the field must go off it. This means gravity must go off the 3-brane (4-spacetime) in the extra dimension but as it leaks out, its past light cone would encompass more and more region outside R and thereby more and more the negative charge. That is it leaks off the brane with diminishing field strength and hence it does not penetrate deep enough. This is the picture quite similar to the Randall-Sundrum braneworld gravity model[7]. If the matter fields remain confined to the 3-brane and only gravity leaks into extra dimensions but not deep enough, then extra dimensions effectively become small (for probing depth of dimension, we need some physical probe that goes there). This is an intuitively very appealing and enlightening classical consideration why gravity cannot remain confined to 4 dimension and at the same time why extra dimension cannot be large?

3. Bianchi derivative

Gravity is the universal force in the sense that it links to everything including the zero mass particles which demands that it can only be described by the curvature of spacetime. This means its dynamics has always to be goverened by the spacetime geometry - a property of the curvature. That is, the gravitational field equation should follow from the Riemann curvature. The Riemann curvature satisfies the Bianchi identity which on contraction yields the second rank symmetric Einstein tensor with vanishing divergence. This provides the appropriate second order differential operator required for deducing the equation of motion for gravitation - the Einstein equation. This should always happen even when higher order terms in curvature are included. It turns out that it is possible to identify the analogue of Riemann tensor which is a homogeneous polynomial in Riemann, and it is the trace of the Bianchi derivative of this polynomial would provide the analogue of the Einstein tensor[2]. This leads to a geometric characterization of the Lovelock polynomial.

In here, we would refer to high energy behavior of gravity, that would ask for inclusion of higher order terms in the curvature. Usually such corrections are evaluated as perturbations against a fixed background spacetime provided by the low energy solution. Since at high energy, spacetime curvature would also be very strong, hence we cannot resort to the usual perturbative analysis but instead have to consider the situation non-perturbatively. That is, include higher order terms in Riemann curvature, derive the equation of motion and then seek its solution. For the next order, we include square of the Riemann curvature and yet we must have a second order quasilinear (linear in the second derivative) equation for unique evolution with a proper initial data. This uniquely identifies a particular combination, $L_{GB} = R^2_{abcd} - 4R^2_{ab} + R^2$, (where $R^2_{ab} = R_{ab}R^{ab}$) known as the Gauss-Bonnet (GB) Lagrangian. This has remarkable property that the squares of the second derivative get canceled out leaving the equation quasilinear. The variation of this as well as the trace of the Bianchi derivative of a fourth rank tensor which is a homogeneous quadratic in Riemann curvature leads to the analogue of G_{ab}, a divergence free H_{ab}.

We define an analogue of R_{abcd} as a homogeneous polynomial[2] as

$$F_{abcd} = P_{abcd} - \frac{n-1}{n(d-1)(d-2)} P(g_{ac}g_{bd} - g_{ad}g_{bc}) \tag{1}$$

where

$$P_{abcd} = Q_{ab}{}^{mn} R_{mncd} \tag{2}$$

and

$$Q^{ab}{}_{cd} = \delta^{aba_1b_1...a_nb_n}_{cdc_1d_1...c_nd_n} R_{a_1b_1}{}^{c_1d_1} ... R_{a_nb_n}{}^{c_nd_n}. \tag{3}$$

Here n is the order of polynomial and $Q^{abcd}{}_{;d} = 0$. For the quadratic case, P_{abcd} reads as

$$P_{abcd} = R_{abmn}R_{cd}{}^{mn} + 4R_{[a}{}^{m}R_{b]mcd} + RR_{abcd} \tag{4}$$

where $P = L_{GB}$. Note that the Bianchi derivative of F_{abcd} does not vanish (that only vanishes for R_{abcd}), however its trace does vanish to give

$$-\frac{n}{2}F^{cd}{}_{[cd;e]} = H_{e;c}^{\ c} = 0 \tag{5}$$

where

$$n(F_{ab} - \frac{1}{2}Fg_{ab}) = H_{ab}. \tag{6}$$

This is an alternative derivation of H_{ab} which results from the variation of the corresponding n-th order term in the Lovelock polynomial which is defined by $Q_{abcd}R^{abcd}$. This is yet another characterization, the trace of the Bianchi derivative yielding the correponding H_{ab}, of the Lovelock gravity. In the GB quadratic case by varying the Lagrangian $L_{GB} = R^2_{abcd} - 4R^2_{ab} + R^2$ we obtain

$$\begin{aligned} H_{ab} = 2(RR_{ab} - 2R_a{}^m R_{bm} - 2R^{mn}R_{ambn} \\ + R_a{}^{mnl}R_{bmnl}) - \frac{1}{2}L_{GB}g_{ab}. \end{aligned} \tag{7}$$

However H_{ab} is non-zero only for $d > 4$ which means GB and higher order Lovelock make non-zero contribution in the equation of motion only in dimension higher than four. This clearly indicates that at high energies gravity cannot remain confined to the four dimensions and hence the consideration of higher dimensions becomes pertinent and relevant. Its dynamics in higher dimensions would be governed by $H_{ab} = -\Lambda g_{ab}$.

Since the corresponding H_{ab} could always be gotten from the trace of the Bianchi derivative for any Lovelock order, gravitational dynamics would thus entirely follow from the geometric properties of spacetime curvature. This is a universal feature which is true for all Lovelock orders and in all dimensions ≥ 4.

4. Lovelock vacuum solutions

Although the Lovelock vacuum solutions are known for long time[8–10] we would here like to probe the universality of their asymptotic large r behaviour[3]. It turns out that this limit is neutral for pure Lovelock solution and its Einstein-Lovelock analogue so long as Λ is non-zero. In particular, the asymptotic limit of the Einstein-Gauss-Bonnet solution with Λ has the same form as the pure Gauss-Bonnet solution and it is the d-dimensional Einstein solution. However their $r \to 0$ limit is radically different, for the former the metric remains regular and finite[5] while for the latter it is singular at $r = 0$. We would also like to draw attention to an interesting property of spherically symmetric vacuum equations that one has ultimately to solve a single first order equation not withstanding the enhanced nonlinearity of the Lovelock gravity.

We shall begin with the GB vacuum equation,

$$H_{ab} = -\Lambda g_{ab} \tag{8}$$

where H_{ab} as given in Eq. (7), for the spherically symmetric metric,

$$ds^2 = e^{\nu} dt^2 - e^{\lambda} dr^2 - r^2 d\Omega_{d-2}^2 \tag{9}$$

where $d\Omega_{d-2}^2$ is the metric on a unit $(d-2)$-sphere. In general ν, λ are functions of both t and r, however as shown in[8] the t-dependence drops out as usual and it then suffices to take them as functions of r alone. To begin with we have $H_t^t = H_r^r$ that immediately determines $\nu = -\lambda$. With this, let us write the non-zero components of H_b^a for the above metric and they read as follows:

$$H_t^t = H_r^r = -\frac{d-2}{2r^4}(1-e^{-\lambda})\left(2re^{-\lambda}\lambda' + (d-5)(1-e^{-\lambda})\right) = -\Lambda \tag{10}$$

$$\begin{aligned} H_\theta^\theta = {} & \frac{1}{2r^4}\Big[r^2 e^{-\lambda}(1-e^{-\lambda})(-2\lambda'' + \lambda'^2) - r^2 e^{-2\lambda}(3-e^{\lambda})\lambda'^2 \\ & + (d-5)(1-e^{-\lambda})\left(-4re^{-\lambda}\lambda' - (d-6)(1-e^{-\lambda})\right)\Big] \\ = {} & -\Lambda \end{aligned} \tag{11}$$

where a prime denotes derivative w.r.t r and all the angular components are equal.

First let us note that the above two equations are not independent and it can easily be seen that the latter is a derivative of the former which was first shown for the usual four dimensional gravity in[11]. It would therefore suffice to integrate the former alone to get the general solution. This is what it should be because there is only one function, λ, to be determined. Eq. (10) could be written as

$$(r^{d-5} f^2)' = \frac{2\Lambda}{d-2} r^{d-2} \tag{12}$$

which readily integrates to give

$$e^{-\lambda} = F = 1 - f, \quad f^2 = \frac{k}{r^{d-5}} + \Lambda_1 r^4 \tag{13}$$

where $2\Lambda/(d-1)(d-2) = \Lambda_1$. This is the general solution of the pure Gauss-Bonnet vacuum which has been obtained by solving the single first order equation. Its large r limit would be

$$F = 1 - \sqrt{\Lambda_1} r^2 - \frac{K}{r^{d-3}} \tag{14}$$

where $K = k/2\sqrt{\Lambda_1}$. This is the Schwarzschild-dS solution for a d-dimensional spacetime. On the other hand let us look at the Einstein-Gauss-Bonnet solution (which is obtained by summing over $n = 1, 2$; i.e. replacing H_{ab} in Eq. (8) by $G_{ab} + \alpha H_{ab}$),[12]

$$F = 1 + \frac{r^2}{2\alpha}[1 - \sqrt{1 + 4\alpha(\frac{M}{r^{d-1}} + \Lambda)}] \tag{15}$$

that would also approximate for large r to

$$F = 1 - \Lambda r^2 - \frac{M}{r^{d-3}}. \tag{16}$$

Thus the two solutions perfectly agree in the large r limit. It should however be noted that for the former the presence of Λ is essential for this limit to exist.

Now we go to the general case and we write $G_{ab}^{(n)}$ for the differential operator resulting from the n-th term in the Lovelock polynomial and in particular, $G_{ab}^{(1)}$ is the Einstein tensor and $G_{ab}^{(2)} = H_{ab}$ of the Gauss-Bonnet. Again $(G_t^t = G_r^r)^{(n)}$ will give $\nu = -\lambda$, and the analogue of Eq. (12) would read as

$$(r^{d-2n-1} f^n)' = \frac{2\Lambda}{d-2} r^{d-2} \tag{17}$$

and it would readily integrate to give

$$f^n = \Lambda_1 r^{2n} + \frac{k}{r^{d-2n-1}}. \tag{18}$$

Again asymptotically it approximates to

$$F = 1 - \Lambda_1^{1/n} r^2 - \frac{K}{r^{d-3}} \tag{19}$$

where now $K = k/n\Lambda_1^{1/n}$. This is the d-dimensional Schwarzschild-dS/AdS black hole (for even n only dS while for odd n it could be dS/AdS with the sign of Λ). The corresponding Einstein-Lovelock solution is simply obtained by summing over n, $\sum_n \alpha_n$ on the left in Eq. (17) with a coupling coefficient for each n. The polynomial on the left becomes difficult to solve for $n \geq 4$. It is however expected that asymptotically the solution should tend to the Einstein solution in d dimensions. We have seen that above for the quadratic $n = 2$ Gauss-Bonnet case and we have also verified it for the cubic $n = 3$ case[13].

Thus we establish that asymptotically the pure Lovelock for a given order n and the Einstein-Lovelock ($\sum_n G^n_{ab}$) tend to the same limit, the Einstein solution in d-dimensional dS/AdS spacetime[3]. This is because higher order Lovelock contributions should wean out asymptotically, however what is rather interesting is the fact that even the order n of the Lovelock polynomial does not matter so long as Λ is included. That is, the large r limit is free of n indicating a universal asymptotic behaviour.

5. Uniform density sphere

We know that the gravitational potential at any point inside a fluid sphere is given by $-M(r)/r^{d-3}$ for $d \geq 4$ dimensional spacetime. Now $M(r) = \int \rho r^{d-2} dr$ which for constant density will go as ρr^{d-1} and then the potential will go as $\rho r^{d-1}/r^{d-3} = \rho r^2$ and is therefore independent of the dimension. This is an interesting general result: for the uniform density sphere, gravity has the universal character that it is independent of the dimension of spacetime and it always goes as r. It is then a natural question to ask, does this result carry over to Einsteinian gravity? In general relativistic language it is equivalent to asking, does the Schwarzschild interior solution always describe the field inside a uniform density sphere in all dimensions $d \geq 4$? Not only that does it carry over to higher order Einstein-Gauss-Bonnet (Lovelock) gravity? The answer is yes[4].

5.1. *Einstein case*

We begin with the general static spherically symmetric metric given by

$$ds^2 = e^{\nu} dt^2 - e^{\lambda} dr^2 - r^2 d\Omega^2_{n-2} \tag{20}$$

where $d\Omega^2_{n-2}$ is the metric on a unit $(n-2)$-sphere (in this Sec. n refers to the dimension of spacetime). For the Einstein equation in the natural units $(8\pi G = c = 1)$,

$$G_{ab} = R_{ab} - \frac{1}{2} R g_{ab} = -T_{ab} \tag{21}$$

and for perfect fluid, $T_a^b = diag(\rho, -p, -p, ..., -p)$, we write

$$e^{-\lambda}(\frac{\lambda'}{r} - \frac{n-3}{r^2}) + \frac{n-3}{r^2} = \frac{2}{n-2}\rho \tag{22}$$

$$e^{-\lambda}(\frac{\nu'}{r} + \frac{n-3}{r^2}) - \frac{n-3}{r^2} = \frac{2}{n-2}p \tag{23}$$

and the pressure isotropy is given by

$$e^{-\lambda}(2\nu'' + \nu'^2 - \lambda'\nu' - 2\frac{\nu'}{r})$$
$$-2(n-3)(\frac{e^{-\lambda}\lambda'}{r} + 2\frac{e^{-\lambda}}{r^2} - \frac{2}{r^2}) = 0. \tag{24}$$

We write this equation in a form that readily yields the universal character of the Schwarzschild interior solution for all $n \geq 4$,

$$e^{-\lambda}(2\nu'' + \nu'^2 - \lambda'\nu' - 2\frac{\nu' + \lambda'}{r} - \frac{4}{r^2}) + \frac{4}{r^2}$$
$$-2(n-4)\Big((n-1)(\frac{e^{-\lambda}}{r^2} - \frac{1}{r^2}) + \frac{2\rho}{n-2}\Big) = 0. \tag{25}$$

We now set the coefficient of $(n-4)$ to zero so that it remains valid for all $n \geq 4$. This then straightway determines $e^{-\lambda}$ without integration and it is given by

$$e^{-\lambda} = 1 - \rho_0 r^2 \tag{26}$$

where $\rho_0 = 2\rho/(n-1)(n-2)$. This when put in Eq. (22) implies constant density. We thus obtain $\rho = const.$ as the neceessary condition for universality of the isotropy equation for all $n \geq 4$. The sufficient condition is obvious from the integration of Eq. (22) for $\rho = const$, giving the same solution as above where a constant of integration is set to zero for regularity at the center. Thus constant density is the necessary and sufficient condition for universality. Alternatively we can say that universality identifies uniform density. The universality is therefore true if and only if density is constant.

As is well known, Eq. (25) on substituting Eq. (26) admits the general solution as given by

$$e^{\nu/2} = A + Be^{-\lambda/2} \tag{27}$$

where A and B are constants of integration to be determined by matching to the exterior solution. This is the Schwarzschild interior solution for a constant density sphere that is independent of the dimension except for a redefinition of the constant density as ρ_0. This proves the universality of the Schwarzschild interior solution for all $n \geq 4$.

The Newtonian result that gravity inside a uniform density sphere is independent of spacetime dimension is thus carried over to general relativity as well despite nonlinearity of the equations. That is, Schwarzschild interior solution is valid for all $n \geq 4$. Since there exist more general actions like Lovelock polynomial and $f(R)$ than the linear Einstein-Hilbert, it would be interesting to see whether this result would carry through there as well. That is what we take up next.

5.2. *Gauss-Bonnet(Lovelock) case*

The gravitational equation will now read as

$$G^a_b + \alpha H^a_b = -T^a_b, \tag{28}$$

where H_{ab} as given in Eq. (7). Now density and pressure would read as follows:

$$\rho = \frac{(n-2)e^{-\lambda}}{2r^2}\Big(r\lambda' - (n-3)(1-e^{\lambda})\Big) + \frac{(n-2)e^{-2\lambda}\tilde{\alpha}}{2r^4}(1-e^{\lambda})\Big(-2r\lambda' + (n-5)(1-e^{\lambda})\Big) \tag{29}$$

$$p = \frac{(n-2)e^{-\lambda}}{2r^2}\Big(r\nu' + (n-3)(1-e^{\lambda})\Big) - \frac{(n-2)e^{-2\lambda}\tilde{\alpha}}{2r^4}(1-e^{\lambda})\Big(2r\nu' + (n-5)(1-e^{\lambda})\Big) \tag{30}$$

The analogue of the isotropy Eq. (25) takes the form

$$I_{GB} \equiv \left(1 + \frac{2\tilde{\alpha}f}{r^2}\right)I_E + \frac{2\tilde{\alpha}}{r}\left(\frac{f}{r^2}\right)'\left[r\psi' + \frac{f}{1-f}\psi\right] = 0 \tag{31}$$

where $\psi = e^{\nu/2}, e^{-\lambda} = 1-f, \tilde{\alpha} = (n-3)(n-4)\alpha$ and I_E is given by LHS of Eq. (24),

$$I_E \equiv \frac{(1-f)}{\psi}\left\{\psi'' - \left(\frac{f'}{2(1-f)} + \frac{1}{r}\right)\psi' - \frac{(n-3)}{2r^2(1-f)}(rf' - 2f)\psi\right\}. \tag{32}$$

From Eq. (29), we write

$$(\tilde{\alpha}r^{n-5}f^2 + r^{n-3}f)' = \frac{2}{n-2}\rho r^{n-2} \tag{33}$$

which integrates for $\rho = const.$ to give

$$\tilde{\alpha}r^{n-5}f^2 + r^{n-3}f = \rho_0 r^{n-1} + k \tag{34}$$

where k is a constant of integration that should be set to zero for regularity at the center and $2\rho/(n-1)(n-2) = \rho_0$ as defined earlier. Solving for f, we get

$$e^{-\lambda} = 1 - f = 1 - \rho_{0GB}r^2 \tag{35}$$

where

$$\rho_{0GB} = \frac{\sqrt{1+4\tilde{\alpha}\rho_0} - 1}{2\tilde{\alpha}}. \tag{36}$$

So the solution is the same as in the Einstein case and the appropriate choice of sign is made so as to admit the limit $\alpha \to 0$ yielding the Einstein ρ_0 (the other choice would imply $\rho_{0GB} < 0$ for positive α). This, when substituted in the pressure isotropy Eq. (31), would lead to $I_E = 0$ in Eq (32) yielding the solution (27) as before. This establishes sufficient condition for universality.

For the necessary condition, we have from Eq. (31) that either

$$\left(\frac{f}{r^2}\right)' = 0 \tag{37}$$

or

$$r\psi' + \frac{f}{1-f}\psi = 0. \tag{38}$$

The former straightway leads with the use of Eq. (33) to the same constant density solution (35) and $I_E = 0$ integrates to Eq. (27) as before. This shows that universality implies constant density as the necessary condition. For the latter case, when Eq. (38) is substituted in Eq. (32) and $I_E = 0$ is now solved for λ, we again obtain the same solution (35). Eq. (33) again implies $\rho = const$ as the necessary condition. Now ψ is determined by Eq. (38), which means the constant A in solution (27) must vanish. Then the solution turns into de Sitter spacetime with $\rho = -p = const.$ which is a particular case of the Schwarzschild solution. This is, however, not a bounded finite distribution.

Thus universality and finiteness of a fluid sphere uniquely identifies the Schwarzschild interior solution for Einstein as well as Einstein-Gauss-Bonnet gravity. That is, gravity inside a fluid sphere of finite radius always goes as r in all dimensions ≥ 4 so long as the density is constant. It is described by the Schwarzschild interior solution. It is only the constant density that gets redefined in terms of ρ_0 and ρ_{0GB}.

Though we have established the universality for EGB, it would in general be true for the Einstein-Lovelock gravity. For that we argue as follows. The entire analysis is based on the two Eqs. (31) and (33). The GB contributions in the former appear in the multiplying factor to the Einstein differential operator, I_E and another term with the factor $\tilde{\alpha}(f/r^2)'$. This indicates that the contributions of higher orders in Lovelock polynomial will obey this pattern to respect quasilinearity of the equation. The higher orders will simply mean inclusion of the corresponding couplings in the multiplying factor as well as in the second term appropriately while the crucial entities, I_E and $(f/r^2)'$ on which the proof of the universality of the Schwarzschild solution hinges remain unchanged. On the other hand,

Eq. (33) is quadratic in f for the quadratic GB action, which indicates that the degree of f is tied to the order of the Lovelock polynomial. It essentially indicates that as ρ_{0GB} is obtained from a quadratic algebraic relation, similarly in higher order its analogue will be determined by the higher degree algebraic relation. However the solution will always be given by Eq. (35). Thus what we have shown above explicitly for EGB will go through for the general Einstein-Lovelock gravity. Thus the Newtonian result of universality of gravity inside a uniform density fluid sphere critically hinges on quasilinearity and would thus only be true for the Lovelock generalization of the Einstein gravity and not for any other. It could in a sense be thought of as yet another identifying feature of the Einstein-Lovelock gravity.

There is also an intriguing and unusual feature of GB(Lovelock) gravity. What happens if the multiplying factor, $1 + 2\tilde{\alpha}f/r^2 = 0$ in Eq. (31)? Then the entire equation becomes vacuous, leaving ψ completely free and undetermined while $e^{-\lambda} = 1 + r^2/2\tilde{\alpha}$. This leads to $p = -\rho = const. = (n-1)(n-2)/8\tilde{\alpha}$, which is an anti-de Sitter distribution for $\alpha \geq 0$. This is a special prescription where density is given by GB coupling α. There is no way to determine ψ, and so we have a case of genuine indeterminacy of the metric. It is because GB(Lovelock) contributes such a multiplying factor involving $(\alpha,\ r,\ f)$ to the second order quasilinear operator, which could be set to zero and thereby annulling the equation altogether. This is a typical feature of Lovelock solutions which have been termed as geometrically free.[8,9] Such a situation also occurs in the Kaluza-Klein split-up of six-dimensional spacetime into the usual M^4 and 2-space of constant curvature in EGB theory [14].

6. Discussion

Since gravity is universal it is always interesting to probe its universal features in various settings. We had set out to establish three universal features of gravity in higher dimensions. The first one was driven by the fact that gravitational equation of motion should always follow from the geometric properties of Riemann curvature even when higher orders in curvature are included. By suitably defining the analogue of Riemann as a homogeneous polynomial in Riemann, we have derived the analogue of the Einstein tensor by the vanishing trace of its Bianchi derivative for the any order in Lovelock polynomial. This is a novel characterization of the Lovelock polynomial[2]. Since it is derived from a geometric property, it is universal.

Next feature is driven by the physical consideration that higher order Lovelock effects should appear as high energy corrections to the Einstein gravity and hence they should evaporate out in the low energy asymptotic limit, $r \to \infty$. That is all the vacuum solutions asymptotically tend to the Einstein solution in general d dimensions. This is true for both pure Lovelock as well as Einstein-GB(Lovelock) solutions. However for the former non-zero Λ is necessary for the asymptotic expansion. Here the universality means that the asymptotic limit is independent of the Lovelock order[3].

Finally it is obvious in the Newtonian theory that gravity inside a uniform density sphere is independent of the spacetime dimension simply because the dimensional change in both mass enclosed in certain radius and gravitational potential square out the spacetime dimension and hence it always goes as r. This is a general result that should be universally true for any good theory of gravity. We establish it for the Einstein as well as for Einstein-Lovelock gravity and its uniqueness. That is it would not be true for other generalizations like $f(R)$ theories. This is yet another characterization of the Lovelock gravity. It turns out that uniform density is necessary and sufficient condition for universality of the Schwarzschild interior solution[4].

Acknowledgments

It is a pleasure to thank the organizers of the 13th Regional Conference on Mathematical Physics held on Oct. 23-31, 2010 at Antalya, Turkey, where this was presented as a plenary lecture for the wonderful hospitality and exciting meeting.

Note added in proof:

Subsequently we have further added and strengthened the universal features[15–17].

References

1. N. Dadhich, *Subtle is the Gravity*, Vaidya-Raychaudhuri Endowment Lecture, gr-qc/0102009.
2. N. Dadhich, *Pramana* **74**, 875 (1010), arxiv:0802.3034.
3. N. Dadhich, *Math Today* (Special issue, A tribute to Prof. P. C. Vaidya) **26**, 37 (2011).
4. N. Dadhich, A. Molina and A. Khugaev, *Phys. Rev.***D81**:104026 (2010).
5. N.Dadhich, *On the Gauss Bonnet Gravity* (Proceedings of the 12th Regional Conference on Mathematical Physics), Eds., M.J. Islam, F. Hussain, A. Qadir, Riazuddin and Hamid Saleem, World Scientific, 331 (hep-th/0509126).
6. R. Arnowitt, S. Deser and C. W. Misner, in *Gravitation: Introduction to current research*, ed. L. Witten (John Wiley, 1962) p. 227.

7. L. Randall and R. Sundrum, *Phys. Rev. Lett.* **83**, 4690 (1999).
8. B. Whitt, *Phys. Rev.* **D38**, 3000 (1988).
9. J. T. Wheeler, *Nucl. Phys.* **B273**, 732 (1986).
10. R. C. Myers and J. Z. Simon, *Phys. Rev.* **38**, 2434 (1988).
11. N. Dadhich, *On the Schwarzschild field*, gr-qc/9704068.
12. D.G. Boulware and S. Deser, *Phys. Rev. Lett.* **55**, 2656 (1985); J. T. Wheeler, *Nucl. Phys.* **B268**, 737 (1986), **B273**, 732 (1986).
13. M. Dehghani and N. Farhangkhah, *Phys. Rev.* **D78**, 064015 (2008).
14. H Maeda and N Dadhich, *Phys. Rev.* **74**, 021501 (2006), **75**, 044007 (2007); N Dadhich and H Maeda, *Int. J. Mod. Phys.*, **D17**, 513 (2008).
15. N Dadhich, SG Ghosh and S Jhingan, *Phys. Lett. B* **711**, 196 (2012); arxiv:1202.4575.
16. N Dadhich, JM Pons and K Prabhu, arxiv:1201.4994.
17. N Dadhich, JM Pons and K Prabhu, arxiv:1110.0673.

SOME INTERESTING CONSEQUENCES OF $f(R)$ THEORY OF GRAVITY

MUHAMMAD SHARIF* and HAFIZA RIZWANA KAUSAR

Department of Mathematics, University of the Punjab,
Quaid-e-Azam Campus, Lahore-54590, Pakistan
**E-mail: msharif@math.pu.edu.pk*

Recently, modified $f(R)$ theory of gravity has attracted many people due to combined motivation coming from high-energy physics, cosmology and astrophysics. The purpose of this paper is two-fold: Firstly, we shall review some exact solutions of the modified field equations. Secondly, we present some new results on plane symmetric gravitational collapse in $f(R)$ theory. Here, we take all higher order curvature terms on the matter side of the field equations and interpret them as a dark source. In this way, we find the effects of dark energy on the dynamics of plane symmetric gravitational collapse.

Keywords: Exact solutions; Dark energy; Gravitational collapse.

1. Introduction

The most fascinating question, which the modern day physics deals with today, is the accelerated expansion of the universe. The explanation for this expansion has both theoretical as well as experimental background. To accommodate this observational data with theoretical predictions, one needs to introduce the dark energy (DE) contributions in the most successful gravitational theory of General Relativity (GR). On the other hand, modified theories of gravity may provide a cosmological accelerating mechanism without introducing any extra DE contribution.

Modified $f(R)$ theory helps in modification of the model to achieve consistency with the experimental tests of solar system. In the last decade, some $f(R)$ models were considered to modify GR at small scales to explain inflation, e.g. $f(R) \propto R^2$ but failed to explain the late time acceleration. A model like $f(R) \propto \frac{1}{R}$ was proposed to explain this acceleration but attained no interest due to conflict with solar system tests[1,2]. A cosmologically viable model needs to satisfy the evolution of big-bang nucleosynthesis, radiation and matter dominated eras. It is hoped that the problem of dark matter can be addressed by using viable $f(R)$ gravity models.

A lot of work has been done in this theory on different issues, for example, see Refs.3-5. Several features including solar system tests[6], Newtonian limit[7], gravitational stability[8] and singularity problems[9] are exhaustively discussed. The vacuum solutions of the field equations has attracted many people. Since the spherical symmetry plays a fundamental role in understanding the nature of gravity, most of the solutions are discussed in this context. Multamäki and Vilja[10] investigated static spherically symmetric vacuum solutions of the field equations. Caramês and Bezerra[11] discussed spherically symmetric vacuum solutions in higher dimensions. Capozziello *et al.*[12] analyzed spherically symmetric solution using Noether symmetry. Azadi *et al.*[13] and Momeni[14] studied constant curvature solution in cylindrical symmetry. Sharif *et al.*[15−20] have studied exact solutions of Bianchi types, plane and spherically symmetric spacetimes both for vacuum and non-vacuum. Multamäki and Vilja[21] investigated non-vacuum solutions by taking perfect fluid. Hollestein and Lobo[22] explored exact solutions of the field equations coupled to non-linear electrodynamics.

Gravitational collapse is one of the basic processes driving evolution within galaxies, assembling giant molecular clouds and producing stars. The study of gravitational collapse is motivated by the fact that it represents one of the few observable phenomena in the universe. The gravitational collapse of a star proceeds to form compact objects as white dwarf, neutron star or black hole primarily depending upon the mass of the star. Since $f(R)$ models also yield black hole solutions, it would be interesting to discuss the features and dynamics of the gravitational collapse in this modified theory. When we take conformal transformation of $f(R)$ action, it is found that Schwarzschild solution is the only static spherically symmetric solution for a model of the form[23] $R + \alpha R^2$.

The phenomenon of gravitational collapse in GR has been examined for spherical, cylindrical, planar and quasi-spherical symmetries[24−29]. Most of the work has been done by using spherically symmetric models[30−33]. Sharif *et al.*[34,35] have worked on many aspects of gravitational collapse in GR as well as in $f(R)$ gravity. In this paper, we extend this work to explore its features in the modified theory for plane symmetry. The paper is organized as follows: In the next section, we review some exact solutions both for vacuum and non-vacuum case. Also, we discuss some geometrical parameters. In section **3**, we formulate the field equations in Einstein frame and consider plane symmetric gravitational collapse. We discuss its dynamics for a complicated dissipative fluid evolution under the effects of dark source terms. In the last section, we summarize the results of the paper.

2. Review of Exact Solutions

In this section, we present the field equations in $f(R)$ theory and their solutions obtained under different assumptions. This theory is the straightforward generalization of GR. The Einstein-Hilbert action in GR is given by

$$S_{EH} = \frac{1}{2\kappa}\int d^4x\sqrt{-g}R, \tag{1}$$

where κ is the coupling constant, R is the Ricci scalar and g is the metric tensor. For $f(R)$ theory of gravity, it is modified as follows

$$S_{f(R)} = \frac{1}{2\kappa}\int d^4x\sqrt{-g}f(R), \tag{2}$$

where $f(R)$ is a general function of the Ricci scalar. Varying this action with respect to the metric tensor, we obtain following set of fourth order partial differential equations

$$F(R)R_{\mu\nu} - \frac{1}{2}f(R)g_{\mu\nu} - \nabla_\mu\nabla_\nu F(R) + g_{\mu\nu}\Box F(R) = \kappa T_{\mu\nu}, \tag{3}$$

where $F(R) \equiv df(R)/dR$ and $T_{\mu\nu}$ is the standard matter energy-momentum tensor. Contraction of the field equations yield

$$F(R)R - 2f(R) + 3\Box F(R) = \kappa T \tag{4}$$

which indicates that the field equations of $f(R)$ gravity will admit a larger variety of solutions than does GR. Further, $T = 0$ does no longer imply $R = 0$ in this theory. In vacuum, the above equation reduces to

$$F(R)R - 2f(R) + 3\Box F(R) = 0. \tag{5}$$

It is obvious from this equation that any metric with constant scalar curvature, say $R = R_0$, is a solution of the contracted equation (5) as long as the following equation holds

$$F(R_0)R_0 - 2f(R_0) = 0. \tag{6}$$

This is called constant curvature condition. Moreover, if we differentiate Eq.(5) with respect to R, we obtain

$$F'(R)R - R'F(R) + 3(\Box F(R))' = 0 \tag{7}$$

which gives a consistency relation for $F(R)$. Using Eq.(5), it follows that

$$f(R) = \frac{-8\pi T + F(R)R + 3\Box F(R)}{2}. \tag{8}$$

Inserting this value of $f(R)$ in the field equations, we obtain

$$\frac{F(R)R-\Box F(R)-8\pi T}{4}=\frac{F(R)R_{\mu\mu}-\nabla_\mu\nabla_\mu F(R)-8\pi T_{\mu\mu}}{g_{\mu\mu}}. \tag{9}$$

In the above equation, the expression on the left hand side is independent of the index μ, so the field equations can be written as

$$A_\mu=\frac{F(R)R_{\mu\mu}-\nabla_\mu\nabla_\mu F(R)-8\pi T_{\mu\mu}}{g_{\mu\mu}}. \tag{10}$$

Notice that A_μ is not a 4-vector rather just a notation for the traced quantity.

2.1. *Constant Curvature Solution*

Constant curvature solutions are directly involved in explaining the accelerating universe[36]. Here we present vacuum solution of plane symmetric spacetime[15] under the assumption of constant Ricci scalar.

2.1.1. *Plane symmetric vacuum solution*

We consider static plane symmetric spacetime of form

$$ds^2=A(x)dt^2-dx^2-B(x)(dy^2+dz^2). \tag{11}$$

The corresponding Ricci scalar becomes

$$R=\frac{1}{2}[\frac{2A''}{A}-(\frac{A'}{A})^2+\frac{2A'B'}{AB}+\frac{4B''}{B}-(\frac{B'}{B})^2], \tag{12}$$

where prime represents derivative with respect to x. Notice that the expression in Eq.(11) is independent of the index μ and hence $A_\mu-A_\nu=0$ for all μ and ν. So, $A_0-A_1=0$ and $A_0-A_2=0$ give

$$[\frac{A'B'}{AB}+(\frac{B'}{B})^2-\frac{2B''}{B}]F-\frac{1}{2A}(A'F'+2F'')=0, \tag{13}$$

$$[\frac{2A''}{A}-(\frac{A'}{A})^2+\frac{A'B'}{AB}-\frac{2B''}{B}]F-\frac{1}{2}(\frac{A'}{A}-\frac{B'}{B})F'=0. \tag{14}$$

Thus we get two non-linear differential equations with three unknowns namely A, B and F. We find solution by using the assumption of constant curvature, $R=R_0$ so that $F'(R_0)=0=F''(R_0)$. Using this condition, Eqs.(13) and (14) reduce to

$$\frac{A'B'}{AB}+(\frac{B'}{B})^2-\frac{2B''}{B}=0, \tag{15}$$

$$\frac{2A''}{A}-(\frac{A'}{A})^2+\frac{A'B'}{AB}-\frac{2B''}{B}=0. \tag{16}$$

These equations are solved by the power law assumption, i.e., $A \propto x^m$ and $B \propto x^n$, where m and n are any real numbers. Thus we use $A = k_1 x^m$ and $B = k_2 x^n$, where k_1 and k_2 are constants of proportionality. It follows that $m = -\frac{2}{3}$ and $n = \frac{4}{3}$, hence the solution becomes

$$ds^2 = k_1 x^{-\frac{2}{3}} dt^2 - dx^2 - k_2 x^{\frac{4}{3}} (dy^2 + dz^2). \tag{17}$$

It can be shown that these values of m and n lead to $R = 0$.

2.2. *Non-constant Curvature Solutions*

Here we discuss both vacuum[17] and non-vacuum solutions[18] of the field equation but without imposing assumption of constant scalar curvature. For this purpose, we take Bianchi type universes.

2.2.1. *Vacuum solution of Bianchi type I universe*

The line element of the Bianchi type I spacetime is given by

$$ds^2 = dt^2 - A^2(t)dx^2 - B^2(t)dy^2 - C^2(t)dz^2. \tag{18}$$

The corresponding Ricci scalar is

$$R = -2[\frac{\ddot{A}}{A} + \frac{\ddot{B}}{B} + \frac{\ddot{C}}{C} + \frac{\dot{A}\dot{B}}{AB} + \frac{\dot{B}\dot{C}}{BC} + \frac{\dot{C}\dot{A}}{CA}], \tag{19}$$

where dot represents derivative with respect to t. The average scale factor a, the volume scale factor v and the mean Hubble parameter H are given as

$$a = \sqrt[3]{ABC}, \quad V = a^3 = ABC, \quad H = \frac{1}{3}(H_1 + H_2 + H_3), \tag{20}$$

where $H_1 = \frac{\dot{A}}{A}$, $H_2 = \frac{\dot{B}}{B}$, $H_3 = \frac{\dot{C}}{C}$ are the directional Hubble parameters in the directions of x, y and z axis respectively. The field equations are

$$\frac{\ddot{A}}{A} - \frac{\ddot{B}}{B} + \frac{\dot{C}}{C}(\frac{\dot{A}}{A} - \frac{\dot{B}}{B}) + \frac{\dot{F}}{F}(\frac{\dot{A}}{A} - \frac{\dot{B}}{B}) = 0, \tag{21}$$

$$\frac{\ddot{B}}{B} - \frac{\ddot{C}}{C} + \frac{\dot{A}}{A}(\frac{\dot{B}}{B} - \frac{\dot{C}}{C}) + \frac{\dot{F}}{F}(\frac{\dot{B}}{B} - \frac{\dot{C}}{C}) = 0, \tag{22}$$

$$\frac{\ddot{A}}{A} - \frac{\ddot{C}}{C} + \frac{\dot{B}}{B}(\frac{\dot{A}}{A} - \frac{\dot{C}}{C}) + \frac{\dot{F}}{F}(\frac{\dot{A}}{A} - \frac{\dot{C}}{C}) = 0. \tag{23}$$

This system yields the following solution[17],

$$A = p_1(nlt + k_1)^{\frac{1}{n}} \exp[\frac{q_1(nlt + k_1)^{\frac{n-1}{n}}}{kl(n-1)}], \quad n \neq 1 \tag{24}$$

$$B = p_2(nlt + k_1)^{\frac{1}{n}} \exp[\frac{q_2(nlt + k_1)^{\frac{n-1}{n}}}{kl(n-1)}], \quad n \neq 1 \tag{25}$$

$$C = p_3(nlt + k_1)^{\frac{1}{n}} \exp[\frac{q_3(nlt + k_1)^{\frac{n-1}{n}}}{kl(n-1)}], \quad n \neq 1, \tag{26}$$

where p_1, p_2, p_3 and q_1, q_2, q_3 satisfy $p_1p_2p_3 = 1$ and $q_1 + q_2 + q_3 = 0$. The cosmological parameters turn out to be

$$H = \frac{l}{nlt + k_1}, \quad V = (nlt + k_1)^{\frac{3}{n}}. \tag{27}$$

Moreover, the function of Ricci scalar for vacuum becomes

$$f(R) = \frac{k}{2}(nlt + k_1)^{\frac{-2}{n}} R + 3kl^2(n-1)(nlt + k_1)^{\frac{-2n-2}{n}}. \tag{28}$$

2.2.2. *Non-vacuum solution of Bianchi type VI_0 universe*

The line element for Bianchi type VI_0 spacetime is given by

$$ds^2 = dt^2 - A^2(t)dx^2 - e^{2\alpha x}B^2(t)dy^2 - e^{-2\alpha x}C^2(t)dz^2, \tag{29}$$

where α is a non-zero constant. The corresponding scalar curvature is

$$R = -2[\frac{\ddot{A}}{A} + \frac{\ddot{B}}{B} + \frac{\ddot{C}}{C} - \frac{\alpha^2}{A^2} + \frac{\dot{A}\dot{B}}{AB} + \frac{\dot{A}\dot{C}}{AC} + \frac{\dot{B}\dot{C}}{BC}]. \tag{30}$$

The cosmological parameters are given as

$$H_x = \frac{\dot{A}}{A}, \quad H_y = \frac{\dot{B}}{B}, \quad H_z = \frac{\dot{C}}{C}, \quad a = (ABC)^{1/3}, \tag{31}$$

$$q = -\frac{a\ddot{a}}{\dot{a}^2}, \quad \Theta = u^a_{;a} = \frac{\dot{A}}{A} + \frac{\dot{B}}{B} + \frac{\dot{C}}{C}, \quad \sigma^2 = \frac{1}{2}\sigma_{ab}\sigma^{ab}, \tag{32}$$

$$H = \frac{1}{3}(\ln V\dot{)} = \frac{1}{3}(\frac{\dot{A}}{A} + \frac{\dot{B}}{B} + \frac{\dot{C}}{C}). \tag{33}$$

Here u^a is the four velocity vector and σ_{ab} is the shear tensor.

The energy-momentum tensor for perfect fluid is

$$T^{\nu}_{\mu} = diag[\rho, -p, -p, -p], \tag{34}$$

where ρ is the density and p is the pressure. The corresponding field equations yield the following solution

$$ds^2 = \left[\frac{(1-n)(m+6n+2mn)\tilde{t}^{2(m+6n+2mn)/3}}{6\alpha^2\tilde{t}^{2(m+6n+2mn)/3}+c_4(m+6n+2mn)}\right]d\tilde{t}^2 - \tilde{t}^2dx^2 - e^{2\alpha x}\tilde{t}^{2n}dy^2 - c_3^2\tilde{t}^{2n}e^{-2\alpha x}dz^2. \quad (35)$$

For this model, the Hubble parameter and the expansion scalar indicate that expansion rate was rapid at initial times of the big bang but it slows down with the passage of time and tends to zero as $\tilde{t} \to \infty$. The ratio $\frac{\sigma}{\Theta}$ indicates that the universe does not achieve isotropy and hence this model represents continuously expanding, shearing universe from the start of the big bang. The scalar curvature and $f(R)$ function turn out to be

$$R = \frac{2\alpha^2(5n+1)}{\tilde{t}^2(n-1)} + 2(5n^2+6n+(2n+1)^2m) \times \frac{-2\alpha^2\tilde{t}^{2(m+6n+2mn)}+c_4(n-1)(m+6n+2mn)}{(n-1)(m+6n+2mn)\tilde{t}^{2(m+6n+2mn+1)}}, \quad (36)$$

$$f(R) = l\tilde{t}^{(2n+1)/3}R + 3(n+1) \times [\frac{-2\alpha^2}{(n-1)(m+6n+2mn)\tilde{t}^2} + \frac{c_4}{\tilde{t}^{2(m+6n+2mn+1)}}]^{1/2}. \quad (37)$$

We know that if $f(R)$ is replaced by $R+\Lambda$, then $f(R)$ theory corresponds to GR, where Λ is interpreted as the energy density of the vacuum[41–44] causing expansion in the universe. Thus $f(R)$ may be used to explain the present cosmological expansion.

3. Dynamics of Gravitational Collapse

In this section, we discuss dynamics of plane symmetric gravitational collapse in $f(R)$ theory using Einstein frame. Notice that Eq.(3) can also be written as

$$G_{\mu\nu} = \frac{\kappa}{F}(T^m_{\mu\nu} + T^{(D)}_{\mu\nu}), \quad (38)$$

where $T^m_{\mu\nu}$ represents the usual matter and

$$T^{(D)}_{\mu\nu} = \frac{1}{\kappa}\left[\frac{f(R)-RF(R)}{2}g_{\mu\nu} + \nabla_\mu\nabla_\nu F(R) - g_{\mu\nu}\Box F(R)\right]. \quad (39)$$

Here "Dark source", $T^{(D)}_{\mu\nu}$, formally plays the role of a source in the field equations and its effect is the same as that of an effective fluid of purely geometrical origin. In fact, this scheme provides all the ingredients needed to tackle the dark side of the universe. Thus Eq.(39) can play the role of DE to explain the effects of DE on cosmological and gravitational phenomena of the universe.

3.1. *Fluid Distribution and the Field Equations*

We consider plane symmetric spacetime by the line element

$$ds^2 = A^2(t,z)dt^2 - B^2(t,z)(dx^2 + dy^2) - C^2(t,z)dz^2. \tag{40}$$

It is assumed that fluid inside the spacetime is locally anisotropic and suffering dissipation in the form of shearing viscosity, heat flow and free streaming radiation. The energy-momentum tensor has the following form

$$T_{\mu\nu} = (\rho+p_\perp)u_\mu u_\nu - p_\perp g_{\mu\nu} + (p_z - p_\perp)\chi_\mu\chi_\nu + u_\mu q_\nu + q_\mu u_\nu + \epsilon l_\mu l_\nu - 2\eta\sigma_{\mu\nu}. \tag{41}$$

Here, we have ρ as the energy density, $p_\perp$ the tangential pressure, p_z the pressure in z-direction, q_μ the heat flux, η the coefficient of shear viscosity, u_μ the four-velocity of the fluid, χ_μ the unit four-vector along the radial direction, l_μ a radial null four-vector and ϵ the energy density of the null fluid describing dissipation in the free streaming approximation. These quantities satisfy the relations

$$\begin{aligned} &u^\mu u_\mu = 1, \quad \chi^\mu\chi_\mu = -1, \quad l^\mu u_\mu = 1, \\ &u^\mu q_\mu = 0, \quad \chi^\mu u_\mu = 0, \quad l^\mu l_\mu = 0 \end{aligned} \tag{42}$$

which are obtained from the following definitions in co-moving coordinates

$$u^\mu = A^{-1}\delta^\mu_0, \quad \chi^\mu = C^{-1}\delta^\mu_3, \quad q^a = qC^{-1}\delta^\mu_3, \quad l^a = A^{-1}\delta^\mu_0 + C^{-1}\delta^\mu_3. \tag{43}$$

Here q is a function of t and r. The shear tensor $\sigma_{\mu\nu}$ is defined by

$$\sigma_{\mu\nu} = u_{(\mu;\nu)} - a_{(\mu}u_{\nu)} - \frac{1}{3}\Theta(g_{\mu\nu} - u_\mu u_\nu), \tag{44}$$

where the acceleration a_μ and the expansion scalar Θ are

$$a_\mu = u_{\mu;\nu}u^\nu, \quad \Theta = u^\mu_{;\mu}. \tag{45}$$

From Eqs.(43) and (44), we obtain

$$\sigma_{11} = -\frac{1}{3}B^2\sigma = \sigma_{22}, \quad \sigma_{33} = \frac{2}{3}C^2\sigma, \quad \sigma = \frac{1}{A}(\frac{\dot{B}}{B} - \frac{\dot{C}}{C}), \tag{46}$$

where σ is the shear scalar. Equations (43) and (45) yield

$$a_3 = -\frac{A'}{A}, \quad \Theta = \frac{1}{A}(\frac{2\dot{B}}{B} + \frac{\dot{C}}{C}), \tag{47}$$

where prime denotes derivative with respect to z.

In Einstein frame, the corresponding field equations become

$$
\begin{aligned}
&(\frac{2\dot{C}}{C}+\frac{\dot{B}}{B})\frac{\dot{B}}{B}+(\frac{A}{C})^2[-\frac{2B''}{B}+(\frac{2C'}{C}-\frac{B'}{B})\frac{B'}{B}] \\
&=\frac{8\pi}{F}[(\rho+\epsilon)A^2+T_{00}^{(D)}], \qquad (48)\\
&-(\frac{B}{A})^2[\frac{\ddot{B}}{B}+\frac{\ddot{C}}{C}-\frac{\dot{A}}{A}(\frac{\dot{B}}{B}+\frac{\dot{C}}{C})+\frac{\dot{B}\dot{C}}{BC}] \\
&+(\frac{B}{C})^2[\frac{A''}{A}+\frac{B''}{B}-\frac{B'C'}{BC}-(\frac{C'}{C}-\frac{B'}{B})\frac{A'}{A}] \\
&=\frac{8\pi}{F}[(p_\perp-\frac{2}{3}\eta\sigma)B^2+T_{11}^{(D)}], \qquad (49)\\
&-(\frac{C}{A})^2[\frac{2\ddot{B}}{B}+(\frac{\dot{B}}{B})^2-\frac{2\dot{A}\dot{B}}{AB}]+(\frac{B'}{B})^2+\frac{2A'B'}{AB} \\
&=\frac{8\pi}{F}[(p_z+\epsilon+\frac{4}{3}\eta\sigma)C^2+T_{33}^{(D)}], \qquad (50)\\
&-\frac{\dot{B}'}{B}+\frac{A'\dot{B}}{AB}+\frac{B'\dot{C}}{BC}=\frac{8\pi}{F}[-(q+\epsilon)AB+T_{03}^{(D)}]. \qquad (51)
\end{aligned}
$$

Here the components of dark fluid are obtained from Eq.(39) as follows

$$
\begin{aligned}
T_{00}^{(D)}&=\frac{A^2}{8\pi}[\frac{f-RF}{2}+\frac{F''}{C^2}-(\frac{2\dot{B}}{B}+\frac{\dot{C}}{C})\frac{\dot{F}}{A^2}-(-\frac{2B'}{B}+\frac{C'}{C})\frac{F'}{C^2}],\\
T_{11}^{(D)}&=\frac{-B^2}{8\pi}[\frac{f-RF}{2}-\frac{\ddot{F}}{A^2}+\frac{F''}{C^2}-(-\frac{\dot{A}}{A}+\frac{\dot{B}}{B}+\frac{\dot{C}}{C})\frac{\dot{F}}{A^2}\\
&+(\frac{A'}{A}+\frac{B'}{B}-\frac{C'}{C})\frac{F'}{B^2}]=T_{22}^{(D)},\\
T_{33}^{(D)}&=\frac{-C^2}{8\pi}[\frac{f-RF}{2}-\frac{\ddot{F}}{A^2}+(\frac{\dot{A}}{A}-\frac{2\dot{B}}{B})\frac{\dot{F}}{A^2}+(\frac{A'}{A}+\frac{2B'}{B})\frac{F'}{B^2}],\\
T_{03}^{(D)}&=\frac{1}{8\pi}(\dot{F}'-\frac{A'}{A}\dot{F}-\frac{\dot{B}}{B}F'). \qquad (52)
\end{aligned}
$$

The Ricci scalar curvature is given by

$$
\begin{aligned}
R=&-2[-\frac{2\ddot{C}}{A^2B}+\frac{2\dot{A}\dot{B}}{A^3BC}+\frac{2A'B'}{ABC^2}+\frac{A''}{AC^2}-\frac{\ddot{C}}{CA^2}+\frac{\dot{A}\dot{C}}{A^2C}\\
&-\frac{A'C'}{AC^3}-\frac{\dot{B}^2}{A^2B^2}+\frac{B'^2}{B^2C^2}+\frac{2B''}{BC^2}-\frac{2\dot{B}\dot{C}}{A^2BC}-\frac{2B'C'}{BC^3}]. \qquad (53)
\end{aligned}
$$

3.2. *Dynamical Equations*

Now, we develop equations that govern the dynamics of plane symmetric collapsing process. For this purpose, we use Misner and Sharp formalism[45].

The proper time derivative and proper derivative in z-direction are defined as[46]

$$D_{\tilde{T}} = \frac{1}{A}\frac{\partial}{\partial t}, \quad D_{\tilde{Z}} = \frac{1}{\tilde{Z}'}\frac{\partial}{\partial z}, \tag{54}$$

where $\tilde{Z} = B$. The velocity of the collapsing fluid is

$$U = D_{\tilde{T}}(\tilde{Z}) = \frac{\dot{B}}{A} \tag{55}$$

which is always negative. The Taub's mass function[47] for plane symmetric spacetime can be written as

$$m(t,z) = \frac{B}{2}(\frac{\dot{B}^2}{A^2} - \frac{B'^2}{C^2}). \tag{56}$$

Using this expression, Eq.(56) implies that

$$E = \frac{B'}{C} = [U^2 - \frac{2m}{B}]^{\frac{1}{2}}. \tag{57}$$

The rate of change of mass with respect to proper time is given by

$$D_{\tilde{T}}m = -\frac{4\pi}{F}\left[(p_z + \epsilon + \frac{4}{3}\eta\sigma + \frac{T_{33}^{(D)}}{B^2})U + E(q + \epsilon - \frac{T_{03}^{(D)}}{AB})\right]\tilde{Z}^2. \tag{58}$$

This shows how mass is varying within the plane hypersurface under the influence of matter variables. The first term shows the contribution of effective pressure in z-direction, radiation density and the DE whose negative effect is balanced with collapsing velocity U. The terms in the second round brackets have negative sign indicating that energy is leaving the system due to heat flux and repulsive effect of curvature fluid. Similarly, we calculate

$$D_{\tilde{Z}}m = \frac{4\pi}{F}\left[\rho + \epsilon + \frac{T_{00}^{(D)}}{A^2} + (q + \epsilon - \frac{T_{03}^{(D)}}{AB})\frac{U}{E}\right]\tilde{Z}^2. \tag{59}$$

This equation describes how different quantities influence the mass between neighboring hypersurfaces in the fluid distribution. It indicates the effects of energy density, radiation density and higher order curvature corrections on adjacent layers of collapsing body. Integration of Eq.(59) yields

$$m = \int_0^{\tilde{Z}} \frac{4\pi}{F}\left[\rho + \epsilon + \frac{T_{00}^{(D)}}{A^2} + (q + \epsilon - \frac{T_{03}^{(D)}}{AB})\frac{U}{E}\right]\tilde{Z}^2 d\tilde{Z}. \tag{60}$$

The dynamical equations can be obtained from the contracted Bianchi identities. Consider the following two equations

$$(\overset{(m)}{T^{\mu\nu}}+\overset{(D)}{T^{\mu\nu}})_{;\nu}u_{\mu}=(\overset{(m)}{T^{0\nu}}{}_{;\nu}+\overset{(D)}{T^{0\nu}}{}_{;\nu})u_0=0, \tag{61}$$

$$(\overset{(m)}{T^{\mu\nu}}+\overset{(D)}{T^{\mu\nu}})_{;\nu}\chi_{\mu}=(\overset{(m)}{T^{3\nu}}{}_{;\nu}+\overset{(D)}{T^{3\nu}}{}_{;\nu})\chi_3=0 \tag{62}$$

which yield

$$\begin{aligned}&\frac{(\mu+\epsilon)^{\cdot}}{A}+(\rho+2\epsilon+p_z+\frac{4}{3}\eta\sigma)\frac{\dot{C}}{AC}+2(\rho+\epsilon+p_{\perp}-\frac{2}{3}\eta\sigma)\frac{\dot{B}}{AB}\\&\frac{(q+\epsilon)'}{AC}+\frac{2A'}{AC}(q+\epsilon)+\frac{2B'}{BC}(q+\epsilon)+D_1=0,\end{aligned} \tag{63}$$

$$\begin{aligned}&\frac{(q+\epsilon)^{\cdot}}{A}+\frac{1}{C}(p_z+\epsilon+\frac{4}{3}\eta\sigma)'+\frac{2(q+\epsilon)(BC)^{\cdot}}{ABC}+(\rho+p_z+2\epsilon\\&+\frac{4}{3}\eta\sigma)\frac{A'}{AC}+2(p_z-p_{\perp}+\epsilon-2\eta\sigma)\frac{B'}{BC}+D_2=0.\end{aligned} \tag{64}$$

Here D_1 and D_2 are the DE contributions given by

$$\begin{aligned}D_1&=\frac{1}{8\pi}[\frac{1}{A^2C^2}(\dot{F}'-\frac{A'}{A}\dot{F}-\frac{\dot{B}}{B}F')]_{,3}+\frac{1}{8\pi A^2}[\frac{f-RF}{2}+\frac{F''}{C^2}-(\frac{2\dot{B}}{B}\\&+\frac{\dot{C}}{C})\frac{\dot{F}}{A^2}-(-\frac{2B'}{B}+\frac{C'}{C})\frac{F'}{C^2}]_{,0}+\frac{1}{8\pi A^2}[\frac{\dot{F}}{A^2}(\frac{3\dot{B}^2}{B^2}+\frac{5\dot{B}\dot{C}}{BC}+\frac{\dot{C}^2}{C^2}\\&-\frac{\dot{A}\dot{B}}{AB}-\frac{\dot{A}\dot{C}}{AC})+\frac{F'}{C^2}(\frac{\dot{B}B'}{B^2}-\frac{A'\dot{B}}{AB}-\frac{A'\dot{C}}{AC}-\frac{\dot{C}^2}{C^2})+\frac{2\ddot{F}}{A^2}(\frac{\dot{B}}{B}+\frac{\dot{C}}{C})\\&+\frac{1}{C^2}(\frac{2A'}{A}+\frac{2B'}{B}+\frac{C'}{C})(\dot{F}'-\frac{A'\dot{F}}{A}-\frac{\dot{C}F'}{C})],\end{aligned} \tag{65}$$

$$\begin{aligned}D_2&=\frac{1}{8\pi}[\frac{1}{A^2C^2}(\dot{F}'-\frac{A'}{A}\dot{F}-\frac{\dot{B}}{B}F')]_{,0}+\frac{1}{8\pi C^2}[\frac{f-RF}{2}-\frac{\ddot{F}}{A^2}+\frac{\dot{F}}{A^2}\\&\times(\frac{\dot{A}}{A}-\frac{2\dot{B}}{B})+(\frac{A'}{A}+\frac{2B'}{B})\frac{F'}{C^2}]_{,3}+\frac{1}{8\pi C^2}[\frac{\dot{F}}{A^2}(\frac{A'\dot{C}}{AC}-\frac{\dot{A}B'}{AB}+\frac{3\dot{B}B'}{B^2}\\&-\frac{B'\dot{C}}{BC}-\frac{\dot{A}A'}{A^2})+\frac{F'}{C^2}(\frac{A'B'}{AB}-\frac{A'C'}{AC}-\frac{3B'^2}{B^2}-\frac{B'C'}{BC}-\frac{A'^2}{A^2})\\&+\frac{\ddot{F}}{A^2}(\frac{A'}{A}+\frac{B'}{B})+\frac{F''}{C^2}(\frac{A'}{A}+\frac{2B'}{B})+\frac{1}{A^2}(\frac{\dot{A}}{A}+\frac{2\dot{B}}{B}+\frac{3\dot{C}}{C})\\&\times(\dot{F}'-\frac{A'\dot{F}}{A}-\frac{\dot{C}F'}{C})].\end{aligned} \tag{66}$$

The acceleration of the collapsing fluid is defined as

$$D_{\tilde{T}}U=\frac{1}{A}\frac{\partial U}{\partial t}=\frac{\ddot{B}}{A^2}-\frac{\dot{A}\dot{B}}{A^3}. \tag{67}$$

Using Eqs.(50), (54) and (56), we have

$$D_{\tilde{T}}U = -4\pi(p_z + \epsilon + \frac{4}{3}\eta\sigma)\tilde{Z} - \frac{m}{\tilde{Z}^2} + \frac{4\pi\tilde{Z}}{C^2}T_{33}^{(D)} + \frac{A'E}{AC}. \qquad (68)$$

Extracting value of $\frac{A'}{A}$ from the above equation and substituting in Eq.(64), it follows that

$$(\rho + p_z + 2\epsilon + \frac{4}{3}\eta\sigma)D_{\tilde{T}}U = -(\rho + p_z + 2\epsilon + \frac{4}{3}\eta\sigma)[\frac{m}{\tilde{Z}^2} + \frac{4\pi\tilde{Z}}{C^2}T_{33}^{(D)}$$
$$+ 4\pi(p_z + \epsilon + \frac{4}{3}\eta\sigma)\tilde{Z}] - E^2[D_{\tilde{Z}}(p_z + \epsilon - \frac{4}{3}\eta\sigma) + \frac{2}{\tilde{Z}}(p_z - p_\perp + \epsilon - 2\eta\sigma)]$$
$$- E[D_{\tilde{T}}(q + \epsilon) + 4(q + \epsilon)\frac{U}{\tilde{Z}} + 2(q + \epsilon)\sigma - D_2]. \qquad (69)$$

This equation yields the effect of different forces on the collapsing process. It can be interpreted in the form of Newton's second law of motion, i.e., Force = mass density × acceleration. Consequently, this equation shows the effect of pressure anisotropy, energy density, density of null fluid, radiation density, viscosity and DE on the rate of gravitational collapse. The terms with negative sign increase the rate of collapse while positive terms create repulsion and slow down the rate.

4. Summary

In this paper, we have presented some interesting consequences of $f(R)$ gravity. It has been shown that DE and dust matter phases can be achieved by exact solutions derived from $f(R)$ cosmological models[48]. Dark energy is considered as one of the causes for expanding universe. In recent papers[15−18], we have discussed expansion of the universe in the context of metric $f(R)$ gravity. Thus we expect that such solutions may provide a gateway towards the solution of DE and dark matter problems.

Firstly, we have reviewed some exact vacuum/non-vacuum solutions with Ricci scalar constant as well as non-constant. Secondly, we have studied the problem of gravitational collapse in $f(R)$ theory. Here the higher order curvature terms are imposed to violate the null energy conditions and thought to be the origin of DE. We have considered the plane symmetric collapsing stars and discussed their dynamics under the influence of $f(R)$ DE. We have formulated the dynamical equations by using contracted Bianchi identities. The dynamical equations help to investigate the evolution of gravitational collapse with time. These equations yield the variation of total energy inside a collapsing body with respect to time and adjacent surfaces. It is worthwhile to mention here that the curvature terms appear to affect the gravitational mass and rate of collapse. Thus $f(R)$ DE slows down the rate of collapse due its repulsive effect.

Acknowledgment

One (MS) of us would like to thank TÜBİTAK for providing the funding to attend this conference.

References

1. T. Chiba, *Phys. Lett.* ***B575***, 1 (2003).
2. A. D. Dolgov and M. Kawasaki, *Phys. Lett.* ***B573***, 1 (2003).
3. S. Capozziello, *Int. J. Mod. Phys.* ***D11***, 483 (2002);
   T. P. Sotiriou, *Class. Quantum Grav.* ***23***, 5117 (2006).
4. S. Nojiri and S.D. Odintsov, *Phys. Lett.* ***B646*** 105 (2007);
   O. Bertolami and J. Pármos, *Class. Quantum Grav.* ***25***, 245017 (2008).
5. S. Capozziello and M. Francaviglia, *Gen. Relativ. Gravit.* ***40***, 357 (2008);
   T. Harko, *Phys. Lett.* ***B669*** 376 (2008).
6. T. Chiba, *Phys. Lett.* ***B575***, 1 (2003);
   L. Amendola and S. Tsujikawa, *Phys. Lett.* ***B660***, 125 (2008).
7. R. Dicke, *Gen. Relativ. Gravit.* ***36***, 217 (2004);
   T. P. Sotiriou, *Phys. Rev.* ***D73***, 063515 (2006).
8. A. D. Dolgov and M. Kawasaki, *Phys. Lett.* ***B573***, 1 (2003);
   V. Faraoni, *Phys. Rev.* ***D74***, 104017 (2006).
9. A. V. Frolov, *Phys. Rev. Lett.* ***101***, 061103 (2008).
10. T. Multamäki and I. Vilja, *Phys. Rev.* ***D74***, 064022 (2006).
11. T. R. P. Caramês and E. R. Bezerra de Mello, *Eur. Phys. J. C* ***64***, 113 (2009).
12. S. Capozziello, A. Stabile and A. Troisi, *Class. Quantum Grav.* ***24***, 2153 (2007).
13. A. Azadi, D. Momeni and M. Nouri-Zonoz: *Phys. Lett.* ***B670***, 210 (2008).
14. D. Momeni, *Int. J. Mod. Phys.* ***D18***, 09 (2009).
15. M. Sharif and M. F. Shamir, *Mod. Phys. Lett.* ***A25***, 1281 (2010).
16. M. Sharif and H. R. Kausar, *J. Japan Phys. Soc.* ***80***, 044004 (2011).
17. M. Sharif and M. F. Shamir, *Class. Quantum Grav.* ***26***, 235020 (2009).
18. M. Sharif and H. R. Kausar, *Astrophys. Space Sci.* ***332***, 463 (2011).
19. M. Sharif and M. F. Shamir, *Gen. Relativ. Grav.* ***42***, 2643 (2010).
20. M. Sharif and H. R. Kausar, *Phys. Lett.* ***B697***, 1 (2011).
21. T. Multamäki and I. Vilja, *Phys. Rev.* ***D76*** 064021 (2007).
22. L. Hollestein and F. S. N. Lobo, *Phys. Rev.* ***D78***, 124007 (2008).
23. B. Whitt, *Phys. Lett.* ***B145***, 176 (1984).
24. A. Di Prisco, et al., *Phys. Rev.* ***D80***, 064031 (2009).
25. K. Nakao and Y. Morisawa, *Class. Quantum Grav.* ***21***, 2101 (2004).
26. M. Sharif and Z. Rehmat, *Gen. Relativ. Grav.* ***42***, 1795 (2010).
27. M. Sharif and Z. Ahmad, *Int. J. Mod. Phys.* ***A23***, 181 (2008).

28. S. Chakraborty and U. Debnath, *Mod. Phys. Lett.* ***A20***, 1451 (2005).
29. S. Chakraborty, S. Chakraborty and U. Debnath, *Int. J. Mod. Phys.* ***D16*** 833 (2007).
30. M. Sharif and K. Iqbal, *Mod. Phys. Lett.* ***A24***, 1533 (2009).
31. M. Sharif and Z. Ahmad, *Mod. Phys. Lett.* ***A22***, 1493 (2007);
M. Sharif and Z. Ahmad, *Mod. Phys. Lett.* ***A22***, 2947 (2007).
32. L. Herrera and N. O. Santos and A. Wang, *Phys. Rev.* ***D78***, 080426 (2008).
33. L. Herrera and N. O. Santos, *Phys. Rev.* ***D70***, 084004 (2004).
34. M. Sharif and A. Siddiqa, *Gen. Relativ. Gravit.* **43**, 73 (2011);
M. Sharif and S. Fatima, *Gen. Relativ. Gravit.* ***43***, 127 (2011);
M. Sharif and G. Abbas, *Mod. Phys. Lett.* ***A24***, 2551 (2009);
M. Sharif and G. Abbas, *Gen. Relativ. Grav.* ***43*** 1179 (2011);
M. Sharif and G. Abbas, *Astrophys. Space Sci.* ***327***, 285 (2010);
M. Sharif and G. Abbas, *J. Korean Phys. Soc.* ***556***, 529 (2010).
35. M. Sharif and H. R. Kausar, *Mod. Phys. Lett.* ***A25***, 3299 (2010);
M. Sharif and H. R. Kausar, *Astrophys. Space Sci.* ***331***, 281 (2011);
M. Sharif and H. R. Kausar, *Int. J. Mod. Phys. D* **20**, 2239 (2011);
M. Sharif and H. R. Kausar, *Astrophys. Space Sci.* ***337*** 805 (2012).
36. S. Nojiri and S. D. Odintsov: *Phys. Lett.* ***B657***, 238 (2007);
S. Nojiri and S. D. Odintsov, *Phys. Rev.* ***D77***, 026007 (2008).
37. K. Kainulainen, V. Reijonen and D. Sunhede, *Phys. Rev.* ***D76***, 043503 (2007).
38. C. G. Böhmer, *Gen. Relativ. Gravit.* ***36***, 1039 (2004).
39. J. P. Singh, R. K., Tiwari and P. Shukla: *Chin. Phys. Lett.* ***24***, 3325 (2007).
40. K. S. Adhav, et al., *Bulg. J. Phys.* ***34***, 260 (2007).
41. Y. B. Zeldovich, *Sov. Phys. JETP Lett.* ***6***, 316 (1967).
42. Y. B. Zeldovich, *Sov. Phys. JETP Lett.* **14**, 1143 (1968).
43. V. L. Ginzburg, D. A. Kirzhnits and A. A. Lyubushin, *Sov. Phys. JETP Lett.* ***33***, 242 (1971).
44. S. A. Fulling, L. Parker and B. L. Hu, *Phys. Rev.* ***D10***, 3905 (1974).
45. C. W. Misner and D. Sharp, *Phys. Rev.* ***136***, B571 (1964).
46. L. Herrera, N.O. Santos and A. Wang, *Phys. Rev.* ***D78***, 080426 (2008).
47. T. Zannias, *Phys. Rev.* ***D41***, 3252 (1990).
48. S. Capozziello, P. Martin-Moruno and C. Rubano, *Phys. Lett.* ***B664***, 12 (2008).

PART C

Quantum Gravity

STRESS-ENERGY CONNECTION: DEGRAVITATING THE VACUUM ENERGY

DURMUŞ ALİ DEMİR
Department of Physics, İzmir Institute of Technology, IZTECH, TR35430, İzmir, Turkey
E-mail: demir@physics.iztech.edu.tr
http://web.iyte.edu.tr/physics/people/durmusalidemir.html

This talk summarizes recent studies on the gravitational properties of vacuum energy in a non-Riemannian geometry formed by the stress-energy tensor of vacuum, matter and radiation. Postulating that the gravitational effects of matter and radiation can be formulated by an appropriate modification of the spacetime connection, we obtain varied geometro-dynamical equations which properly comprise the usual gravitational field equations with, however, Planck-suppressed, non-local, higher-dimensional additional terms. The prime novelty brought about by the formalism is that, the vacuum energy does act not as the cosmological constant but as the source of the gravitational constant. The formalism thus deafens the cosmological constant problem by channeling vacuum energy to gravitational constant. Nevertheless, quantum gravitational effects, if any, restore the problem via the graviton and graviton-matter loops, and the mechanism proposed here falls short of taming such contributions to cosmological constant.

Keywords: Stress-energy connection; Cosmological constant; Gravitational constant.

1. Introduction

In four-dimensional spacetime, there exist three symmetric, valency-two, divergence-free tensor fields. They are the metric tensor $g_{\alpha\beta}$, the Einstein tensor $G_{\alpha\beta}(g,\Gamma)$, and the stress-energy tensor $T_{\alpha\beta}$ of matter, radiation and vacuum. Some linear combination of them

$$aG_{\alpha\beta} + bg_{\alpha\beta} + cT_{\alpha\beta} \tag{1}$$

must vanish as there is no other tensor field exhibiting the same properties. Therefore, by an appropriate choice of the constants a, b and c one is led to the Einstein field equations of gravitation[1]

$$G_{\alpha\beta}\left(g,\Gamma\right) = -\Lambda_0 g_{\alpha\beta} + \left(8\pi G_N\right) T_{\alpha\beta} \tag{2}$$

where Λ_0 – Einstein's cosmological constant (CC)[2] – represents the intrinsic curvature of spacetime, and G_N stands for the gravitational constant. The tensor fields in (1) are simultaneously divergence-free if the connection Γ is compatible with metric, that is, if it is the Levi-Civita connection

$$\Gamma^{\lambda}_{\alpha\beta} = \frac{1}{2} g^{\lambda\rho}\left(\partial_\alpha g_{\beta\rho} + \partial_\beta g_{\rho\alpha} - \partial_\rho g_{\alpha\beta}\right) . \tag{3}$$

It is this connection that generates the Einstein tensor $G_{\alpha\beta}(g,\Gamma)$ (further details: Conventions[a]).

In general, $T_{\alpha\beta}$ involves all the matter and force fields as well as the metric tensor. Indeed, $T_{\alpha\beta}$ is computed from the quantum effective action which encodes quantum fluctuations of entire matter and all forces but gravity in the background geometry determined by $g_{\alpha\beta}$. Quantum theoretic structure ensures that

$$T_{\alpha\beta} = -\mathtt{E}\, g_{\alpha\beta} + \mathtt{t}_{\alpha\beta} \tag{6}$$

where $\mathtt{E}$ is the energy density of the vacuum, and $\mathtt{t}_{\alpha\beta}$ is the stress-energy tensor of everything but the vacuum. Putting $T_{\alpha\beta}$ into Eq. (2) gives rise to an effective CC

$$\Lambda_{\mathtt{eff}} = \Lambda_0 + 8\pi G_N \mathtt{E} \tag{7}$$

which must nearly saturate the present expansion rate of the Universe

$$\Lambda_{\mathtt{eff}} \lesssim H_0^2 \tag{8}$$

where $H_0 \simeq 73.2\,\mathrm{Mpc}^{-1}\,\mathrm{s}^{-1}\,\mathrm{km}$ according to the WMAP seven-year mean[3].

If Λ_0 not $\Lambda_{\mathtt{eff}}$ were used, the bound (8) would furnish, through the observational value of H_0 quoted above, an empirical determination of Λ_0, as for any other fundamental constant of Nature. The same does not apply

---

[a]Conventions: In general, a given connection $\text{𐰉}^{\lambda}_{\alpha\beta}$ generates the Einstein tensor

$$G_{\alpha\beta}\left(g,\text{𐰉}\right) = R_{\alpha\beta}\left(\text{𐰉}\right) - \frac{1}{2} g_{\alpha\beta} R\left(g,\text{𐰉}\right) \tag{4}$$

where the metric tensor $g_{\alpha\beta}$ forms the Ricci scalar $R\left(g,\text{𐰉}\right) \equiv g^{\mu\nu} R_{\mu\nu}\left(\text{𐰉}\right)$ from the Ricci tensor $R_{\alpha\beta}\left(\text{𐰉}\right) \equiv R^{\mu}_{\alpha\mu\beta}\left(\text{𐰉}\right)$ with

$$R^{\mu}_{\alpha\nu\beta}\left(\text{𐰉}\right) = \partial_\nu\, \text{𐰉}^{\mu}_{\beta\alpha} - \partial_\beta\, \text{𐰉}^{\mu}_{\nu\alpha} + \text{𐰉}^{\mu}_{\nu\lambda}\text{𐰉}^{\lambda}_{\beta\alpha} - \text{𐰉}^{\mu}_{\beta\lambda}\text{𐰉}^{\lambda}_{\nu\alpha} \tag{5}$$

being the Riemann tensor. By the way, the symbol 𐰉, the letter *b* in Turkic runic, is short-hand for *bagh* meaning 'connection' in Turkish.

to $\Lambda_{\mathtt{eff}}$, however. This is because the vacuum energy density E, equaling the zero-point energies of quantum fields plus enthalpy released by various phase transitions, is much larger than $\Lambda_{\mathtt{eff}}^{\mathtt{exp}}/8\pi G_N$. Therefore, previously determined, experimentally confirmed matter and forces down to the terascale $M_W \sim$ TeV, are expected to induce a vacuum energy density of order M_W^4. This is enormous compared to $\Lambda_{\mathtt{eff}}^{\mathtt{exp}}/8\pi G_N$, and hence, enforcement of $\Lambda_{\mathtt{eff}}$ to respect the bound (8) introduces a severe tuning of Λ_0 and $8\pi G_N$E up to at least sixty decimal places. This immense tuning becomes incrementally worse as the electroweak theory is extended to higher and higher energies. As a result, we face the biggest naturalness problem – the cosmological constant problem (CCP) – which plagues both particle physics and cosmology.

Over the decades, since its first solidification[4], the CCP has been approached by various proposals and interpretations, as reviewed and critically discussed by certain works[5,6]. Each proposal involves a certain degree of speculation in regard to going beyond (2) by postulating novel symmetry arguments, relaxation mechanisms, modified gravitational dynamics and statistical interpretations[5,6]. Except for the nonlocal, acausal modification of gravity[7] and the anthropic approach[8], most of the solutions proposed for the CCP seem to overlook the already-existing vacuum energy density $\mathcal{O}\left[\mathrm{TeV}^4\right]$ induced by known physics down to the terascale[9]. However, any resolution of the CCP, irrespective of how speculative it might be, must, in the first place, provide an understanding of how this existing energy component is to be tamed.

The CCP is a highly inextricable problem. Therefore, its crystallization proves highly critical in determining the correct solution strategy. The discussions above make it clear that *problem stems from covariantly constant parts* in equation (2). This means that the solution strategy, if any, must, first of all, look for ways of nullifying those terms involving constant multiples of $g_{\alpha\beta}$ in the fundamental structure (1)[10]. Clearly, this would be accomplished if metric tensor is prohibited to have vanishing divergence. This observation can be taken to indicate that a resolution of the CCP might be found in non-Riemannian geometries in which metric tensor is not convariantly constant or, equivalently, the non-metricity tensor does not vanish. One can thus put forth the interpretation that *the CCP is actually the problem of finding the correct method for incorporating the stress-energy tensor $T_{\alpha\beta}$ into the matter-free gravitational field equations so that the vacuum energy* E, *however large it might be, does not contribute to the effective CC.* Depending on how this incorporation is made, the gravitational field equations can admit variant interpretations and maneuvers for the vacuum energy, and it might then be possible to achieve a resolution for the CCP.

The material of this talk was covered by author's previous papers[11,12]. The talk is largely based on the paper 12, and repeats its contents. Given in Sec. 2 below is a detailed discussion of the method. The novel concept of 'stress-energy connection' will be introduced therein. Sec. III discusses certain questions concerning the workings of the mechanism. Sec. IV concludes.

2. Stress-Energy Connection and Cosmological Constant

In accord with the reasonings above, a formalism will be proposed and analyzed in this section. The key ingredient of the formalism shall be non-Riemannian geometries based on the stress-energy tensor of matter, radiation and vacuum.

In regions of spacetime devoid of energy, momentum, stress or pressure distribution, curving of the spacetime fabric is governed by the matter-free gravitational field equations

$$G_{\alpha\beta}\left(\between_V, V\right) = V_{\alpha\beta} \tag{9}$$

written purposefully in a slightly different form by utilizing the 'metric tensor'

$$V_{\alpha\beta} = -\Lambda_0 g_{\alpha\beta} \tag{10}$$

which is nothing but the stress-energy tensor of nothingness. It generates the connection

$$\left(\between_V\right)^{\lambda}_{\alpha\beta} = \frac{1}{2}\left(V^{-1}\right)^{\lambda\mu}\left(\partial_\alpha V_{\beta\mu} + \partial_\beta V_{\mu\alpha} - \partial_\mu V_{\alpha\beta}\right) \tag{11}$$

which is identical to the Levi-Civita connection in (3). This equivalence between $\between_V$ and Γ holds for any value of Λ_0 provided that it is strictly constant.

If the region of spacetime under concern is endowed with an energy, momentum, stress or pressure distribution, which are collectively encapsulated in the stress-energy tensor $T_{\alpha\beta}$, the matter-free gravitational field equations (9) change to

$$G_{\alpha\beta}\left(\between_V, V\right) = V_{\alpha\beta} + 8\pi G_N T_{\alpha\beta} \tag{12}$$

wherein the two sources are seen to directly add up[1]. It is this additive structure that is responsible for the CCP.

In search for a resolution for the CCP, certain clues are provided by the scaling properties of gravitational field equations. Under a rigid Weyl rescaling[13]

$$g_{\alpha\beta} \to a^2 g_{\alpha\beta} \tag{13}$$

the gravitational field equations (12) take the form

$$G_{\alpha\beta}\left(\Gamma_V, V\right) = a^2 V_{\alpha\beta} + 8\pi\left(G_N a^{-2}\right) T_{\alpha\beta}\left(a^d \mu_{(d)}\right) \tag{14}$$

where $\mu_{(d)}$ is a mass dimension-d coupling in the matter sector. The geometrodynamical variables $\left(\Gamma_V\right)^{\lambda}_{\alpha\beta}$ and $G_{\alpha\beta}\left(\Gamma_V, V\right)$ are strictly invariant under the global rescaling (13). However, sources $V_{\alpha\beta}$ and $G_N T_{\alpha\beta}$, containing fixed scales corresponding to masses, dimensionful couplings and renormalization scale, do not exhibit any invariance as such. Notably, however, even if the bare CC Λ_0 vanishes completely or if the matter sector possesses exact scale invariance ($T_{\alpha\beta} \to a^{-2} T_{\alpha\beta}$), gravitational field equations are never Weyl invariant simply because Newton's constant is there to scale as a^{-2}.

A short glance at (14) reveals that, the combination

$$G_{\alpha\beta}\left(\Gamma_V, V\right) - 8\pi\left(G_N a^{-2}\right) T_{\alpha\beta}\left(a^d \mu_{(d)}\right) \tag{15}$$

owns the transformation property of the Einstein tensor pertaining to a non-Riemannian geometry..This is readily seen by noting that, a general connection Γ can always decomposed as

$$\Gamma^{\lambda}_{\alpha\beta} = \left(\Gamma_V\right)^{\lambda}_{\alpha\beta} + \Delta^{\lambda}_{\alpha\beta} \tag{16}$$

where Δ is a rank (1,2) tensor field. In response to this split structure, the Einstein tensor of Γ breaks up into two

$$\mathbb{G}_{\alpha\beta}\left(\Gamma, V\right) = G_{\alpha\beta}\left(\Gamma_V, V\right) + \mathcal{G}_{\alpha\beta}\left(\Delta, V\right) \tag{17}$$

where $\mathcal{G}_{\alpha\beta}\left(\Delta, V\right)$, not found in GR, reads as

$$\mathcal{G}_{\alpha\beta}\left(\Delta, V\right) = \mathcal{R}_{\alpha\beta}(\Delta) - \frac{1}{2} V_{\alpha\beta}\left(V^{-1}\right)^{\mu\nu} \mathcal{R}_{\mu\nu}(\Delta) \tag{18}$$

with

$$\mathcal{R}_{\alpha\beta}\left(\Delta\right) = \nabla_{\mu}\Delta^{\mu}_{\alpha\beta} - \nabla_{\beta}\Delta^{\mu}_{\mu\alpha} + \Delta^{\mu}_{\mu\nu}\Delta^{\nu}_{\alpha\beta} - \Delta^{\mu}_{\beta\nu}\Delta^{\nu}_{\alpha\mu} \,. \tag{19}$$

Under the global scaling in (13), $G_{\alpha\beta}\left(\Gamma_V, V\right)$ stays at its original value yet $\mathcal{G}_{\alpha\beta}\left(\Delta, V\right)$ exhibits modifications contingent on how $\Delta^{\lambda}_{\alpha\beta}$ depends on the metric tensor. Formally, the Einstein tensor in (17) changes to

$$G_{\alpha\beta}\left(\Gamma_V, V\right) + \mathcal{G}_{\alpha\beta}\left(\Delta(a), V\right) \tag{20}$$

which obtains the same structure as the combination in (15) as far as the scaling properties of individual terms are concerned.

At this point, there arises a crucial question as to whether their formal similarity under scaling can ever promote (20) to a novel formulation alternative to (15). In other words, can part of (15) involving the stress-energy tensor arise, partly or wholly, from $\mathcal{G}_{\alpha\beta}(\Delta, V)$? Can matter and radiation be put in interaction with gravity by enveloping $T_{\alpha\beta}$ into connection instead of adding it to $V_{\alpha\beta}$ as in (12)? These questions, which are at the heart of the novel formulation being constructed, cannot be answered without a proper understanding of the tensorial connection Δ. To this end, one observes that generating $T_{\alpha\beta}$ from $\mathcal{G}_{\alpha\beta}(\Delta, V)$ can be a quite intricate process since while $T_{\alpha\beta}$ is divergence-free $\mathcal{G}_{\alpha\beta}(\Delta, V)$ is not

$$\nabla^{\alpha}\mathcal{G}_{\alpha\beta}(\Delta, V) \neq 0 \tag{21}$$

because $\mathcal{R}_{\alpha\beta}(\Delta)$, as it is not generated by commutators of $\nabla^{\between_V}$ or $\nabla^{\between}$, is not necessarily a true curvature tensor to obey the Bianchi identities. For relating $\Delta^{\lambda}_{\alpha\beta}$ to $T_{\alpha\beta}$, it proves facilitative to introduce a symmetric tensor field

$$\mathbb{T}_{\alpha\beta} = -\Lambda g_{\alpha\beta} + \Theta_{\alpha\beta} \tag{22}$$

which will be related to $T_{\alpha\beta}$ in the sequel. For definiteness, $\mathbb{T}_{\alpha\beta}$, similar to the stress-energy tensor $T_{\alpha\beta}$, is split into a covariantly-constant part which is its first term (Λ is strictly constant), and a generic symmetric tensor field $\Theta_{\alpha\beta}$ which does, by construction, not contain any covariantly-constant structure. As an obvious way of incorporating $T_{\alpha\beta}$ into (9) via $\between$, one can write

$$\between^{\lambda}_{\alpha\beta} = \left(\between_{V+\mathbb{T}}\right)^{\lambda}_{\alpha\beta} \tag{23}$$

which follows from (11) by replacing $V_{\alpha\beta}$ therein with $V_{\alpha\beta} + \mathbb{T}_{\alpha\beta}$. As a result, Δ becomes

$$\begin{aligned}\Delta^{\lambda}_{\alpha\beta} &= \frac{1}{2}\left((V+\mathbb{T})^{-1}\right)^{\lambda\nu}\left(\nabla_{\alpha}\mathbb{T}_{\beta\nu} + \nabla_{\beta}\mathbb{T}_{\nu\alpha} - \nabla_{\nu}\mathbb{T}_{\alpha\beta}\right)\\ &= \frac{1}{2}\left((V+\mathbb{T})^{-1}\right)^{\lambda\nu}\left(\nabla_{\alpha}\Theta_{\beta\nu} + \nabla_{\beta}\Theta_{\nu\alpha} - \nabla_{\nu}\Theta_{\alpha\beta}\right)\end{aligned} \tag{24}$$

which is seen to be a sensitive probe of $\Theta_{\alpha\beta}$ since it vanishes identically as $\Theta_{\alpha\beta} \to 0$.

By way of (23), the Einstein tensor in (20) takes the form

$$G_{\alpha\beta}\left(\between_V, V\right) + \mathcal{G}_{\alpha\beta}\left((\Lambda_0 + \Lambda)a^2, \Theta(a)\right) \tag{25}$$

whose comparison with (14) reveals the following features:

(1) The parameter Λ in (22) must be related to the gravitational constant G_N. Actually, a relation of the form

$$\Lambda + \Lambda_0 = (8\pi G_N)^{-1} \tag{26}$$

is expected on general grounds.

(2) In the limit $T_{\alpha\beta} \to 0$, the gravitational field equations (12) uniquely reduce to the matter-free field equations (9). Likewise, the gravitational field equations to be obtained here, as suggested by (23), must smoothly reduce to (9) as $\mathbb{T} \to 0$. Therefore, any functional relation $\mathbb{T}_{\alpha\beta} = \mathbb{T}_{\alpha\beta}[T]$ between $\mathbb{T}$ and T should exhibit the correspondence

$$T_{\alpha\beta} = 0 \Longleftrightarrow \mathbb{T}_{\alpha\beta} = 0\,. \tag{27}$$

In addition, as $T_{\alpha\beta} \to -\mathtt{E} g_{\alpha\beta}$, the right-hand side of (9) changes to $(1 + \mathtt{E}/\Lambda_0)\, V_{\alpha\beta}$, which clearly signals the CCP. In contrast, however, as $\mathbb{T}_{\alpha\beta} \to -\Lambda g_{\alpha\beta}$, $\mathbb{G}_{\alpha\beta}\left(\between_{V+\mathbb{T}}, V\right)$ reduces to the matter-free form $\mathbb{G}_{\alpha\beta}\left(\between_{V}, V\right)$. In other words, even if matter and radiation are discarded, that is, $T_{\alpha\beta} = -\Lambda g_{\alpha\beta}$ ($\mathtt{t}_{\alpha\beta} = 0$), the gravitational field equations (12) suffer from the CCP. However, when $\mathbb{T}_{\alpha\beta} = -\Lambda g_{\alpha\beta}$ ($\Theta_{\alpha\beta} = 0$), $\mathbb{G}_{\alpha\beta}\left(\between_{V+\mathbb{T}}, V\right)$ remains unchanged at $\mathbb{G}_{\alpha\beta}\left(\between_{V}, V\right)$ with complete immunity to Λ.

These observations evidently reveal the physical and CCP-wise relevance of the method.

As a matter of course, the dynamical equation

$$\mathbb{G}_{\alpha\beta}\left(\between_{V+\mathbb{T}}, V\right) = V_{\alpha\beta}\,, \tag{28}$$

as directly follows from (9) via the replacement $\between_{V} \to \between_{V+\mathbb{T}}$, forms the germ of the CCP-free gravitational dynamics under attempt. Under (17), it gives

$$G_{\alpha\beta}\left(\between_{V}, V\right) = V_{\alpha\beta} - \mathcal{G}_{\alpha\beta}\left(\Delta, V\right) \tag{29}$$

which refines the germinal equation (28). To proceed further, it is necessary to establish the relation between $\mathbb{T}_{\alpha\beta}$ and $T_{\alpha\beta}$ so that (28) reduces to (12), at least approximately. This reduction does of course not affect the value of CC; it stays put at Λ_0. On the other hand, with (26) relating Λ to G_N, on physical grounds, one expects $|\Lambda| \gg |\Theta|$. Then, all quantities, in particular, $\Delta^{\lambda}_{\alpha\beta}$ can be expanded in powers of Θ/Λ such that (29), at the leading order, is to return the gravitational field equations (12). As a matter of fact, the dynamical equation (29), after using

$$\left(\left(V + \mathbb{T}\right)^{-1}\right)_{\alpha\beta} = (8\pi G_N) g_{\alpha\beta} - (8\pi G_N)^2 \Theta_{\alpha\beta} + (8\pi G_N)^3 \Theta^{\mu}_{\alpha} \Theta_{\mu\beta} - \ldots$$

takes the form

$$G_{\alpha\beta}\left(\between_V, V\right) = \mathtt{C}^{(0)}_{\alpha\beta} + (8\pi G_N)\mathtt{C}^{(1)}_{\alpha\beta} + (8\pi G_N)^2\mathtt{C}^{(2)}_{\alpha\beta} + \dots \tag{30}$$

where $\mathtt{C}^{(n)}_{\alpha\beta}$ are valency-two symmetric tensor fields encapsulating all the terms of order $(8\pi G_N)^n$. For $n = 0$, the tensorial connection $\Delta^{\lambda}_{\alpha\beta}$ vanishes identically, and hence,

$$\mathtt{C}^{(0)}_{\alpha\beta} = V_{\alpha\beta} \tag{31}$$

so that (28) directly reduces to the matter-free gravitational field equations (9) for $\mathbb{T}_{\alpha\beta} = 0$ as well as $\mathbb{T}_{\alpha\beta} = -\Lambda g_{\alpha\beta}$.

For $n = 1$,

$$\Delta^{\lambda}_{\alpha\beta} = 4\pi G_N(\nabla_\alpha \Theta^{\lambda}_{\beta} + \nabla_\beta \Theta^{\lambda}_{\alpha} - \nabla^{\lambda}\Theta_{\alpha\beta}) \tag{32}$$

is linear in $\Theta_{\alpha\beta}$, and so is the derivative part of $\mathcal{R}_{\alpha\beta}\left(\Delta\right)$. Then, $\mathcal{G}_{\alpha\beta}\left(\Delta, V\right)$ defined in (18) yields

$$\mathtt{C}^{(1)}_{\alpha\beta} = -2\left[\mathtt{K}^{-1}\left(\nabla\right)\right]^{\mu\nu}_{\alpha\beta}\Theta_{\mu\nu} \tag{33}$$

where

$$\begin{aligned}\left[\mathtt{K}^{-1}\right]_{\alpha\beta\mu\nu}\left(\nabla\right) = {} & \frac{1}{8}\left(\nabla_\mu\nabla_\alpha g_{\nu\beta} + \nabla_\mu\nabla_\beta g_{\alpha\nu}\right) + \frac{1}{8}\left(\nabla_\nu\nabla_\alpha g_{\mu\beta} + \nabla_\nu\nabla_\beta g_{\alpha\mu}\right) \\ & - \frac{1}{8}\left(\nabla_\alpha\nabla_\beta + \nabla_\beta\nabla_\alpha\right) g_{\mu\nu} - \frac{1}{8}\left(\nabla_\mu\nabla_\nu + \nabla_\nu\nabla_\mu\right) g_{\alpha\beta} \\ & - \frac{1}{8}\Box\left(g_{\alpha\mu}g_{\beta\nu} + g_{\alpha\nu}g_{\mu\beta} - 2g_{\alpha\beta}g_{\mu\nu}\right)\end{aligned} \tag{34}$$

is nothing but the inverse propagator for a 'massless spin-2 field' in the background geometry generated by $g_{\alpha\beta}$. To reproduce the gravitational field equations (12) correctly, one must impose

$$-2\left[\mathtt{K}^{-1}\left(\nabla\right)\right]^{\mu\nu}_{\alpha\beta}\Theta_{\mu\nu} = -2\left[\mathtt{K}^{-1}\left(\nabla\right)\right]^{\mu\nu}_{\alpha\beta}\mathbb{T}_{\mu\nu} \equiv \mathtt{t}_{\alpha\beta} \tag{35}$$

where "$\mathtt{t}_{\alpha\beta}$" was defined in (6) to involve 'no covariantly-constant part'. This equality lies at the heart of the mechanism being proposed, and therefore, its analysis and examination prove vital for further progress. The main question is this: Can the right-hand side of (35) ever involve a covariantly-constant part (of the form $c_1 g_{\alpha\beta}$ with c_1 constant) added to $\mathtt{t}_{\alpha\beta}$? If the answer turns out to be affirmative then whole mechanism collapses down since c_1/M^2, unless guaranteed to lie near Λ_0 by some reason, brings back the CCP. In examining, one first notes that the equality (35) works fine for both $\Theta_{\alpha\beta}$ and $\mathbb{T}_{\alpha\beta}$ since a covariantly-constant part (like $\Lambda g_{\alpha\beta}$) is automatically nullified by $\left[\mathtt{K}^{-1}\left(\nabla\right)\right]^{\mu\nu}_{\alpha\beta}$. Therefore, if $\mathtt{t}_{\alpha\beta}$ in (35) is to change to $\mathtt{t}_{\alpha\beta} + c_1 g_{\alpha\beta}$ there has to be an appropriate structure within $\mathbb{T}_{\alpha\beta}$. The requisite structure is found to be

$$\delta\mathbb{T}_{\alpha\beta} = \left[\mathtt{K}\left(\nabla\right)\right]_{\alpha\beta}^{\mu\nu}\left(c_1 g_{\mu\nu}\right) \equiv k_1 g_{\alpha\beta} \tag{36}$$

where structure of the spin-2 propagator $\left[\mathtt{K}\left(\nabla\right)\right]_{\alpha\beta}^{\mu\nu}$ guarantees that $k_1 = \pm\infty$ independent of the value of c_1. This result implies that the covariantly-constant part of $\mathbb{T}_{\alpha\beta}$ in (22) changes to $(\Lambda + k)g_{\alpha\beta} \equiv \Lambda_{\mathtt{eff}} g_{\alpha\beta}$ with $\Lambda_{\mathtt{eff}} = \pm\infty$. In other words, an infinite Λ in $\mathbb{T}_{\alpha\beta}$ corresponds to a covariantly-constant part of the form $c_1 g_{\alpha\beta}$ in (35). However, $\Lambda \to \pm\infty$ in $\mathbb{T}_{\alpha\beta}$ causes the tensorial connection $\Delta_{\alpha\beta}^{\lambda}$ in (24) to vanish, and hence, the germinal equation (28) to reduce to the original matter-free gravitational field equations (9). This implies that an infinite Λ prohibits the incorporation of matter and radiation into (9). These observations and findings should provide enough evidence that "$\mathtt{t}_{\alpha\beta}$" in (35) is the stress-energy tensor of everything but vacuum; it cannot have a covariantly-constant part. It is precisely what was meant in writing (22), and hence, everything but vacuum gravitates precisely as in the GR. Obviously, $\Theta_{\alpha\beta}$ is related to $\mathtt{t}_{\alpha\beta}$ non-locally yet causally since $\Theta_{\alpha\beta}$ involves values of $\mathtt{t}_{\alpha\beta}$ in every place and time as propagated by the 'massless spin-2 propagator' $\mathtt{K}_{\alpha\beta\mu\nu}\left(\nabla\right)$. By inverting (35) one finds

$$\mathbb{T}_{\alpha\beta} = \Theta_{\alpha\beta}^{0} - \frac{1}{2}\left[\mathtt{K}\left(\nabla\right)\right]_{\alpha\beta}^{\mu\nu}\mathtt{t}_{\mu\nu} \tag{37}$$

where $\Theta_{\alpha\beta}^{0} \equiv -\Lambda g_{\alpha\beta}$ is covariantly-constant. In fact, it must be proportional to the vacuum energy density in (6), that is, $\Theta_{\alpha\beta}^{0} \propto \mathtt{E} g_{\alpha\beta}$. Consequently,

$$\mathbb{T}_{\alpha\beta} = -\mathtt{L}^2\mathtt{E} g_{\alpha\beta} - \frac{1}{2}\left[\mathtt{K}\left(\nabla\right)\right]_{\alpha\beta}^{\mu\nu}\mathtt{t}_{\mu\nu} \tag{38}$$

wherein $\Lambda = \mathtt{L}^2\mathtt{E}$, and $\mathtt{L}^2$, having the dimension of area, arises for dimensionality reasons. This expression establishes a direct relationship between $\mathbb{T}_{\alpha\beta}$ and $T_{\alpha\beta}$ so that $\mathbb{T}_{\alpha\beta} = 0 \Longleftrightarrow T_{\alpha\beta} = 0$, as was discussed in detail in relation to (27). Actually, it is possible to interpret the result (38) in a more general setting by generalizing the propagator (34) to massive case

$$\begin{aligned}\left[\mathcal{K}^{-1}\right]_{\alpha\beta\mu\nu}\left(\nabla, \mathtt{L}^2\right) &= \left[\mathtt{K}^{-1}\right]_{\alpha\beta\mu\nu}\left(\nabla\right) \\ &\quad - \frac{f\left(\mathtt{L}^2\Box\right)}{4\mathtt{L}^2}\left(g_{\alpha\beta}g_{\mu\nu} - g_{\alpha\mu}g_{\beta\nu} - g_{\alpha\nu}g_{\mu\beta}\right)\end{aligned} \tag{39}$$

where the operator $f\left(\mathtt{L}^2\Box\right)/\mathtt{L}^2$ serves as the 'mass-squared' parameter with the distributional structure

$$f(x) = \begin{cases} 1, & x = 0 \\ 0, & x \neq 0 \end{cases} \tag{40}$$

similar to the one used in 7. In (39), care is needed in interpreting the 'mass term' in that there is actually no 'spin-2 mass term' to speak about: It vanishes for non-uniform sources like $\mathtt{t}_{\alpha\beta}$ and stays constant for uniform sources like $\Lambda g_{\alpha\beta}$. Clearly, the 'massive propagator' above automatically reproduces the result in (38)

$$\mathbb{T}_{\alpha\beta} = [\mathcal{K}]^{\mu\nu}_{\alpha\beta}\left(\nabla, \mathtt{L}^2\right) T_{\mu\nu} = -\mathtt{L}^2 \mathtt{E} g_{\alpha\beta} - (1/2)\mathtt{K}\left(\nabla\right)^{\mu\nu}_{\alpha\beta} \mathtt{t}_{\mu\nu} \tag{41}$$

thanks to the distributional structure of $f(x)$.

For $n = 2$ and higher, the tensorial connection $\Delta^{\lambda}_{\alpha\beta}$ goes like Θ^{n-1} times $\nabla\Theta$, and is always proportional to $\Delta(n = 1)$. More explicitly,

$$\Delta^{\lambda}_{\alpha\beta}(n) = \left[\Pi^{n-1}_{k=1}(-8\pi G_N)^k \Theta^{\lambda}_{\mu_k}\right] \Delta^{\mu_1}_{\alpha\beta}(1) \tag{42}$$

where each Θ factor is expressed in terms of $\mathtt{t}$ via (41). Gradients of $\Delta^{\lambda}_{\alpha\beta}(n)$ and bilinears $\left[\Delta(n-k) \otimes \Delta(k)\right]_{\alpha\beta}$ $(k = 1, 2, \dots, n-1)$ add up to form $\mathtt{C}^{(n)}_{\alpha\beta}$ in accord with the structure of $\mathcal{G}_{\alpha\beta}(\Delta)$ in (18). In contrast to the three tensor fields $G_{\alpha\beta}(\between_V, V)$, $\mathtt{C}^{(0)}_{\alpha\beta}$ and $\mathtt{C}^{(1)}_{\alpha\beta}$, it is not clear if $\mathtt{C}^{(n\geq 2)}_{\alpha\beta}$ acquires vanishing divergence, in general. Therefore, the gravitational field equations

$$G_{\alpha\beta} = -\Lambda_0 g_{\alpha\beta} + (8\pi G_N)\mathtt{t}_{\alpha\beta} + \mathcal{O}\left[(8\pi G_N \nabla\Theta)^2, (8\pi G_N)^2 \Theta\nabla\nabla\Theta\right] \tag{43}$$

distilled from the germinal dynamics in (28), are insensitive to vacuum energy density $\mathtt{E}$ yet suffer from a serious inconsistency that the divergence of $\mathtt{C}^{(n\geq 2)}_{\alpha\beta}$ may not vanish at all. The next section will give a critique of the formalism, as developed so far.

3. More on the Formalism

Comparison of (43) with (12) raises certain questions pertaining to the consistency of the elicited gravitational dynamics. There are mainly three questions:

Question 1. What precludes $\mathcal{G}_{\alpha\beta}(\Delta, V)$ from developing a covariantly-constant part that can act as the CC?

Question 2. What must be the structure of $\mathbb{T}_{\alpha\beta}$ such that, despite Eq.(21), $\nabla^{\alpha} G_{\alpha\beta}(\Delta, V)$ is nullified to make both sides of (29) divergence-free?

Question 3. What is the status of CCP under the formalism developed here?

Answers to these questions will disclose the physical meaning, scope and reach of the gravitational field equations (43).

3.1. *Answer to Question 1*

It is of prime importance to determine if the quasi Einstein tensor $\mathcal{G}_{\alpha\beta}(\Delta, V)$ can develop a covariantly-constant part since this type of contribution can cause the CCP.

As the definition of $\Delta^{\lambda}_{\alpha\beta}$ in (24) manifestly shows, Λ, in whatever way it might be related to $\mathtt{E}$, does not provide any contribution to CC. In fact, a nontrivial $\Delta^{\lambda}_{\alpha\beta}$ originates from $\Theta_{\alpha\beta}$ only. Though it vanishes identically for $\Theta_{\alpha\beta} = 0$, it remains nonvanishing even for $\Lambda = 0$. Therefore, $\mathcal{G}_{\alpha\beta}(\Delta, V)$ depends critically on $\Theta_{\alpha\beta}$, and any value it takes, covariantly-constant or otherwise, is governed by $\Theta_{\alpha\beta}$. There is no such sensitivity to Λ.

As dictated by the structure of the quasi curvature tensor $\mathcal{R}_{\alpha\beta}$ in (19), for $G_{\alpha\beta}(\Delta, V)$ to develop a covariantly-constant part, at least one of

$$\nabla_{\mu}\Delta^{\mu}_{\alpha\beta}\,,\ \Delta^{\mu}_{\mu\nu}\Delta^{\nu}_{\alpha\beta}\,,\ \nabla_{\beta}\Delta^{\mu}_{\mu\alpha}\,,\ \Delta^{\mu}_{\beta\nu}\Delta^{\nu}_{\alpha\mu} \tag{44}$$

must be partly proportional to the metric tensor $g_{\alpha\beta}$ or must partly take a constant value when contracted with the metric tensor. Concerning the first and second structures above, a reasonable ansatz is $\Delta^{\lambda}_{\alpha\beta} \ni U^{\lambda}g_{\alpha\beta}$ where U^{α} is a vector field. With this structure for $\Delta^{\lambda}_{\alpha\beta}$, all one needs is to set $\nabla_{\mu}U^{\mu} = c_1$ for $\nabla_{\mu}\Delta^{\mu}_{\alpha\beta} \ni c_1 g_{\alpha\beta}$, and $U_{\mu}U^{\mu} = c_2$ for $\Delta^{\mu}_{\mu\nu}\Delta^{\nu}_{\alpha\beta} \ni c_2 g_{\alpha\beta}$, where c_1 and c_2 are constants. With the same ansatz for $\Delta^{\lambda}_{\alpha\beta}$, the remaining terms in (44) give rise to a covariantly-constant part in $\mathcal{G}_{\alpha\beta}(\Delta, V)$ not by themselves but via $V_{\alpha\beta}\left(V^{-1}\right)^{\mu\nu}\mathcal{R}_{\mu\nu}(\Delta)$. Indeed, $\nabla_{\beta}\Delta^{\mu}_{\mu\alpha} \ni \nabla_{\beta}U_{\alpha}$ and $\Delta^{\mu}_{\beta\nu}\Delta^{\nu}_{\alpha\mu} \ni U_{\alpha}U_{\beta}$, and they contract to c_1 and c_2 for $\nabla_{\mu}U^{\mu} = c_1$ and $U_{\mu}U^{\mu} = c_2$, respectively. A more accurate ansatz for a symmetric tensorial connection would be

$$\widetilde{\Delta}^{\lambda}_{\alpha\beta} = aU^{\lambda}g_{\alpha\beta} + b\left(\delta^{\lambda}_{\alpha}U_{\beta} + U_{\alpha}\delta^{\lambda}_{\beta}\right)\,. \tag{45}$$

As follows from (19), the Ricci tensor $\widetilde{\mathcal{R}}_{\alpha\beta}$ for this particular connection becomes symmetric for $a = -5b$, and the Einstein tensor

$$\widetilde{\mathcal{G}}_{\alpha\beta} = b\left(\nabla_{\alpha}U_{\beta} + \nabla_{\beta}U_{\alpha}\right) - 22b^2U_{\alpha}U_{\beta} + b\nabla\cdot U g_{\alpha\beta} + b^2 U\cdot U g_{\alpha\beta} \tag{46}$$

contributes to the CC by its third term in an amount $\delta\Lambda_0 = 4bc_1$ if $\nabla_{\mu}U^{\mu} = c_1$, and by its fourth term in an amount $\delta\Lambda_0 = -b^2c_2$ if $U_{\mu}U^{\mu} = c_2$. These results ensure that, at least for a connection in the form of (45), the CCP could be resurrected depending on how the contribution of U^{μ} compares with the bare term Λ_0. To this end, being a symmetric tensorial connection with symmetric Ricci tensor, $\widetilde{\Delta}^{\lambda}_{\alpha\beta}$ in (45) can be directly compared to $\Delta^{\lambda}_{\alpha\beta}$ in (24) to find

$$\frac{1}{2}\nabla_{\alpha}\log\left(\text{Det}\left[\mathbb{T}\right]\right) = \widetilde{\Delta}^{\mu}_{\mu\alpha} = 0 \tag{47}$$

and

$$\frac{1}{2}\left(\mathbb{T}^{-1}\right)^{\lambda\rho}\left(2\nabla^{\alpha}\mathbb{T}_{\alpha\rho} - \nabla_{\rho}\mathbb{T}^{\alpha}_{\alpha}\right) = g^{\alpha\beta}\widetilde{\Delta}^{\lambda}_{\alpha\beta} = -18bU^{\lambda}\,. \tag{48}$$

The first condition, namely the one in (47), requires $\mathbb{T}_{\alpha\beta} = \widetilde{c}\, g_{\alpha\beta}$ where $\widetilde{c}$ is a constant. In other words, (47) enforces $\Theta_{\alpha\beta} = 0$, and its replacement in (48) consistently gives $b = 0$. Therefore, at least for connections structured like (45), there does not exist a $\Theta_{\alpha\beta}$ to equip $\mathcal{G}_{\alpha\beta}\left(\Delta, V\right)$ with a covariantly-constant part.

Despite the firmness of this result, one notices that, it is actually not necessary to force $\Delta^{\lambda}_{\alpha\beta}$ to be wholly equal to $\widetilde{\Delta}^{\lambda}_{\alpha\beta}$ since it is sufficient to have only part of $\mathcal{G}_{\alpha\beta}\left(\Delta, V\right)$ be covariantly-constant. Thus, in general, one can write

$$\Delta^{\lambda}_{\alpha\beta} = \widetilde{\Delta}^{\lambda}_{\alpha\beta} + \mathcal{D}^{\lambda}_{\alpha\beta} \tag{49}$$

where $\mathcal{D}^{\lambda}_{\alpha\beta} = \mathcal{D}^{\lambda}_{\beta\alpha}$, and $\nabla_{\beta}\mathcal{D}^{\mu}_{\mu\alpha} = \nabla_{\alpha}\mathcal{D}^{\mu}_{\mu\beta}$ for $\mathcal{R}_{\alpha\beta}\left(\mathcal{D}\right) = \mathcal{R}_{\beta\alpha}\left(\mathcal{D}\right)$. This condition enforces either $\mathcal{D}^{\mu}_{\mu\alpha} = 0$ or $\mathcal{D}^{\mu}_{\mu\alpha} = \nabla_{\alpha}\Phi$, Φ being a scalar. The former, which was used for $\widetilde{\Delta}^{\lambda}_{\alpha\beta}$ in (45), does not change the present conclusion. The latter, which was used for $\Delta^{\lambda}_{\alpha\beta}$ in (24), guarantees that $\Delta^{\lambda}_{\alpha\beta}$ and $\mathcal{D}^{\lambda}_{\alpha\beta}$ are identical up to some determinant-preserving transformations. More accurately, while $\Delta^{\lambda}_{\alpha\beta}$ makes use of $\mathbb{T}_{\alpha\beta}$, $\mathcal{D}^{\lambda}_{\alpha\beta}$ involves $\mathcal{T}_{\alpha\beta}$ which must equal $\mathbb{M}^{\mu}_{\alpha}\mathbb{T}_{\mu\nu}\left(\mathbb{M}^{-1}\right)^{\nu}_{\beta}$ with $\mathbb{M}_{\alpha\beta}$ being a generic tensor field. All these results ensure that, $\Delta^{\lambda}_{\alpha\beta}$ cannot cause $\mathcal{G}_{\alpha\beta}\left(\Delta, V\right)$ to develop a covariantly-constant part, at least for tensorial connections of the form (45).

3.2. *Answer to Question 2*

The left-hand side of (43) is divergence-free by the Bianchi identities; however, its right-hand side exhibits no such property for $n \geq 2$. Indeed, unlike GR wherein the right-hand side obtains vanishing divergence by the conservation of matter and radiation flow, the right-hand side of (43) lacks such a property because the quasi curvature tensor $\mathcal{R}^{\mu}_{\alpha\nu\beta}\left(\Delta\right)$ does not obey the Bianchi identities. A remedy to this conservation problem, an aspect that the initiator work[11] was lacking, comes via the expansion

$$\mathbb{T}_{\alpha\beta} = -\Lambda\sum_{n=0}^{\infty}\left(-8\pi G_N\right)^n\Theta^{(n)}_{\alpha\beta} = -\Lambda g_{\alpha\beta} + \Theta^{(1)}_{\alpha\beta} - \left(8\pi G_N\right)\Theta^{(2)}_{\alpha\beta} + \ldots \tag{50}$$

over a set of tensor fields $\{\Theta^{(0)}_{\alpha\beta} \equiv g_{\alpha\beta}, \Theta^{(1)}_{\alpha\beta}, \Theta^{(2)}_{\alpha\beta}, \dots\}$, and requiring terms at the n–th order to give, through the dynamics of $\Theta^{(n)}_{\alpha\beta}$, a conserved tensor field $\mathtt{C}^{(n)}_{\alpha\beta}$. In (50), use has been made of $\Lambda \simeq (8\pi G_N)^{-1}$ as follows from (26) thanks to the extreme smallness of $|\Lambda_0|$. Clearly, $\Theta^{(1)}_{\alpha\beta}$ in (50) corresponds to Θ in (22), and $\Theta^{(n\geq 2)}_{\alpha\beta}$ represent the added features for achieving consistency in (43).

Despite the structure (50), $\mathtt{C}^{(0)}_{\alpha\beta}$ and $\mathtt{C}^{(1)}_{\alpha\beta}$ both stay put at their previous values in (31) and (33), respectively. The only difference is that Θ in (35) is replaced by $\Theta^{(1)}_{\alpha\beta}$, and hence, what appears in (38) are the first two terms of (50). Consequently, at levels of $n = 0$ and $n = 1$, gravitational dynamics in (43) stay intact to the serial structure of $\mathbb{T}$ introduced in (50). At the higher orders, $n \geq 2$, the situation changes due to the introduction of $\Theta^{(n\geq 2)}_{\alpha\beta}$. For example, if $n = 2$, the tensorial connection $\Delta^{\lambda}_{\alpha\beta}$ is quadratic in $\Theta^{(1)}{}_{\alpha\beta}$ and linear in $\Theta^{(2)}{}_{\alpha\beta}$

$$\Delta^{\lambda}_{\alpha\beta}(2) = 8\pi G_N \left(-\Theta^{(1)}{}^{\lambda}{}_{\rho} \Delta^{\rho}_{\alpha\beta}(1) + 4\pi G_N \left(\delta^{(2)}\right)^{\lambda}_{\alpha\beta} \right) \tag{51}$$

which differs from (42) by the presence of

$$\left(\delta^{(2)}\right)^{\lambda}_{\alpha\beta} = \nabla_{\alpha}\Theta^{(2)}{}^{\lambda}_{\beta} + \nabla_{\beta}\Theta^{(2)}{}^{\lambda}_{\alpha} - \nabla^{\lambda}\Theta^{(2)}{}_{\alpha\beta} \tag{52}$$

induced by $\Theta^{(2)}{}_{\alpha\beta}$ alone. Replacement of (51) in (29) yields $\mathcal{O}\left[(8\pi G_N)^2\right]$ terms which involve both $\Theta^{(2)}{}_{\alpha\beta}$ and $\Theta^{(1)}{}_{\alpha\beta}$, where the latter is related to $\mathtt{t}_{\alpha\beta}$ via Eq. (35).

The Bianchi-wise consistency and completeness of Einstein field equations are based on the feature that the three tensor fields, $G_{\alpha\beta}(\between_V, V)$, $\mathtt{C}^{(0)}_{\alpha\beta}$ and $\mathtt{C}^{(1)}_{\alpha\beta}$, are the only divergence-free symmetric tensor fields in 4-dimensional spacetime[14]. There exist no other divergence-free, symmetric tensor fields with which $\mathtt{C}^{(n\geq 2)}_{\alpha\beta}$ can be identified. In fact, there is no analogue of Huggins tensor in curved space[14,15]. Consequently, instead of strict vanishing of the divergences of $\mathtt{C}^{(n\geq 2)}_{\alpha\beta}$, which cannot be achieved, one must be content with non-vanishing yet higher order remnants to be canceled by divergences of higher orders. More accurately, if divergence of $\mathtt{C}^{(n)}_{\alpha\beta}$, in the equation of motion (30), gives a remnant at order of $(n+1)$-st and higher then divergence at the n-th level is effectively nullified. At the $n = 2$ level, for instance, one can consider the tensor field

$$\mathtt{C}^{(2)}_{\alpha\beta} = \Big(- \boxminus_{\alpha\beta} g_{\mu\nu} + \boxminus_{\alpha\mu} g_{\beta\nu} + \boxminus_{\beta\mu} g_{\alpha\nu} - \nabla_\mu \nabla_\nu g_{\alpha\beta} - 2G_{\alpha\mu\beta\nu} + \frac{1}{2}\Box \left(2g_{\mu\nu}g_{\alpha\beta} - g_{\alpha\mu}g_{\beta\nu} - g_{\alpha\nu}g_{\beta\mu}\right)\Big)\Omega^{\mu\nu} \tag{53}$$

where $\boxminus_{\alpha\beta} \equiv \nabla_\alpha \nabla_\beta - G_{\alpha\beta}$, $G_{\alpha\mu\beta\nu} \equiv R_{\alpha\mu\beta\nu} - \frac{1}{2} g_{\alpha\beta} R_{\mu\nu}$, and

$$\Omega_{\alpha\beta} = c_1 {\Theta^{(1)}}^\mu_{\ \alpha} {\Theta^{(1)}}_{\mu\beta} + c_2 {\Theta^{(1)}}^\mu_{\ \mu} {\Theta^{(1)}}_{\alpha\beta} + c_3 {\Theta^{(1)}}^\mu_{\ \mu} {\Theta^{(1)}}^\nu_{\ \nu} g_{\alpha\beta} + c_4 \mathtt{t}_{\alpha\beta} \tag{54}$$

with $c_{1,\dots,4}$ being dimensionless constants. Obviously, divergence of $\Omega_{\alpha\beta}$ does not vanish, and it is non-local due to its dependence on ${\Theta^{(1)}}_{\alpha\beta}$. Expectedly, divergence of $\mathtt{C}^{(2)}_{\alpha\beta}$ does not vanish yet it is $\mathcal{O}\left[(8\pi G_N)\mathtt{t}\nabla\Omega\right]$ on the equation of motion (30). It is sufficiently suppressed since it falls at the $n = 4$ order, and may be made to cancel with the divergence of $n = 4$ term. This progressive, systematic cancellation works well as long as divergence of $\mathtt{C}^{(n)}_{\alpha\beta}$ produces terms at the n–th and $(n+1)$–st orders so that the n–th order term cancels the non-vanishing divergence coming from the $(n-1)$–st order. This procedure, order by order in $(8\pi G_N)$, adjusts $\mathbb{T}_{\alpha\beta}$, more correctly its $\Theta_{\alpha\beta}$ part, to guarantee the conservation of matter and radiation flow.

In general, the mechanism proposed involves higher powers of G_N associated with higher powers of $\Theta^{(n)}$ encoding the matter sector. Accordingly, the dynamical equations are expected to involve higher powers of curvature tensors. These higher order contributions from either sector are constrained by the Bianchi identities. In fact, $\mathtt{C}^{(n)}_{\alpha\beta}$ encode nothing but these mutual contributions from material and gravitational sectors. This is best illustrated by $\mathtt{C}^{(2)}_{\alpha\beta}$ in (53): Curvature tensors and covariant derivatives acting on $\Omega_{\alpha\beta}$ are collected together to make the divergence of $\mathtt{C}^{(2)}_{\alpha\beta}$ higher order.

Also, one notes that the expression of $\mathtt{C}^{(2)}_{\alpha\beta}$ in (53) serves only as an illustration. It is obviously not exhaustive, as $\mathtt{C}^{(2)}_{\alpha\beta}$ cannot be guaranteed to depend on $\Theta^{(1)}$ through only Ω. It may well involve structures like $\nabla\Theta^{(1)}\nabla\Theta^{(1)}$ or $\Theta^{(1)}\nabla\nabla\Theta^{(1)}$. One also notes that, however it is composed of ${\Theta^{(1)}}_{\alpha\beta}$ and $\mathtt{t}_{\alpha\beta}$, $\Omega_{\alpha\beta}$ originates from ${\Theta^{(2)}}_{\alpha\beta}$ as the remnant of competing $\Theta^{(1)}$– and $\Theta^{(2)}$–dependent parts of (51). Essentially, what is happening is that ${\Theta^{(2)}}_{\alpha\beta}$ gets expressed in terms of ${\Theta^{(1)}}_{\alpha\beta}$ via $\Omega_{\alpha\beta}$ so that the divergence of $\mathtt{C}^{(2)}_{\alpha\beta}$ jumps to $n = 4$ level.

3.3. *Answer to Question 3*

Having arrived at the gravitational field equations (43), it is clear that Λ_0 stands out as the only dark energy source to account for the observational value of the CC[3]. In other words, one is left with the identification

$$\Lambda_{\mathtt{eff}} = \Lambda_0 \lesssim H_0^2 \tag{55}$$

to be constrasted with (7) in GR. It is manifest that this result involves no fine or coarse tuning of distinct curvature sources. The vacuum energy $\mathtt{E}$, instead of gravitating, generates the gravitational constant G_N via

$$\left(8\pi G_N\right)^{-1} \simeq \mathtt{L}^2\mathtt{E} \tag{56}$$

where $\mathtt{L}^2$ is an area parameter which converts the vacuum energy into Newton's constant. This parameter is not fixed by the model. Essentially, it adjusts itself against possible variations in vacuum energy density $\mathtt{E}$ so that G_N is correctly generated. If $\mathtt{E} \sim \left(M_{EW}\right)^4$ then $\mathtt{L}^2 \sim m_\nu^{-2}$. In this scenario, contributions to vacuum energy from quantal matter whose loops smaller than the electroweak scale are canceled by some symmetry principle. Low-energy supersymmetry is this sort of symmetry. On the other hand, if $\mathtt{E} \sim \left(8\pi G_N\right)^{-2}$ then $\mathtt{L} \sim \ell_{Pl}$. In this case vacuum energy stays uncut up to the Planck scale, and $\mathtt{E}$ and $\mathtt{L}^2$ happen to be determined by a single scale. Therefore, this case turns out to be the most natural one compared to cases where the vacuum energy falls to an intermediate scale. In a sense, the worst case of GR translates into the best case of the present scenario.

As was also noted in Ref. 11, the result (56) guarantees that matter and radiation are prohibited from causing the CCP. In spite of this, one must keep in mind that quantum gravitational effects can restore the CCP by shifting Λ_0 by quartically-divergent contribution of the graviton and graviton-matter loops. If gravity is classical, however, the mechanism successfully avoids the CCP by canalizing the vacuum energy deposited by quantal matter into the generation of the gravitational constant. Namely, stress-energy connection alters the role and meaning of the vacuum energy in a striking way. Newton's constant is the outlet of the vacuum energy.

A critical aspect of the mechanism, which has not been mentioned so far, is that the seed dynamical equations (28) do not follow from an action principle. Indeed, the germ of the mechanism rests entirely upon the matter-free gravitational field equations in GR, and it is not obvious if it can ever follow from an action principle. Though one can argue for the Einstein-Hilbert action at the linear level in (43), the non-local, higher-order terms do not fit into this picture. Thus, one concludes that, gravitational field equations at finis involve non-local, Planck-suppressed higher-order effects, and they are difficult, if not impossible, to derive from an action principle.

4. Conclusion

The CCP is too perplexing to admit a resolution within the GR or quantum field theory. Any attempt at adjudicating the problem is immediately faced with the conundrum that the fundamental equations are to be processed to offer a resolution for the CCP by maintaining all the successes of quantum field theory and GR.

In the present work, gravity is taken classical yet matter and radiation are interpreted as quantal. The vacuum energy deposited by quantal matter and its gravitational consequences are explored in complete generality by erecting a non-Riemannian geometry on the stress-energy tensor. By using the scaling properties of gravitational field equations in GR as a guide, it has been inferred that stress-energy tensor can be incorporated into gravitational dynamics by modifying the connection. This observation gives rise to a novel framework in which the gravitational constant G_N derives from the vacuum energy. In fact, vacuum energy, instead of curving the spacetime, happens to generate the gravitational constant. Indeed, contrary to GR, the vacuum energy induced by quantal matter is not 'cosmological constant'; it just sources the 'gravitational constant'. The CC stays put at its bare value, and its identification with the observational value involves no tuning of distinct quantities as long as gravity is classical. Quantum gravitational effects bring back the CCP by adding to Λ_0 quartically-divergent contributions of the graviton and graviton-matter loops.

In spite of these observations, the model is in want of certain rectifications for a number of vague aspects. One of them is the absence of an action principle. Another aspect concerns a complete analysis of the quantum gravitational effects. Another point to note is the parameter $\mathtt{L}^2$ whose dynamical origin is obscure. Finally, the case $|\Theta| \lesssim |\Lambda|$ must be studied in depth to determine strong gravitational effects. All these points and many not mentioned here are topics of further analyses of the model.

The literature consists of numerous attempts at solving the CCP. The proposals conceptually and practically vary in a rather wide range (See the long list of references in the review volumes[5,6,9] as well as the paper 11. Recent work based on extended gravity theories are listed below[16]). The mechanism proposed in this work, which significantly improves and expands Ref. 11, differs from those in the literature by its ability to tame the vacuum energy induced by already known physics down to the terascale, by its immunity to any symmetry principle beyond general covariance, and by its originality in canalizing the vacuum energy to generation of the gravitational constant.

Acknowledgements

The author is grateful to Organizers of the 13th Regional Conference on Mathematical Physics, October 27-31, 2010, Antalya, Turkey. For interesting questions and comments, he is indebted to the audience, in particular, N. Dadhich, C. Germani, V. Husain and S. Randjbar-Daemi. He thanks Anne Frary for editing the text.

References

1. A. Einstein, *Annalen Phys.* **49**, 769 (1916) [*Annalen Phys.* **14**, 517 (2005)].
2. A. Einstein, Sitzungsber. Preuss. Akad. Wiss. Berlin (*Math. Phys.*) **1917**, 142 (1917).
3. E. Komatsu *et al.* [WMAP Collaboration], *Astrophys. J. Suppl.* **192**, 18 (2011). [arXiv:1001.4538 [astro-ph.CO]].
4. Y. B. Zeldovich, *JETP Lett.* **6**, 316 (1967) [*Pisma Zh. Eksp. Teor. Fiz.* **6**, 883 (1967)].
5. S. Weinberg, *Rev. Mod. Phys.* **61**, 1 (1989).
6. S. Nobbenhuis, *Found. Phys.* **36**, 613-680 (2006). [gr-qc/0411093]; P. J. E. Peebles, B. Ratra, *Rev. Mod. Phys.* **75**, 559-606 (2003). [astro-ph/0207347].
7. N. Arkani-Hamed, S. Dimopoulos, G. Dvali *et al.*, [hep-th/0209227]; G. Dvali, S. Hofmann, J. Khoury, *Phys. Rev.* **D76**, 084006 (2007). [hep-th/0703027 [HEP-TH]].
8. S. Weinberg, *Phys. Rev. Lett.* **59**, 2607 (1987).
9. S. Weinberg, arXiv:astro-ph/9610044; S. M. Carroll, *AIP Conf. Proc.* **743**, 16 (2005). [astro-ph/0310342]
10. N. Dadhich, Talk in these proceedings. See also: arXiv:1006.1552 [gr-qc].
11. D. A. Demir, *Found. Phys.* **39**, 1407-1425 (2009). [arXiv:0910.2730 [hep-th]].
12. D. A. Demir, arXiv:1102.2276 [hep-th]. (to appear in *Phys. Lett. B*)
13. A. Iorio, L. O'Raifeartaigh, I. Sachs, C. Wiesendanger, *Nucl. Phys.* **B495**, 433-450 (1997). [hep-th/9607110]; D. A. Demir, *Phys. Lett.* **B584**, 133-140 (2004). [hep-ph/0401163].
14. D. Lovelock, *J. Math. Phys.* **12**, 498 (1971).
15. D. Lovelock, *Lett. Nouvo Cimento* **10**, 581 (1974); B. Kerrighan, *Gen. Rel. Grav.* **13**, 283 (1981).
16. Y. Bisabr, *Gen. Rel. Grav.* **42**, 1211 (2010). [arXiv:0910.2169 [gr-qc]]; R. A. Porto, A. Zee, *Class. Quant. Grav.* **27**, 065006 (2010). [arXiv:0910.3716 [hep-th]]; I. L. Shapiro, J. Sola, *Phys. Lett.* **B682**, 105-113 (2009). [arXiv:0910.4925 [hep-th]]; F. Bauer, J. Sola, H. Stefancic,

Phys. Lett. **B688**, 269-272 (2010). [arXiv:0912.0677 [hep-th]]; E. Alvarez, R. Vidal, *Phys. Rev.* **D81**, 084057 (2010). [arXiv:1001.4458 [hep-th]]; P. Chen, [arXiv:1002.4275 [gr-qc]]; N. J. Poplawski, [arXiv:1005.0893 [gr-qc]]; P. D. Mannheim, [arXiv:1005.5108 [hep-th]]; S. F. Hassan, S. Hofmann, M. von Strauss, *JCAP* **1101**, 020 (2011). [arXiv:1007.1263 [hep-th]]; H. Azri, A. Bounames, [arXiv:1007.1948 [gr-qc]]; D. Metaxas, [arXiv:1010.0246 [hep-th]]; P. Jain, S. Mitra, S. Panda *et al.*, [arXiv:1010.3483 [hep-ph]]; G. A. M. Angel, [arXiv:1011.4334 [gr-qc]].

THE INFORMATION LOSS PARADOX AND THE HOLOGRAPHIC PRINCIPLE

ASGHAR QADIR
Centre for Advanced Mathematics & Physics,
National University of Sciences & Technology,
H-12, Islamabad, Pakistan
E-mail: aqadirmath@yahoo.com
www.camp.nust.edu.pk

Great interest was generated when it was announced that Hawking had admitted that he had lost a bet with Susskind. The bet had been about the "information loss paradox" in which Hawking had tried to show that quantum theory, in its merger with gravity, is incomplete and *both* quantum theory and relativity need to be modified. Susskind had refuted the argument using quantum arguments why the information would *not* be lost. In the process he formulated a "black hole complementarity principle". It turns out that work of Wheeler and I leads to the same conclusions qualitatively and there is reason to believe that introducing the uncertainty principle might give a quantitative result with no further quantum input.

Keywords: Black holes; Complementarity; Hawking radiation; Holographic principle; Information loss.

1. Introduction

There has been a running battle between Relativity and Quantum Theory since their inception. Einstein, who had been one of the founders of both theories, believed that the latter was only an "effective theory" in which the lack of precise knowledge was covered up by using Statistics — in much the same way as is done in Statistical Mechanics. Bohr, on the other hand, argued that Nature is inherently uncertain and there are limits on what is "knowable". The debate between Einstein and Bohr is discussed in great detail in Ref. 1. This debate dating from the start of the twentieth century was revived at the end of that century with Hawking playing the role of Einstein and Susskind and 't Hooft playing the role of Bohr[2]. The debate made the headlines when it was announced that Hawking admitted to losing a bet that he and Susskind had made. Here I will briefly review the debate and then show how work of John Wheeler[3–5] and I has a bearing on that debate. Before that, for completeness I will mention the terminology and concepts that will be used.

One of the most dramatic predictions of Relativity is that there must be regions of space where the curvature of spacetime due to gravitational collapse cannot be withstood and there will be unending collapse leading to regions of infinite curvature. Though the region is commonly talked of as a point in space, that would not be true in general. In fact, for the simplest result of gravitational collapse, called a Schwarzschild black hole, a "line singularity" results. If the region of infinite curvature, called the *singularity*, is totally enclosed by a surface from which nothing can escape, it is called a *black hole*. The name was given by Wheeler[6] because not even light can escape, thus rendering the object totally black and other objects can fall into it as if it were a hole. The first candidate black hole (Cygnus X-1) was identified by Remo Ruffini[7]. There is overwhelming evidence[8] that there is a multi-million solar mass black hole at the centre of our galaxy (Sgr A*). Though pedants continue to debate whether the objects mentioned are "really" black holes the astrophysical community continues to deal with them as if they are.

The idealized description of a black hole is described by a geometry that is unchanging with time. Since there is time-translational invariance, energy is conserved. The radius of a black hole is proportional to its mass and hence the area of the black hole to the square of the mass. Since nothing can come out, the mass must not decrease with time and hence the area will be non-decreasing. These have been called the laws of *black hole dynamics* in analogy with the laws of thermodynamics[9]. The significance of the analogy is that the first law of thermodynamics also gives energy conservation and the second law gives *entropy* as non-decreasing. However, it may have been regarded as just an amusing similarity if there had not been more to it. I will come to that shortly.

Whereas Relativity changes the classical way of thinking drastically, it does not totally overturn it like Quantum Theory does. In the latter the scientific certainties are lost. The World becomes a matter of happenstance and chance that is not predictable. Objects are not clearly either waves or particles but their nature depends on how they are observed. Humpty Dumpty said "Words mean what I choose them to mean"[10]. In the Humpty Dumpty Quantum World "Objects are what we choose them to be". However, this loss of the classical certainties is not total but is limited to some extent. While *precise* predictions are not possible, approximate predictions are. Though objects can be either waves or particles, by Bohr's complementarity principle they *cannot be both at the same time.* Depending on the choice of experiment, they will either appear as particles or as waves. Till the observation is made they are not both they are neither, existing in a state of limbo as a wave function that is to be "collapsed" by an observation.

The problem of the "collapse of the wave function" lies at the heart of the difference between Relativity and Quantum Theory. As shown by Einstein, Poldolsky and Rosen[11] this collapse leads to some effects that seem to violate causality. Though it turns out that no actual signal is sent faster than light, there are effects at a distance that do not admit of a simple spacetime description in terms of point particles. One has a choice of either accepting the loss of causality or adopting non-locality. Most currently opt for the latter. This problem also lies at the heart of the modern version of the old Bohr-Einstein debate.

2. Black Hole Thermodynamics and Hawking Radiation

Penrose propounded a paradox involving black holes. Though I heard him talk of it in late 1968, he might have constructed the paradox earlier. He considered a civilization that lives around a black hole. Their means of producing power is to fill a box with thermal radiation taken from the surroundings, lower it to the vicinity of the black hole by a spring and throw the radiation in. Since the radiation has an equivalent mass, a lighter box is brought up than went down. Consequently, the spring stores energy that can be used. This seems to violate the second law of thermodynamics, which can be paraphrased as "There is no such thing as a free lunch". Here we not only seem to get free energy out, but we reduce the thermal pollution around. *Not only do we have a free lunch, we get paid for eating it!* Penrose argued that the only way to save the second law of thermodynamics is to require that the black hole has an entropy that rises more in the above process, than is lost by the surroundings. Thus, the civilization only gains over a short duration and later generations will have to pay for the lunch of their ancestors (apparently a common fate for the later generations). However, his suggestion was that the entropy would be measured somehow by the conformal curvature tensor.

The next development was when a student of John Wheeler, Jakob Bekenstein, gave an argument that provided a measure of the entropy[12]. To explain the argument I need to first state a point originally made by Planck[13]. He noted that Special Relativity reduces to Classical Mechanics if we can take $1/c$ to be negligible, Quantum Theory reduces to Classical Mechanics if we can take $\hbar$ negligible and General Relativity reduces to Special Relativity if we can take G to be negligible. Thus Relativity and Quantum Theory would both be significant in a regime when none of these are negligible. Thus, in such a situation we would need a theory of Quantum Gravity. To see this in perspective he defined units of length mass and time that are constructed from these universal constants. These are $l_P = \sqrt{G\hbar/c^3}$, $m_P = \sqrt{\hbar c/G}$ and $t_P = \sqrt{G\hbar/c^5}$. The Planck energy is then $E_P = \sqrt{\hbar c^5/G}$. The values come out to be $l_P \sim 10^{-33}$, $m_P \sim 10^{-5} gm$, $t_P \sim 10^{-42} s$, $E_P \sim 10^{19} GeV$.

It is well known that entropy can be thought of as a measure of the disorder of a system. Further, order in a system can be thought of as information stored in it. As such, if the information content of a system is increased the number of states in the system should go up and hence the entropy should increase. Bekenstein considered the change in a black hole if a 'bit' of information is put into it. How do we decide what a single 'bit' of information is? He suggested that it should be the smallest amount of energy that could be put in. Since the least energy would be in a photon, the proposal was to use the least energy photon, which means the *largest wave-length photon* that could enter it. Bear in mind that if the photon's wave-length is larger than the black hole diameter it would go around the hole. Thus the energy put in would be $E = hc/\lambda$, where $\lambda = 2r_s = GM/c^2$, m being the mass of the black hole and r_S being the Schwarzschild radius. (We have assumed that the black hole is a simple one with no spin or electric charge.) The net result is that the increase in the area of the black hole is $\delta A = 16\pi^2 l_P^2$, i.e. the increase in area is simply the Planck area (times $16\pi^2$). Thus the entropy should come out to be proportional to the black hole area, *independent of the black hole area*!

The next step in the story comes from a more direct inclusion of the Quantum Theory. Fulling[14] had tried to quantize scalar fields in a linearly accelerating frame. The result had been odd. On computing the expectation value of the number operator defined in the Schwarzschild background in the Minkowski space vacuum states, he obtained a fractional value. Consequently he concluded that the procedure was not meaningful. In fact the title of his paper was about the "non-uniqueness of quantization in curved spacetimes". Of course, he had not taken a genuinely *curved* space. Hawking tried to repeat the procedure in a curved space, namely for a Schwarzschild black hole. The result gave, of all things, *a Planck spectrum.* He tried taking into account the process of collapse and doing the calculation using

Feynman path integrals instead of the Bogolioubov transformations used by Fulling. He concluded that there must be something to it as he still got a Planck spectrum. The temperature came out to be $T = \hbar c/8\pi kGm$. He interpreted this as radiation coming from the black hole. This is now known as Hawking radiation and the temperature is known as the Hawking temperature.

With this development a very clear identification of the laws of thermodynamics and those for black holes emerged. For the identification of the area of the black hole with entropy to make sense there needed to be a temperature associated with the black hole. This would need to be an *intensive* thermodynamic variable. The temperature appearing is the surface gravity, which is intensive. In fact, now one had the zeroth law also provided for black holes as there is a temperature defined. The third law is not apparent. It may seem strange to have a black hole radiate. Classically there could be no radiation but quantum mechanically one can think of the energy as tunnelling out of the black hole. What now seems to occur is that energy comes out of the black hole and thus the mass decreases and hence the temperature increases. This would lead to enhanced radiation and hence a sharper decrease in mass. The end result must be that the black hole will radiate entirely away.

3. The Information Loss Paradox

I am now in a position to explain the information loss paradox. Hawking conceived of a pure quantum state thrown into the black hole. Had the black hole been classical the pure state would go into the black hole and that would be the end of the story. While it would not be seen outside ever again, it would continue to exist inside. The information would be lost to the outside world but would not be lost to the Universe. However, if the black hole radiates, whatever was inside the hole would emerge out and no information would stay inside the hole. Since it comes out as thermal radiation it is totally mixed. Thus information (the pure state) was thrown into the black hole and was lost. This is not allowed. Put another way, elementary quantum evolution is *unitary* and unitary evolution of a pure state should yield a pure state. However, as it comes out as a mixed state, unitarity has been lost. Since the theory, in the presence of a strong gravitational field, contradicts itself, *it is incomplete*. Einstein would have loved it! Not only is quantum theory proved to be incomplete, the demonstration is in Quantum terms and not, as Einstein used to try to prove, in non-quantum terms. This way it should avoid the rebuttals of the Copenhagen interpretation.

G. 't Hooft and L. Susskind[2] felt that there must be a flaw in the argument and tried to demonstrate it. However, they were not able to prove their point by the next morning (as Bohr used to do with the arguments of Einstein). Nevertheless, Susskind and Hawking entered into a bet, Hawking maintaining that his argument would not be answerable and Susskind that it would. Finally, the counter-argument came in two parts: *the black hole complementarity principle*; and *the holographic principle.* I will not go into these in any great detail but will just try to give the essence of the arguments.

For the former, the idea is analogous to Bohr's complementarity principle. Since no information from inside can come out, there is no experiment that can be conducted from the outside that would take into account the view of the Universe from within. If, from the inside, *the black hole is not seen to evaporate* there would be no paradox constructed. Bohr would have loved it! The rebuttal is quintessentially quantum. However, the question of information loss posed from the outside remains unanswered. To answer it we need to go to the latter principle.

Noting that Bekenstein's argument only needs to measure the *area* of the hole and not its volume, one would like to argue that the information is entirely stored on the *surface of the hole* and not in the volume. On the face of it, it seems obvious that more information could have got stored in the volume than on the surface. However, it was conjectured that the information was stored like a holograph stores a 3-d picture of an object and that an image of the 3-d object can then be reconstructed in the same way. Thus, one would suppose, that the maximum information can be stored on the surface of the black hole. This is known as the *Maldacena conjecture*[15]. Using a (2 + 1)-dimensional black hole, the so-called BTZ black hole[16], the conjecture was proved in the context of superstring (or rather D-brane) theory. The spacetime for this black hole is hyperbolic of the "anti-de Sitter" (AdS) type. In such a spacetime, it can be shown[2] that the information can equally well be thought of as stored at the outer horizon of the Universe. Consequently, *the information is never lost.*

Hawking accepted that the rebuttal was valid and paid the bet. It appeared that the Quantum Theory was not only vindicated but came out triumphant – stronger than ever.

4. The Qadir-Wheeler Suture Model and Information Loss

It appears that a "purely" classical argument could give support to the black hole complementarity idea. This would support the Susskind claim in one way but would temper the triumph of Quantum Theory, as the argument *is* classical. To explain this idea I need to quickly explain the "Qadir-Wheeler suture model".

This is not meant to be an actual model of the Universe but a "toy model" designed to prove a point of principle. Penrose[17] had conjectured that the black hole singularity would be simultaneous with the cosmological singularity (if the Universe is closed) in some appropriately defined frame. That would imply that the end of the Universe would not be causally *after* the formation of the black hole in any frame. Wheeler and I wanted to prove this point of principle. We first tried to prove it by using the "York time"[9] and a black hole in existence from the beginning to the end of the Universe[4]. The spacetime is broken into a sequence of spacelike hypersurfaces. Writing the unit timelike vector normal to the (spacelike) hypersurface as $\mathbf{n}$, the extrinsic curvature tensor is $\mathbf{K} = \nabla\mathbf{n}$. The requirement is that its trace, K, be constant over the entire hypersurface and that it increase as we go along the sequence, so that no two hypersurfaces have the same value of K. Then K is proportional to the York time.

For our purpose we first took a Schwarzschild lattice universe[9]. Though the simultaneity was established, it proved nothing as the black hole never formed. To address this issue we had the black hole form at the phase of maximum expansion of a Friedman closed model universe. Though it worked, the problem with this attempt was that one had to limit one's attention to times *after* the phase of maximum expansion. If one followed the evolution backward, the black hole would "stick out of the Universe" at the beginning. (This work was never published.) We therefore had to (as George W. Bush used to say) "do more".

We considered two closed Friedman model universes of different densities at the phase of maximum expansion. The rarer it is the larger it will be, the denser it is the smaller it will be. The evolution of the two would proceed differently as well. We now cut out a portion of the rarer model and replace it by a matched portion of the denser one. It is so arranged that the two match perfectly at the start of the Universe. Then, from the point of view of the rarer portion, the denser part looks like a density fluctuation. Since the two portions have different densities they will evolve differently. As such, they will not continue to match perfectly. We join the two portions by a Schwarzschild geometry. This cut-and-paste toy model was called "the suture model"[4,5]. As we go backwards to the Big Bang the Schwarzschild region will vanish. We define simultaneity by timelike hypersurfaces of constant mean extrinsic curvature. Then, at some stage of evolution, from the point of view of the rarer portion, the denser one will be hidden behind the event horizon and will, therefore, be a black hole. In terms of these hypersurfaces of simultaneity we can then ask whether the black hole singularity forms first or the Big Crunch singularity. We found that the two form singularities form simultaneously. I then checked that the result is robust and can be generalized to more than one hole[3].

As time proceeds from the Bang to the phase of maximum expansion, the Schwarzschild region must expand, but after that what happens? Does the Schwarzschild region expand or shrink? For that matter, is it necessary that the phases of maximum expansion lie on the same hypersurface of simultaneity? The answer to this question is "No". As may have been expected, the denser region passes the phase of maximum expansion *before* the rarer one does. That is how the black hole can form in the rarer region. As such, for the question of what happens to the Schwarzschild region after the phase of maximum expansion, one needs to ask "which one?" It is clearly inconsistent to have it shrink with the denser and yet expand with the rarer. The only thing that could happen would be for the Schwarzschild region to keep on growing. What looked like a suture in the early phase starts looking like a corridor in the later phases. Once both phases of maximum expansion are passed, the corridor must become continuously narrower at both ends and hence in between. Thus the *volume* can decrease while the *length* of the corridor goes on increasing.

Of course, the entire model is based on a closed Friedman Universe. The information loss paradox and the black hole complementarity principle are based on there being no time that the "inside" and the "outside" of the black hole come together. As such, one should be thinking of an open Universe. The model can be extended to an open Universe by taking the rarer part of the model to be sufficiently rare to be open. The foliation procedure would go through exactly in the same way for an open Universe as for a closed. The point of principle that was proved by the original suture model would, however, no longer apply. Instead, in that case, we would have to conclude that in terms of the York time, *the black hole singularity never forms*! The point is that if the "outside Universe" never ends, there can never be enough (York) time for the black hole singularity to form. It is, as it were, "held up" by the rest of the Universe. One could see this result in another way. One could perform a conformal transformation on the "outside

Universe" to make it compact. Then the earlier work on the suture model would prove that the end of the Universe is simultaneous with the formation of the singularity. Reversing the conformal transformation, there is no end of the Universe, and hence (since the end is to be simultaneous with the formation of the singularity) the singularity never forms.

I mentioned this point to John Wheeler when we had just constructed and foliated the suture model in 1987, long before Hawking's argument about information loss. He was not interested in it as he had a religious belief that the Universe is closed. I use the term "religious" literally and not figuratively. Wheeler held that the purpose of human existence is to ponder the origins of the Universe and by observing them (on the basis of his belief in the Copenhagen interpretation of the quantum theory), *create the Universe.*

The question arises as to *how* the length increases and the volume decreases with York time. This was answered by foliating the Schwarzschild spacetime by spacelike hypersurfaces of constant mean extrinsic curvature[18] and then approximating the length of the corridor by a Taylor series in the inverse powers of the York time[19]. It turns out that the proper length of the Schwarzschild region increases as $K^{4/3}$, while the area decreases as K^2. Thus the volume decreases as $K^{-2/3}$. As the foliating hypersurface radius shrinks, the number of Planck area units that can be fitted onto a unit proper length reduce. However, they cannot reduce below the number 1 in any case. As such, there will be a stage at which we have to stop the foliation. Now Relativity is time-reversal invariant. Hence, once the lowest value is reached the only allowable next step would be for the hypersurfaces to start growing again. This means that the black hole would, at that point of York time, convert into a *white hole* and finally spew all the material out. *Never would any information that had gone in be lost.*

Note the remarkable similarity of this view of the final outcome of the black hole and the information it contained, with the black hole complementarity principle of Susskind. Yet, there is a profound difference. While that principle ends up saying that the outcome depends on the view of the observer, here we come to a definite answer for all observers. Note, also, the remarkable similarity of this picture to the Hawking radiation picture. Both predict that the black hole will ultimately cease to exist. In both views, the larger the hole is, initially, the longer it will take for it to end. Yet, once again, there is a profound difference. The Hawking argument depends on a different mechanism than this one.

I have been at pains to distinguish between this approach and that in the Hawking and Susskind arguments on the basis of use of the quantum theory. However, it must be admitted that the quantum theory *is* used in making the smallest building blocks for the area of the foliating hypersurfaces. This minimal use is in line with Bekenstein's arguments and the development of statistical physics. However, no appeal is made to the foundational aspects of quantum theory or to quantum field theory. The former remains somewhat controversial. The latter requires that there should be a genuine quantum theory of gravity. Due to the nonlinearity of Relativity, any conclusion that neglects the quantization of the spacetime itself, must be suspect.

5. Conclusion

I must admit that I have glibly glossed over various difficulties of actually constructing the open suture model and foliating it. The process would not be difficult but would be tedious and time-consuming. With our current understanding of the model and the current computing power, there is no doubt that it could all be implemented. I have also glossed over the actual calculation of the limit beyond which the foliation would be stopped by the Planck limit for the foliating hypersurface. Thinking in terms of phase space, it is easy enough to see it in principle but would require actual calculations to do consistently in Relativity. This work is planned for the future. Again, while it is clear that the results to be obtained from the suture model are qualitatively similar to the predictions of Hawking radiation, the details are likely to be quite different. That, also, needs to be investigated further.

One might wonder about the asymptotic behaviour of the proper length of the Schwarzschild singularity by itself, without reference to the suture model, in terms of York time. It turns out that it picks up an extra factor of $\ln K$ (see Ref. 20). This fact is of interest in itself and has an implication for lower dimensional theories of gravity. In an n-dimensional space (of an $n+1$ dimensional spacetime), since the hypersurface "area" decreases as $K^{(n-1)}$, the general change in the spatial volume would $\sim K^{(1-n)/3} \ln K$. But this means that a black hole in a (2+1)-dimensional theory would have a slow reduction in volume, but in a (1+1)-dimensional theory *it would not be a black hole*, as the volume would not tend to zero but tend (logarithmically) to infinity! Since the claims to explain black hole entropy in superstring terms need (1+1)-dimensional gravity, the significance of those claims should be re-considered.

It could be hoped that by following through the minimal requirement of quantum theory in a gravitational context we may get insight into the desperately needed theory of quantum gravity.

Acknowledgments

I am grateful to the AS-ICTP for travel support to attend the Thirteenth Regional Conference, where this work was presented. I am also grateful to the organizers of the Conference for local hospitality in Antalya, Turkey. Most of all, I would like to thank Prof. Leonard Susskind, who encouraged me to "resurrect the idea" that Wheeler had ignored.

References

1. M. Jammer, *The Conceptual Development of Quantum Mechanics*, (McGraw-Hill, New York, 1966).
2. L. Susskind, *The Black Hole War: My Battle with Stephen Hawking to Make the World Safe for Quantum Mechanics*, (Black Bay Books / Little Brown and Company 2008).
3. A. Qadir, *Proc. Fifth Marcel Grossmann Meeting*, eds. D. G. Blair and M. J. Buckingham (World Scientific, Singapore, 1989).
4. A. Qadir and J. A. Wheeler, *From SU(3) to Gravity: Yuval Ne'eman Festschrift*, eds. E. S. Gotsman and G. Tauber, (Cambridge University Press, Cambridge, 1985).
5. A. Qadir and J. A. Wheeler, *Nucl. Phys. B, Proc. Suppl.*, eds. Y. S. Kim and W. W. Zachary, **6**, 345 (1989).
6. R. Ruffini and J. A. Wheeler, *Physics Today*, 30 Jan. 1971.
7. R. Ruffini, *Physics and Contemporary Needs, Vol. 1*, ed. Riazuddin, (Plenum Press, New York, 1977).
8. F. De Paolis, G. Ingrosso, A. A. Nucita, A. Qadir and A. F. Zakharov, *Gen. Rel. & Gravit.* **43** 977 (2011).
9. C. W. Misner, K. S. Thorne and J. A. Wheeler, *Gravitation*, (W. H. Freeman, New York, 1973).
10. Lewis Carol, *Alice's Adventures in Wonderland*, (Macmillan and Company, London, 1865).
11. A. Einstein, B. Podolsky and N. Rosen, *Phys. Rev.* **47**, 777 (1935).
12. J. D. Bekenstein, *Phys. Rev. D* **7**, 2333 (1973).
13. M. Planck, *Sitzungsberichte der Königlich Preußischen Akademie der Wissenschaften zu Berlin* **5**, 440 (1899).
14. S. Fulling, *Phys. Rev. D* **7**, 2860 (1973).
15. J. M. Maldacena, *Adv. Theor. Math. Phys.* **2**, 231 (1998).
16. M. Banados, C. Teitelboim, J. Zanelli, *Phys. Rev. Lett.* **69**, 1849 (1992).
17. R. Penrose, *Road to Reality: A Complete Guide to the Laws of the Universe*, (A. Knopf, London, 2004).
18. A. Pervez, A. Qadir and A. A. Siddiqui, *Phys. Rev. D* **51**, 4598 (1995).
19. A. t-Hussain and A. Qadir, *Phys. Rev. D* **63**, 083502 (2001).
20. A. Qadir and A. A. Siddiqui, *Int. J. Mod. Phys. D* **18**, 397 (2009).

PART D

Quantum Field Theory

GENERALIZED NONCOMMUTATIVE GAUGE THEORY*

F. ARDALAN

Department of Physics, Sharif University of Technology
&
Institute for Research in Fundamental Sciences (IPM),
P. O. Box 19395-5746, Tehran, Iran
E-mail: ardalan@ipm.ir

The structure of the translationally invariant noncommutative field theory in determined and the corresponding gauge theory is presented.

1. Introduction

Noncommutative gauge theories naturally appear as the low energy limit of String theory in the presence of the antisymmetric background field B with the action of the string,

$$S = \frac{1}{4\pi\alpha'} \int_{\Sigma} (g_{ij}\partial_a x^i \partial^a x^j - 2\pi i\alpha' B_{ij}\epsilon^{a,b}\partial_a x^i \partial_b x^j).$$

Canonical quantization of the string then forces noncommutativity of space[2],

$$[x^i, x^j]_\star = i\Theta^{ij}.$$

Open string amplitudes are then shown to give a vertex leading to a noncommutative gauge theory[3]

$$F_{\mu\nu}(x) = \partial_\mu A_\nu(x) - \partial_\nu A_\mu(x) + ig[A_\mu(x), A_\nu(x)]_\star. \tag{1}$$

The Lagrangian density of the resulting gauge theory is

$$\mathcal{L} = \Psi \star (i\partial - gA) \star \psi\psi - \frac{1}{4}F_{\mu\nu} \star F^{\mu\nu},$$

*Talk based on work done in collaboration with N. Sadooghi[1].

where

$$f(x) \star g(x) \equiv e^{\frac{i\theta_{\mu\nu}}{2}\frac{\partial}{\partial\xi_\mu}\frac{\partial}{\partial\zeta_\nu}} f(x+\xi)g(x+\zeta)|_{\xi=\zeta=0}.$$

This star product is associative but not commutative.

This theory gives similar Feynman diagrams as in ordinary gauge theories, with the same propagators

$$\overset{\mu \quad k \quad \nu}{\bullet\!\!\sim\!\sim\!\sim\!\sim\!\!\bullet} \qquad D_{\mu\nu}(k) = -\frac{ig_{\mu\nu}}{k^2}, \tag{2}$$

because

$$\int dx^d f \star M g = \int dx^d fg;$$

but modified vertices which have a crucial extra phase factor

$$V_\mu(p_1, p_2; k_1) = ig(2\pi)^4\delta^4(p_1 + p_2 + k_1)\gamma_\mu exp(\frac{-i\Theta_{\eta\sigma}}{2}p_1^\eta p_2^\sigma);$$

leading to better UV behavior.

Planar loop diagrams are easily seen not to involve these phases, whereas the nonplanar diagrams involve them, which improve the high momentum behavior of the integrals as long as external lines are on shell. However, if the external lines are part of a larger diagram and thus off shell, they will give an IR divergent contribution, leading to nonrenormalizability. This is serious problem, the so called UV/IR problem.

In an explicit example, the one loop expression in scalar theory,

$$S = \int d^4x(\frac{1}{2}(\partial_\mu\phi)^2 + \frac{1}{2}m^2\phi^2) = \frac{1}{4!}g^2\phi \star \phi \star \phi \star \phi,$$

for the two point functions give,

$$\Gamma^{(2)}_{1,\text{planar}} = \frac{g^2}{3(2\pi)^4}\int \frac{d^4k}{k^2 + m^2},$$

$$\Gamma^{(2)}_{1,\text{nonplanar}} = \frac{g^2}{6(2\pi)^4}\int \frac{d^4k}{k^2 + m^2}e^{ikXp};$$

which asymptotically behave as

$$\Gamma^{(2)}_{1,\text{planar}} = \frac{g^2}{48\pi^2}(\bigwedge^2 - m^2\ln(\frac{\Lambda^2}{m^2} + 0(1)),$$

$$\Gamma^{(2)}_{1,\text{nonplanar}} = \frac{g^2}{96\pi^2}(\bigwedge^2_{\text{eff}} - m^2\ln(\frac{\Lambda^2_{\text{eff}}}{m^2} + 0(1)),$$

where

$$\bigwedge_{\text{eff}}^{2} = \frac{1}{1/\bigwedge^{2} + pop},$$

and

$$pop \equiv -p.\theta^2.p.$$

The crucial point is the different behaviours for the two limits,

a) $pop \ll \frac{1}{\Lambda^2}$, which is the commutative theory result, and

b) $pop \gg \frac{1}{\Lambda^2}$, where a finite result appears.

Note that the two limits do not commute.

2. Generalization of the Star Product

Recently a translationally invariant generalization of the star product has been suggested[4],

$$f \star g = \frac{1}{(2\pi)^{\frac{d}{2}}} \int dp^d dq^d dk^d e^{ip.x} \tilde{f}(q)\tilde{g}(k)K(p,q,k).$$

Translation invariance,

$$\mathcal{T}_a(f) \star \mathcal{T}_a(g) = \mathcal{T}_a\left(f \star g\right), \tag{3}$$

with

$$\mathcal{T}_a(f)(x) = f(x+a), \tag{4}$$

then implies,

$$K(p,q,k) = e^{\alpha(p,q)}\delta(k-p+q);$$

consequently

$$f \star g = \frac{1}{(2\pi)^{\frac{d}{2}}} \int dp^d dq^d e^{ip.x} \tilde{f}(q)\tilde{g}(p-q)e^{\alpha(p,q)}.$$

Associativity of the new star product imposes the constraint equation on the structure functions,

$$\alpha(p,q) + \alpha(q,r) = \alpha(p,r) + \alpha(p-r, q-r). \tag{5}$$

We will now proceed to solve this equation. Generally, decomposing α into the real and imaginary parts,

$$\alpha(p,q) = \alpha_1(p,q) + i\alpha_2(p,q), \tag{6}$$

it is readily verified that[1]

$$\alpha_1(p,q) = \eta_1(q) - \eta_1(p) + \eta_1(p-q), \tag{7}$$

where $\eta_1(p) \equiv \frac{1}{2}\ \alpha(0,p)$, is an arbitrary function of momenta with

$$\eta_1(-p) = \eta_1(p). \tag{8}$$

$$\eta_1(0) = 0. \tag{9}$$

This function shows up as a weight in the integrals,

$$\int d^dx\ f(x) \star g(x) = \int d^dx\ g(x) \star f(x) = \int \frac{d^dp}{(2\pi)^d}\ e^{2\eta_1(p)} f(p)g(p). \tag{10}$$

In Ref. 4, it was noted that loop diagrams involve a particular combination, ω, of the structure functions

$$\omega(p,q) \equiv -\frac{i}{2}[\alpha(p+q,p) - \alpha(p+q,q)].$$

If we now write down the imaginary part of α as

$$\alpha_2(p,q) = \omega(p,q) + \xi(p,q), \tag{11}$$

we can easily see that ω and ξ satisfy the following relations:

$$\omega(p,q) \ \text{ is real}, \tag{12}$$

$$\omega(p,q) = -\omega(q,p), \tag{13}$$

$$\omega(-p,-q) = \omega(p,q), \tag{14}$$

$$\xi(p,q) = \xi(q,p), \tag{15}$$

$$\xi(-p,-q) = -\xi(p,q). \tag{16}$$

The way to find the explicit form of ω and ξ is to expand them in power series of p and q components. (here we limit ourselves to two dimensions),

$$\zeta(P,Q) = \sum_{\substack{\{i_1,i_2,j_1,j_2\} \\ i_1+i_2+j_1+j_2=N}} a_{i_1,i_2;j_1,j_2} p_1^{i_1} p_2^{i_2} q_1^{j_1} q_2^{j_2}, \tag{17}$$

where ζ signifies either ω or ξ, which satisfy

$$\zeta(\vec{p},\vec{q}) + \zeta(\vec{q},\vec{r}) = \zeta(\vec{p},\vec{r}) + \zeta(\vec{p}-\vec{r},\vec{q}-\vec{r}).$$

In the obvious shorthand notation,

$$\zeta(\vec{p},\vec{q}) = \sum_{\vec{i},\vec{j}} a_{\vec{i},\vec{j}}\ \mathbf{p}^i\ \mathbf{q}^j \tag{18}$$

we find

$$\binom{i+k}{k} a_{\vec{i}+\vec{k},\vec{j}-\vec{k}} = (-1)^k \binom{j}{k} a_{\vec{i},\vec{j}}, \qquad \text{with} \qquad \vec{k} \leq \vec{j}. \tag{19}$$

This equation implies both

$$a_{\vec{j},\vec{i}} = -a_{\vec{i},\vec{j}}, \tag{20}$$

and

$$a_{\vec{j},\vec{i}} = (-1)^{j-i} a_{\vec{i},\vec{j}}, \tag{21}$$

where in the equation i and j in the exponent are $i \equiv i_1 + i_2$ and $j \equiv j_1 = j_2$. This implies that

$$(j_1 + j_2) - (i_1 + i_2) \text{ is odd}, \tag{22}$$

which means that the function has odd parity which is only satisfied by $\xi(\vec{p}, \vec{q})$. However, there is an exception when

$$\vec{i} = (0,1), \quad \vec{j} = (1,0), \text{ and } \vec{k} = (0,0). \tag{23}$$

The recurrence relation then yields only an identity

$$a_{0,1,1,0} = a_{0,1,1,0}. \tag{24}$$

There is a single solution possible for ω, as argued in Ref. 4,

$$\omega(\vec{p}, \vec{q}) = \vec{p} \bigwedge \vec{q}.$$

For the other function ξ, we find the solution

$$\xi_{j_1 j_2}(\vec{p}, \vec{q}) = (q_1 - p_1)^{j_1+1}(q_2 - p_2)^{j_2} + p_1^{j_1+1} p_2^{j_2} - q_1^{j_1+1} q_2^{j_2} \tag{25}$$

with

$$j_1 + j_2 = 2n,$$

leading to

$$\xi_N(\vec{p}, \vec{q}) = \sum_{n_1, n_2} C_{n_1 n_2} \xi_{n_1 n_2}(\vec{p}, \vec{q}) \tag{26}$$

with arbitrary real numbers C_{n1n2}. Clearly, the most general solution of ξ in two dimensions is

$$\xi_N(\vec{p}, \vec{q}) = \eta_2(\vec{q}) - \eta_2(\vec{p}) + \eta_2(\vec{p} - \vec{q}), \tag{27}$$

with $\eta_2(\vec{p})$ an arbitrary odd function of $(\vec{p})$. In arbitrary dimensions, the form of (27) is seen to satisfy the associativity equation.

To summarize, we have found that the most general form of the structure function α of the generalized star product is of the form

$$\alpha(\vec{p},\vec{q}) = \sigma(\vec{p},\vec{q}) + i\omega(\vec{p},\vec{q}), \tag{28}$$

with ω real and in the form of

$$\omega(\vec{p},\vec{q}) = \vec{p} \wedge \vec{q}$$

and σ in terms of a complex function

$$\sigma(\vec{p},\vec{q}) = \eta(\vec{q}) - \eta(\vec{p}) + \eta(\vec{p}-\vec{q}), \tag{29}$$

$$\eta(\vec{p}) \equiv \eta_1(\vec{p}) + i\eta_2(\vec{p}), \tag{30}$$

with $\eta_1(\vec{p})$ an arbitrary even function of and $\eta_2(\vec{p})$ an arbitrary odd function of $\vec{p}$, satisfying $\eta_1(\vec{0}) = \eta_2(\vec{0}) = 0$; thus $\eta(-\vec{p}) = \eta^*(\vec{p})$.

3. New NC Gauge Theory

The gauge theory based on the new star product is a straightforward generalization of the usual Moyal gauge theory with certain modifications.

The diagrams, which we briefly present, are now more complicated:

Photon propagator (in Feynman gauge $\xi = 1$):

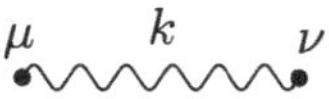

μ k ν

$$D_{\mu\nu}(k) = -\frac{ig_{\mu\nu}}{k^2} e^{-2\eta_1(k)}. \tag{31}$$

Ghost propagator:

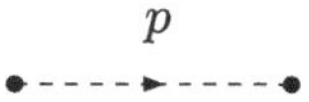

p

$$G(p) = \frac{i}{p^2} e^{-2\eta_1(p)}. \tag{32}$$

$\bar{\psi}_\alpha A_\mu \psi_\beta$-Vertex:

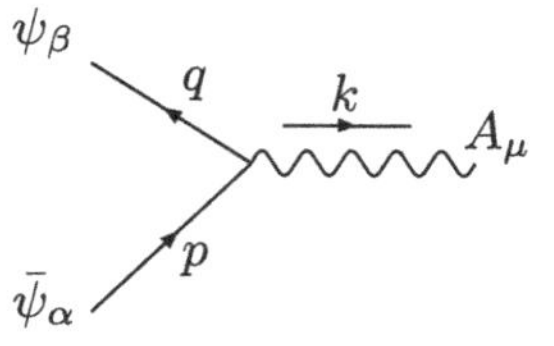

ψ_β q k A_μ p $\bar{\psi}_\alpha$

$$V_{\mu,\alpha\beta}(p,q;k)$$

$$= ig(2\pi)^4\delta^4(p-k-q)(\gamma_\mu)_{\alpha\beta} e^{\alpha(0,-q)} e^{\alpha(-q,-p)}$$

$$= ig(2\pi)^4\delta^4(p-k-q)(\gamma_\mu)_{\alpha\beta} e^{[\eta(-p)+\eta(q)+\eta(k)]}\, e^{-i\omega(p,q)}. \tag{33}$$

$A_{\mu_1}A_{\mu_2}A_{\mu_3}$*-Vertex:*

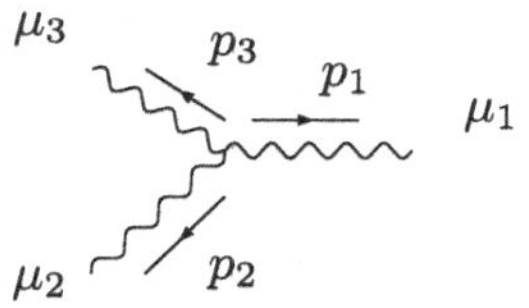

$$V_{\mu_1\mu_2\mu_3}(p_1, p_2, p_3)$$

$$= 2g(2\pi)^4\delta^4(p_1 + p_2 + p_3) \times e^{[\eta(p_1)+\eta(p_2)+\eta(p_3)]} \sin(\omega(p_1, p_2))$$
$$\times [g_{\mu_1\mu_2}(p_1 - p_2)_{\mu_3} + g_{\mu_1\mu_3}(p_3 - p_1)_{\mu_2} + g_{\mu_3\mu_2}(p_2 - p_3)_{\mu_1}]. \quad (34)$$

$A_{\mu_1}A_{\mu_2}A_{\mu_3}A_{\mu_4}$*-Vertex:*

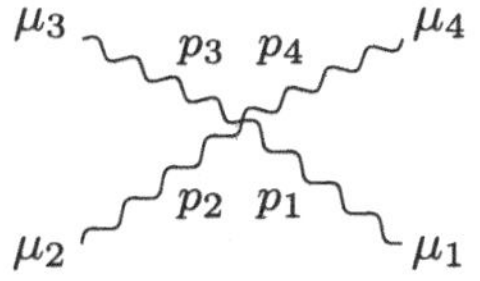

$$V_{\mu_1\mu_2\mu_3\mu_4}(p_1, p_2, p_3, p_4)$$

$$= -4ig^2(2\pi)^4\delta^4(p_1 + p_2 + p_3 + p_4)e^{[\eta(p_1)+\eta(p_2)+\eta(p_3)+\eta(p_4)]}$$
$$\times \{\sin[\omega(p_1, p_2)] \sin[\omega(p_3, p_4)] (g_{\mu_1\mu_3}g_{\mu_2\mu_4} - g_{\mu_1\mu_4}g_{\mu_2\mu_3})$$
$$+ \sin[\omega(p_1, p_3)] \sin[\omega(p_2, p_4)] (g_{\mu_1\mu_2}g_{\mu_3\mu_4} - g_{\mu_1\mu_4}g_{\mu_2\mu_3})$$
$$+ \sin[\omega(p_1, p_4)] \sin[\omega(p_2, p_3)] (g_{\mu_1\mu_2}g_{\mu_3\mu_4} - g_{\mu_1\mu_3}g_{\mu_2\mu_4})\}. \quad (35)$$

$\bar{c}cA_\mu$*-Vertex:*

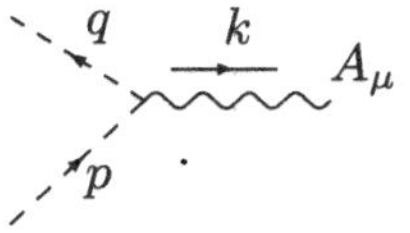

$$G_\mu(p, q; k)$$

$$= 2ig(2\pi)^4\delta^4(p - k - q)p_\mu e^{[\eta(-p)+\eta(k)+\eta(q)]} \sin(\omega(p, q)). \quad (36)$$

One-loop fermion self-energy:

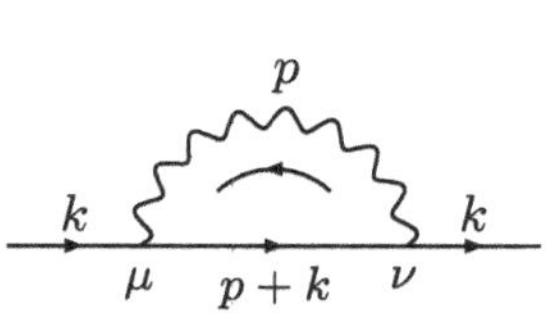

$$-i\Sigma(k)$$

$$= -g^2\mu^\epsilon\, e^{2\eta_1(k)} \int \frac{d^dp}{(2\pi)^d} \frac{\gamma_\mu[\gamma \cdot (k + p) + m]\gamma^\mu}{p^2[(p + k)^2 - m^2]}. \quad (37)$$

References

1. F. Ardalan and N. Sadooghi *Translational-invariant noncommutative gauge theory*, Phys. Rev. D83 **025014**, (2011), arXiv: hep-th/0805064.
2. F. Ardalan, H. Arfaei and M. M. Sheikh-Jabbari, *Noncommutative geometry from strings and branes*, JHEP **9902**, 016 (1999), arXiv: hep-th/9810072.
3. N. Seiberg and E. Witten, *String theory and noncommutative geometry*, JHEP **9909**, 032 (1999), arXiv: hep-th/9908142.
4. S. Galluccio, F. Lizzi and P. Vitale, *Translation invariance, commutation relations and ultraviolet/infrared mixing*, JHEP **0909**, 054 (2009), arXiv: 0907.3640 [hep-th].

CP VIOLATION AND FLAVOR PHYSICS IN GAUGE-HIGGS UNIFICATION SCENARIO

C. S. LIM

Department of Physics, Kobe University, Kobe, 657-8501, Japan
E-mail: lim@kobe-u.ac.jp

We discuss new types of mechanisms to break CP symmetry and to realize flavor mixing in the scenario of gauge-Higgs unification, which guarantees a finite Higgs mass (from the viewpoint of hierarchy problem) and is an attractive scenario for the physics beyond the standard model. In the scenario the Higgs interactions are governed by gauge principle and it may shed some light on the arbitrariness problem of the Higgs interactions. On the other hand, since the Yukawa couplings are real to start with CP violation is a challenging issue. Two kinds of new mechanism of "spontaneous" CP violation are discussed to be possible, one by the complex structure of compactified extra space and one by the VEV of the Higgs. Also the Yukawa couplings are originally universal and hierarchical fermion mass spectrum and flavor mixing are other challenging issues. The hierarchical fermion masses are known to be naturally realized by the exponential suppression factor without fine tuning. Concerning the violation of flavor symmetry, non-degenerate bulk masses are new source of flavor violation beyond the argument of Glashow-Weinberg, and turns out to lead to FCNC at the tree level, though we find the existence of "GIM-like" mechanism in FCNC processes.

Keywords: Gauge-Higgs unification; Extra dimension; CP violation.

1. Introduction

The standard model of elementary particles seems to have unsettled problems in its Higgs sector as follows:

(1) The hierarchy problem:

The standard model is regarded as an effective theory which is valid up to the cutoff Λ. The Higgs mass gets "quadratic divergence" in its quantum correction. Then the problem is how to guarantee the hierarchy $M_W \ll \Lambda$ naturally under the correction (the problem of "quadratic divergence").

(2) CP violation:

In spite of the great success of the Kobayashi-Maskawa (K-M) model, the origin of CP violation still seems to be not conclusive yet. In particular, it has been argued that the CP violation of K-M model is not enough to generate observed amount of matter in the universe.

(3) Fermion masses and mixings:

The origin of hierarchical fermion masses and flavor mixings is still mysterious.

Though the problem 2 stated above seems to have no relation to the Higgs sector, let us recall that in K-M model, CP violating observables appear only through the violation of flavor symmetry (through flavor changing neutral current (FCNC) processes), whose origin is in the Higgs sector.

These problems stem from the fact that there is no guiding principle (or symmetry) to restrict the structure of Higgs interactions. Gauge symmetry (in ordinary 4 dimensional space-time), which completely determines the property of gauge interactions, cannot constrain the structure of Yukawa couplings and Higgs self-interaction.

We discuss "Gauge-Higgs Unification (GHU)" scenario as an interesting attractive candidate of New Physics (Physics Beyond the Standard Model), which is expected to shed some lights on these problems relying on higher dimensional gauge symmetry. The GHU is a unified theory of gauge and Higgs interactions, mediated by bosons whose spins differ by 1 unit ($s = 1,\ 0$). The scenario is realized in the framework of higher dimensional gauge theory. The extra space component A_y of the higher dimensional gauge field $A_M = (A_\mu, A_y)$ ($\mu = 0 - 3$) behaves as a Lorentz scalar from the 4-dimensional (4D) point of view and its "zero-mode" (the mode with zero 4D mass) is identified with our Higgs field

$$A_y^{(0)}(x) = H(x). \tag{1}$$

Thus Higgs is originally a gauge field ! In 4D space, gauge fields possess spin 1 and cannot be identified with Higgs with spin 0. Such apparent contradiction is evaded once space-time is made higher dimensional. As is well-known, photon never gets mass even in quantum level because of local gauge symmetry. Thus we expect that as long as Higgs is originally gauge field its mass does not suffer from "quadratic divergence". In fact the local operator corresponding to the Higgs mass-squared $m_A^2 A_y^2$ is not allowed since it is not invariant under "higher dimensional" local gauge transformation. By higher dimensional gauge transformation we mean a transformation where gauge parameter λ depends not on 4D coordinates

x^μ but on the extra-space coordinate y. Let us note under such gauge transformation A_y transforms inhomogeneously, thus making the local operator forbidden:

$$A_y \to A'_y = A_y + \partial_y \lambda(y). \tag{2}$$

The idea of Gauge-Higgs unification itself is not new[1,2]. In particular, Hosotani proposed a mechanism to break gauge symmetry dynamically by the VEV of A_y, meaningful for non-Abelian gauge theories, "Hosotani mechanism". Later the scenario was (a sort of) revived in a paper[3]. Namely, we showed that the quantum correction to the Higgs mass m_H is finite because of the higher dimensional gauge symmetry and that the GHU provides us with a new avenue to solve the hierarchy problem without invoking SUSY. The scenario has been actively studied later on as a new candidate of the theory of New Physics (see also ref 4).

If the A_y gets some sort of mass term without any derivatives, it should survive even for constant A_y. Now a natural question to ask is whether such constant A_y or the VEV $\langle A_y \rangle$ ever has some physical meaning. It is usually claimed that such constant gauge field has zero field strength and therefore equivalent to empty gauge field ("pure gauge" configuration). It is interesting to note that in the case of Aharonov-Bohm (A-B) effect, constant vector potential yields the A-B phase (or Wilson loop), which corresponds to the magnetic flux penetrating the closed contour. The same thing happens in GHU scenario when the extra space is compactified on a circle S^1. The Wilson-loop along the circle

$$W = \mathrm{Tr}(P \exp(ig \oint A_y dy)) \tag{3}$$

is clearly the function of A_y without any derivatives. W is obviously gauge invariant. The point here is that the Wilson-loop is a non-local gauge invariant operator of A_y, which has a physical meaning since the circle is non-simply-connected space. Thus the function of W may appear at quantum level and it is nothing but the potential of the Higgs H. Though the Higgs generally obtain its mass at quantum level it never has UV-divergence, just because the operator relevant for the Higgs mass is non-local and does not suffer from the UV-divergence, which originates from local interactions.

Interestingly, other attractive attempts of New Physics have some close relations with the scenario of GHU.

(i) Dimensional deconstruction

Arkani-Hamed, Cohen and Georgi have proposed an impressive scenario in ordinary 4-dimensional (4D) space-time[5], where the quadratically divergent quantum correction to the Higgs mass disappears without relying on supersymmetry. The key ingredient to eliminate the UV divergence is N-times repetition of gauge symmetry: $SU(m) \times SU(m) \times \ldots \times SU(m)$. They claim that when $N \geq 3$ the UV-divergence disappears in the Higgs potential. Actually, this scenario can be regarded as a latticized 5D gauge theory, where the extra dimension has N lattice sites. In fact, at the limit $N \to \infty$ the effective potential they got for the Higgs is confirmed to coincide with what we obtain in GHU (see for instance Ref. 6). This close relation is not surprising, since the N independent gauge transformation parameters $\lambda_i(x^\mu)$ $(i = 1 - N)$ correspond to the gauge parameter in 5D space-time if the index i is identified with the extra space coordinate y. Thus, fifth dimension is generated by the strong dynamics in 4D in their scenario !

(ii) Little Higgs model

Another interesting candidate of New Physics, now under active studies, is the scenario of "Little Higgs". This is a 4D scenario where the broken generators G/H of some global symmetry G provide Higgs as a (pseudo) Nambu-Goldstone boson. This may be regarded to be "dual" (through "holographic principle") to the 5D GHU, where A_y associated with the broken generators G/H of higher dimensional local gauge symmetry G provide Higgs. Also there are many circumstantial evidences that these two scenarios are closely related. One example is that in Little Higgs there is a shift symmetry, i.e. the symmetry under the global gauge transformation $G \to G + c$ (G : NG boson, c : constant), which is similar to the local gauge transformation $A_y \to A_y + \partial_y \lambda$ in the GHU scenario.

(iii) Superstring theory

The low energy (point particle) limit of superstring theory, i.e. 10 dimensional SUSY Yang-Mills theory, is a sort of GHU. In fact, it has been known that the extra space component of E_8 gauge field provides Higgs field.

(iv) Application to cosmology

An interesting possibility to utilize the extra space component of the gauge field A_y as an inflaton also has been pointed out[7]. The claim is that $A_y^{(0)}$ may be a natural candidate for the inflaton, since the local gauge symmetry can stabilize the inflaton potential even under the quantum gravity corrections, which is anticipated to break global symmetries: the scenario is thus called "ultra natural inflation".

2. Minimal GHU model

For the scenario of GHU to become viable, we have to present a minimal model based on the scenario, just as Minimal Supersymmetric Standard Model (MSSM) in the case of SUSY scenario to solve the hierarchy problem. In the GHU, it turns out that the gauge group should be enlarged. This is because the Higgs, being gauge field, belongs to adjoint repr., while the Higgs is the fundamental repr. of SU(2). Recall that in the heterotic string theory, Higgs belonging to the fundamental repr. of E_6 comes from the adjoint repr. of E_8 as the part of broken generator of E_8. Thus a natural way out of this problem is to enlarge the gauge symmetry.

As the minimal model of GHU, we have proposed SU(3) unified electroweak model formulated on 5D space-time where the compact extra space is an orbifold S^1/Z_2 (Ref. 6). Orbifolding is a convenient way to get a chiral theory and by assigning non-trivial eigenvalues (Z_2 -parity) for each of the members of the fundamental repr. (Ref. 8), the necessary breaking $SU(3) \rightarrow SU(2) \times U(1)$ is realized:

$$\Psi(-y) = \mathcal{P}\gamma^5\Psi(y) \quad (\mathcal{P} = \text{diag}(+,+,-)) \tag{4}$$

where fermionic field Ψ belongs to the triplet of SU(3). The gauge symmetry breaking is realized in the form that the part of the generators corresponding to those of SU(2)×U(1) have Kaluza-Klein (KK) zero modes and therefore remain massless. Namely the zero mode sector of gauge-Higgs bosons is given as

$$A_\mu = \frac{1}{2}\begin{pmatrix} W_\mu^3 + \frac{B_\mu}{\sqrt{3}} & \sqrt{2}W_\mu^+ & 0 \\ \sqrt{2}W_\mu^- & -W_\mu^3 + \frac{B_\mu}{\sqrt{3}} & 0 \\ 0 & 0 & -\frac{2}{\sqrt{3}}B_\mu \end{pmatrix}, \tag{5}$$

$$A_y = \frac{1}{\sqrt{2}}\begin{pmatrix} 0 & 0 & \phi^+ \\ 0 & 0 & \phi^0 \\ \phi^- & \phi^{0*} & 0 \end{pmatrix}. \tag{6}$$

As we expected in the part of broken generators of A_y we get ordinary Higgs doublet $(\phi^+,\ \phi^0)^t$. Remarkably, we have got exactly what we need in the SU(2)×U(1) Standard Model: $8 \rightarrow 1 + 3 + 2 \times 2$.

Let us note that in GHU the following issues are non-trivial:

- to break CP,
- to realize fermion mass hierarchy,
- to accommodate flavor mixing,

since the Yukawa coupling, being originally gauge coupling, is real and universal among generations, to start with. On the other hand, once these issues are successfully solved, the scenario provides new types of mechanisms for CP and flavor violation, which we expect should have enough predictive power.

3. CP violation in GHU

We now discuss the violation of CP symmetry in the scenario of GHU. Note that as long as higher dimensional gauge theory itself is CP invariant with real gauge coupling, without any CP phase, CP violation, if realized, should be a sort of "spontaneous" breaking. We have proposed two possibilities for such spontaneous breaking, which will be discussed successively below.

3.1. *CP violation due to compactification*

One of a few possibilities to break CP symmetry is to invoke the manner of compactification, which determines the vacuum state of the theory[9]. The key ingredient is the fact that higher dimensional C, P transformations defined as usual by, e.g. C matrix satisfying $C^{\dagger}\Gamma_M C = -(\Gamma_M)^t$, do not reduce to ordinary 4D transformations, in general, since they act on the "internal space" of fermions as well. Namely they cause the unitary transformations between 4D fermions (e.g. in 6D case one 6D fermion decomposes into two 4D fermions). Thus some modification is necessary to get ordinary C, P transformations. Interestingly, the modified CP transformation turns out to act on the extra space coordinates non-trivially: it acts as a complex conjugation of the complex homogeneous coordinates for the extra space (in even dimensions). Let us take 6D case for an illustrative purpose. In the base, where 6D spinor decomposes into two 4D spinors,

$$\Psi_6 = \begin{pmatrix} \psi \\ \Psi \end{pmatrix}, \tag{7}$$

gamma matrices are given as

$$\Gamma^{\mu} = \gamma^{\mu} \otimes I_2 = \begin{pmatrix} \gamma^{\mu} & 0 \\ 0 & \gamma^{\mu} \end{pmatrix}, \quad \Gamma^{y} = \gamma^5 \otimes i\sigma_1 = \begin{pmatrix} 0 & i\gamma^5 \\ i\gamma^5 & 0 \end{pmatrix},$$
$$\Gamma^{z} = \gamma^5 \otimes i\sigma_2 = \begin{pmatrix} 0 & \gamma^5 \\ -\gamma^5 & 0 \end{pmatrix}. \tag{8}$$

The C matrix satisfying $C^{\dagger}\Gamma_M C = -(\Gamma_M)^t$ is easily known to be

$$C = C_4 \otimes \sigma_2 \quad (C_4 = i\gamma_0\gamma_2), \tag{9}$$

as σ_2 acts as a "charge conjugation" in the internal space (ψ and Ψ forms a doublet in the internal space with SU(2) symmetry), having a property $\sigma_2^\dagger \sigma_i \sigma_2 = -(\sigma_i)^t$. We, however, realize that such defined charge conjugation does not reduce to ordinary 4D transformation, since σ_2 causes SU(2) rotation in the internal space. We thus modify the P and C transformations as

$$\begin{aligned} P: \quad \Psi_6 \quad &\rightarrow \quad (\gamma^0 \otimes \sigma_3)\Psi_6, \\ C: \quad \Psi_6 \quad &\rightarrow \quad (C_4 \otimes \sigma_3)\bar{\Psi}_6^t, \end{aligned} \tag{10}$$

where in the internal space we have taken σ_3 instead of the unit matrix, just for convenience. Accordingly the transformation properties of a vector $V^M = \bar{\Psi}_6 \Gamma^M \Psi_6$ is uniquely determined and we find (with y, z being two coordinates of extra space):

$$\begin{aligned} P: \ (y, z) &\rightarrow (y, z), \\ C, \ CP: \ (y, z) &\rightarrow (y, -z). \end{aligned} \tag{11}$$

Thus, introducing a complex coordinate $\omega = y + iz$, the CP transformation is nothing but a complex conjugation:

$$CP: \quad \omega \quad \rightarrow \quad \omega^*. \tag{12}$$

It is interesting to note that though it is well-known that the CP transformation acts on a field with complex conjugation, it also acts as complex conjugation in the space of extra space as well. Now CP is a sort of reflection symmetry in the extra space and therefore is a discrete symmetry as we naturally expect.

As an example, consider Type-I superstring theory with 6-dimensional Calabi-Yau manifold as the extra space defined by

$$\sum_{a=1}^{5} (\omega^a)^5 - C(\omega^1 \omega^2 \cdots \omega^5) = 0 \tag{13}$$

imposed on CP^4. The above "defining equation" is not invariant under $\omega \rightarrow \omega^*$ if the coefficient C is complex and CP is broken by the "complex structure" of the compactification.

We have proposed a mechanism in higher dimensional gauge theory where CP symmetry is broken by orbifold compactification[10]. The main purpose of our work was to realize CP violation in the framework of higher dimensional gauge theories such as GHU or the low energy theory of superstring, but not string theory itself, with much simpler compact spaces, such as orbifold. We discussed the CP violation in the 6-dimensional U(1)

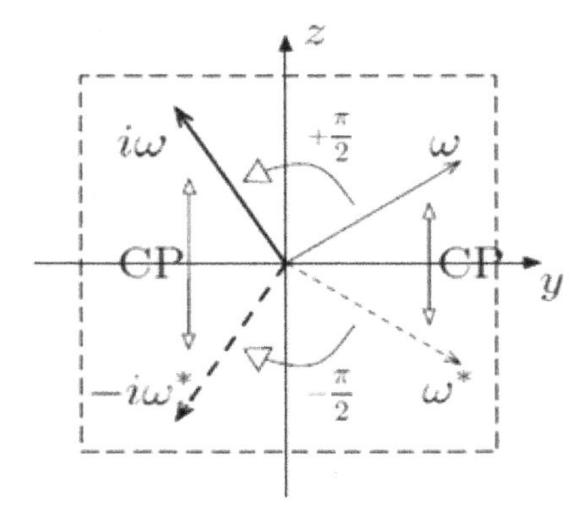

Fig. 1. CP transformation on the complex plane of ω and the orbifording condition

gauge theory due to the compactification on the orbifold T^2/Z_4. It is easy to understand the origin of CP violation without performing explicit calculations: we easily know that CP transformation is not compatible with the condition of orbifolding as follows. In terms of the complex coordinate $\omega = y + iz$ (y, z : extra space coordinates), the Z_4 orbifolding condition is nothing but the identification

$$i\omega \sim \omega. \tag{14}$$

After the CP transformation, the condition reads as

$$(-i)\omega^* \sim \omega^*. \tag{15}$$

Thus the identification of the points connected by the rotation of angle $\frac{\pi}{2}$ is changed to the identification due to the rotation of $-\frac{\pi}{2}$: CP is an "orientation changing operator" (see Fig. 1). Thus, CP transformation is not compatible with orbifolding condition, and therefore CP symmetry is expected to be broken.

The Lagrangian of our model is given as

$$\begin{aligned}\mathcal{L}_{QED} &= \overline{\Psi_6}\left\{\Gamma^M(i\partial_M + gA_M) - m_B\right\}\Psi_6 \\ &\quad -\frac{1}{4}(\partial_M A_N - \partial_N A_M)^2 \ \ (M, N = 0-3, y, z),\end{aligned} \tag{16}$$

where the 6D Dirac fermion is decomposed into two 4D Dirac fermions

$$\Psi_6 = \begin{pmatrix}\psi \\ \Psi\end{pmatrix}. \tag{17}$$

Z_4 eigenvalue t may take $t = \pm 1, \pm i$ with $t^4 = 1$. The assignment of t for each field is as follows

$$\begin{aligned}
&A_\mu(x, i\omega) = A_\mu(x, \omega) \ (t = 1, \ \omega \equiv \frac{y + iz}{\sqrt{2}}), \\
&A_\omega(x, i\omega) = (-i) A_\omega(x, \omega) \ (t = -i, \ A_\omega = \frac{A_y - iA_z}{\sqrt{2}}), \\
&\psi(x, i\omega) = (-i)\psi(x, \omega) \ (t = -i), \\
&\Psi(x, i\omega) = \Psi(x, \omega) \ (t = 1).
\end{aligned} \tag{18}$$

The presence of eigenvalues $t = -i$ signals the possibility of CP violation. We have confirmed by explicit computation that the interaction vertices for non-zero KK gauge bosons ("photons") do have CP violating phases[10]. Such obtained CP violating phases have been confirmed to remain even after the re-phasing of the fields. We also have identified "Jarlskog-type" re-phasing invariant parameters in our model.

A remarkable thing of our mechanism is that CP violation is realized in "QED" with only 1 generation of fermion. It is therefore a new type of mechanism of CP violation, different from that in the K-M model.

3.2. *CP violation due to the VEV of the Higgs*

Another possibility to break CP is due to the VEV of some field which has odd CP eigenvalue. We argue that the VEV $\langle A_y \rangle$ of the Higgs in the GHU scenario, or the VEV of Wilson-loop plays such a role.

We have shown that neutron electric dipole moment (EDM) gets contribution already at 1-loop level in the model, although we assume the presence of only 1 generation[11]. The model we took is the minimal 5D SU(3) GHU model explained above.

Let us note that to get EDM, both P and CP have to be broken. P symmetry, however, is broken anyway by the orbifolding, S^1/Z_2. In 5D space-time, CP transformation can be defined just as in the 4D case, since the spinor has 4 components as in 4D space-time:

$$CP: \quad \Psi(x^\mu, y) \to i\gamma^0\gamma^2\Psi(x_\mu, y)^*. \tag{19}$$

Correspondingly, the transformations of space-time and fields are fixed as

$$\begin{aligned}
CP: \ &x^\mu \to x_\mu, \ y \to y, \\
&A_\mu(x^\mu, y) \to -A^\mu(x_\mu, y)^t, \ A_y(x^\mu, y) \to -A_y(x_\mu, y)^t.
\end{aligned} \tag{20}$$

Thus we realize that the VEV of A_y has an odd CP-eigenvalue and leads to CP violation.

We have confirmed by explicit calculation of Feynman diagrams that the EDM appears already at 1-loop level as the result of CP violation (for the detail see Ref. 11). Again, CP violation is realized with only 1 generation and therefore the mechanism provides a new type of CP violation, that is only possible in higher dimensional gauge theory and different from that of K-M model.

4. Flavor Physics in GHU[a]

To achieve flavor violation is another non-trivial issue in GHU scenario, since Yukawa couplings are originated from gauge coupling, which is universal for all flavors. As a new feature of higher dimensional model with Z_2-orbifolding, Z_2-odd bulk masses

$$\epsilon(y) M_i \bar{\psi}_i \psi_i \quad (\epsilon(y) : \text{sign function}) \tag{21}$$

are allowed, with M_i being different depending on each flavor. This bulk mass term can be new source of the violation of flavor symmetry, specific to higher dimensional model.

The bulk masses lead to the localization of Weyl fermions at different fixed points depending on its chirality. This localization results in small overlap integral of mode functions of left- and right-handed Weyl fermions (see Fig. 2) and therefore leads to exponentially suppressed Yukawa couplings,

$$\sim g e^{-\pi M_i R} \ (R : \text{the radius in the case of } S^1). \tag{22}$$

Note that in the scenario of GHU, the observed hierarchical fermion masses are understood as the result of originally universal mass of weak scale, with the hierarchical structure of the masses being naturally realized by the exponential factor without fine tuning (though to get the top quark mass is a non-trivial problem).

4.1. *Flavor changing neutral current (FCNC) processes*

Once flavor mixing due to the violation of flavor symmetry is realized, it leads to flavor changing neutral current (FCNC) processes, such as the "rare processes"

$$K^0 \leftrightarrow \bar{K}^0, \quad K_L \to \mu\bar{\mu}, \quad B^0 \leftrightarrow \bar{B}^0, \ldots. \tag{23}$$

[a]The material of this section is based on Ref. 12.

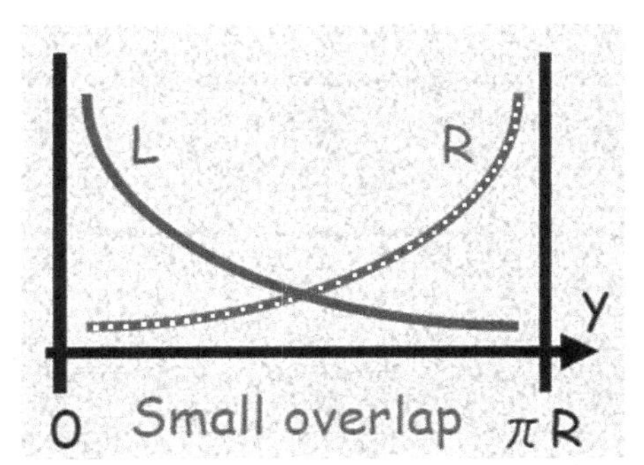

Fig. 2. The localization of mode functions of fermion at fixed points depending on the chirality

Natural flavor conservation in GHU

FCNC had played central roles for the foundation of the Standard Model (introduction of charm (GIM) and 3$^{\rm rd}$ generation (K-M)). Still, it has been playing a crucial role in order to check the viability of New Physics scenarios (SUSY, Technicolor, etc.). We ask if the condition of "natural flavor conservation" proposed by Glashow and Weinberg[13] is met in GHU, i.e. if FCNC processes at the tree level are naturally forbidden (in order to make them "rare"). We find that the condition by Glashow-Weinberg to guarantee natural flavor conservation

"Fermions with the same electric charge and the same chirality should possess the same quantum numbers"

is satisfied in our model. But, the new sauce of flavor violation, characteristic to higher dimensional theory, i.e. the non-degenerate bulk masses, is known to lead to FCNC due to the exchanges of non-zero Kaluza-Klein modes of gauge bosons already at the tree level. As a remark, FCNC processes in pure zero-mode sector are still forbidden, since the mode functions for zero-mode (4D massless) gauge bosons are "flat" in the direction of extra-dimension.

$$
\begin{aligned}
& g \int \text{const.} \times |f_{iL}(y)|^2 dy \\
& \propto \int |f_{iL}(y)|^2 dy = 1 : \quad \text{universal for all } i.
\end{aligned}
$$

Thus no FCNC processes due to the exchange of zero-mode gauge bosons appear at the tree level, even after the flavor mixing. Or, we may say that at low energies the theory reduces to the Standard Model, and concerning the

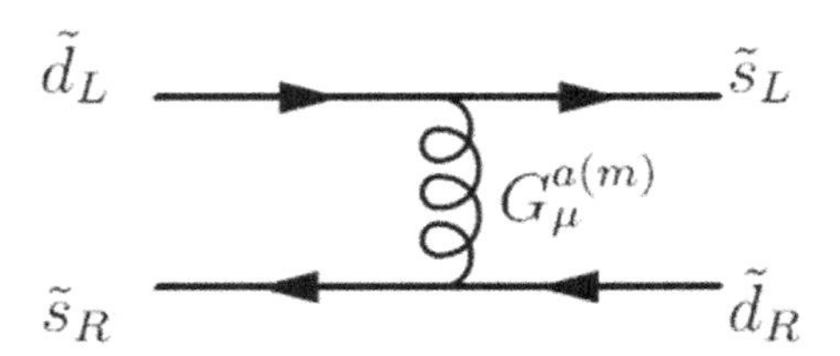

Fig. 3. $K^0 \leftrightarrow \bar{K}^0$ mixing due to the exchange of non-zero KK modes of gluon

zero-mode sector, the condition by Glashow-Weinberg is enough to suppress FCNC.

4.2. *GIM-like mechanism*

The above argument breaks down for the interaction of (massive) gauge bosons of non-zero KK modes, and FCNC appears already at the tree level, since the mode functions of the non-zero KK modes are no longer flat. Interestingly, however, in our analysis mentioned below, "GIM-like mechanism" is found to operate: for large bulk mass M_i, fermions almost perfectly localize at the fixed points, and for them the mode functions of KK gauge bosons look like constant. Hence gauge coupling becomes almost universal.

In fact, we found exponential suppression factor of FCNC[12]:

$$-\frac{\pi^2}{2}\left(e^{-2\pi M^1 R} + e^{-2\pi M^2 R}\right) - \frac{\pi}{2R}\frac{(M^1)^2 - M^1 M^2 + (M^2)^2}{M^1 M^2 (M^1 - M^2)} \times \left(e^{-2\pi M^1 R} - e^{-2\pi M^2 R}\right).$$

Note that

$$(\frac{m_{q_i}}{M_W})^2 \sim \exp(-2\pi M_i R), \tag{24}$$

and the suppression is similar to that of GIM mechanism by

$$\frac{m_c^2 - m_u^2}{M_W^2}. \tag{25}$$

4.3. *FCNC process: $K^0 \leftrightarrow \bar{K}^0$ mixing*

We considered the dominant contribution to the typical FCNC process, $K^0 \leftrightarrow \bar{K}^0$ mixing, due to the exchange of non-zero KK modes of gluon (see Fig.3).

We found the following:

(1) Though it appears at tree level, the rate of the mixing is suppressed by GIM-like mechanism and the inverse powers of the compactification scale (the decoupling of KK particles): Namely FCNC processes are under control in GHU scenario.
(2) A lower bound of the compactification scale is obtained by comparing our prediction with the room left for the discrepancy between the Standard Model prediction[14] and the data:

$$\frac{1}{R} > \mathcal{O}(10)\text{TeV}. \tag{26}$$

5. Summary

We summarize what we have discussed and found below.

(1) The gauge-Higgs unification (GHU) scenario guarantees a finite Higgs mass (in the viewpoint of hierarchy problem) and is an attractive scenario for the physics beyond the standard model.
(2) In the GHU scenario the Higgs interactions are governed by gauge principle and it may shed some light on the arbitrariness problem of the Higgs interactions.
(3) In GHU, as the Yukawa couplings are real to start with, CP violation is a challenging issue. Two new mechanisms of "spontaneous" CP violation were proposed, one by the complex structure of compactified extra space and one by the VEV of the Higgs. These mechanisms provide CP violation beyond that of Kobayashi-Maskawa model and "unitarity triangle".
(4) As the Yukawa couplings are universal in GHU, to start with, fermion mass hierarchy and flavor mixing are also challenging issues.
(5) In this scenario, hierarchical fermion masses are the result of (originally) universal mass of weak scale, with the hierarchy being naturally realized by the exponential factor $e^{-\pi M_i R}$ without fine tuning .
(6) The non-degenerated bulk mass M_i is a new source of flavor violation beyond the argument of Glashow-Weinberg, and leads to FCNC at the tree level.
(7) We found that "GIM-like" mechanism is operative in $K^0 \leftrightarrow \bar{K}^0$ mixing.
(8) The exchange of non-zero KK modes of gluon at the tree level yields the amplitude of $K^0 \leftrightarrow \bar{K}^0$, suppressed by GIM-like mechanism and the compactification scale (decoupling), and the data put a lower bound on the compactification scale:

$$\frac{1}{R} > \mathcal{O}(10)\text{TeV}. \tag{27}$$

References

1. N. S. Manton, *Nucl. Phys. B* **158**, 141 (1979).
2. Y. Hosotani, *Phys. Lett. B* **126**, 309 (1983); *Phys. Lett. B* **129**, 193 (1983); *Annals Phys.* **190**, 233 (1989).
3. H. Hatanaka, T. Inami and C. S. Lim, *Mod. Phys. Lett. A* **13**, 2601 (1998).
4. G.R. Dvali, S. Randjbar-Daemi and R. Tabbash, *Phys. Rev. D* **65**, 064021 (2002).
5. N. Arkani-Hamed, A.G. Cohen and H. Georgi, *Phys. Rev. Lett.* 864757 (2001); *Phys. Lett. B* **513**, 232 (2001).
6. M. Kubo, C. S. Lim and H. Yamashita, *Mod. Phys. Lett. A* **17**, 2249 (2002).
7. N. Arkani-Hamed, H.-C. Cheng, P. Creminelli and L. Randall, *Phys. Rev. Lett.* **90**, 221302 (2003); T. Inami, Y. Koyama, C.S. Lim and S. Minakami, *Progr. Theor. Phys.* **122**, 543(2009).
8. Y. Kawamura, *Prog. Theor. Phys.* **105**, 999 (2001).
9. C.S. Lim, *Phys. Lett. B* **256**, 233 (1991); A. Strominger and E. Witten, *Commun. Math. Phys.* **101**, 231 (1985).
10. C.S. Lim, N. Maru and K. Nishiwaki, *Phys. Rev. D* **81**, 076006 (2010).
11. Y. Adachi, C.S. Lim and N. Maru, *Phys. Rev. D* **80**, 055025 (2009).
12. Y. Adachi, N. Kurahashi, C.S. Lim and N. Maru, *Journ. of High Energy Phys.* **1011**, 150 (2010).
13. S. L. Glashow and S. Weinberg, Phys. Rev. D 15, 1958 (1977).
14. T. Inami and C.S. Lim, *Prog. Theor. Phys.* **65**, 297 (1981).
15. B. W. Bestbury, *J. Phys. A* **36**, 1947 (2003).
16. L. Lamport, *LaTeX, A Document Preparation System*, 2nd edition (Addison-Wesley, Reading, Massachusetts, 1994).
17. C. Jarlskog, in *CP Violation* (World Scientific, Singapore, 1988).

HIDDEN QUANTUM-MECHANICAL SUPERSYMMETRY IN EXTRA DIMENSIONS

MAKOTO SAKAMOTO

Department of Physics, Kobe University,
Rokkodai, Nada, Kobe 657-8501, Japan,
E-mail: dragon@kobe-u.ac.jp

We study higher dimensional field theories with extra dimensions from a 4d spectrum point of view. It is shown that 4d mass spectra of spinor, gauge and gravity field theories are governed by quantum-mechanical supersymmetry. The 4d massless modes turn out to correspond to zero energy vacuum states of the supersymmetry. Allowed boundary conditions on extra dimensions compatible with the supersymmetry are found to be severely restricted.

Keywords: Extra dimension; Supersymmetry; 4d spectrum.

1. Introduction

Gauge theories in higher dimensions are a promising candidate beyond the Standard Model. Such theories turn out to possess unexpectedly rich properties that shed new light and give a deep understanding on high energy physics. In fact, it has been shown that new mechanisms of gauge symmetry breaking[1–5], spontaneous supersymmetry breaking[6], and breaking of translational invariance[7,8] can occur, and that various phase structures arise in field theoretical models on certain topological manifolds[9–11]. Furthermore, new diverse scenarios of solving the hierarchy problem have been proposed[12–15].

Higher dimensional field theories will be described by 4d effective theories at low energies. Since we could not directly see extra dimensions, in particular, higher dimensional symmetries such as higher dimensional gauge symmetry and general covariance symmetry, one might ask what are remnants of the symmetries which originate from extra dimensions. They have to be hidden in the 4d effective theories. This is our motivation to investigate higher dimensional field theories from a 4d mass spectrum point of view. Our results show that the 4d mass spectrum is governed by quantum-mechanical supersymmetry (QM SUSY). Especially, the 4d massless spectrum is closely related to zero energy vacuum states of the supersymmetry and depends crucially on boundary conditions of extra dimensions, which are severely restricted by compatibility with QM SUSY.

In higher dimensional scalar theories, QM SUSY would appear in 4d spectrum but its appearance is found to be accidental. In higher dimensional spinor, gauge and gravity theories, QM SUSY always appears in 4d mass spectrum. Its origin turns out to be chiral symmetry, higher dimensional gauge symmetry and higher dimensional general covariance symmetry for spinor, gauge and gravity theories, respectively. It is interesting to note that all the symmetries guarantee the masslessness of the fields.

The paper is organized as follows: In section 2, we summarize the characteristic properties of QM SUSY. In the subsequent sections, we examine the 4d mass spectrum of higher dimensional scalar, spinor, gauge and gravity theories, separately and show that QM SUSY always appears in the 4d mass spectrum except for scalar theories. The section 7 is devoted to conclusions.

2. Minimal Supersymmetry Algebra

In any higher dimensional spinor/gauge/gravity theories with extra dimensions, quantum-mechanical supersymmetry turns out to be hidden in 4d spectrum and to play an important role to determine the spectrum of massless 4d fields, which are crucial ingredients in constructing low energy effective theories. The supersymmetric structure is found to be summarized in the minimal supersymmetry algebra, which consists of the hermitian operators H, Q and F, defined by

$$H = Q^2, \tag{1}$$

$$(-1)^F Q = -Q\,(-1)^F, \tag{2}$$

$$(-1)^F = \begin{cases} +1 & \text{for “bosonic” states,} \\ -1 & \text{for “fermionic” states,} \end{cases} \tag{3}$$

where H and Q are the Hamiltonian and the supercharge. The operator F is called a “fermion” number operator and the eigenvalues of $(-1)^F$ are given by $+1$ for “bosonic” states and -1 for “fermionic” ones, although the words, boson and fermion, have nothing to do with particles of integer spins and half-odd integer spins. The readers should not confuse the quantum-mechanical supersymmetry with supersymmetry in quantum field theory, which implies a symmetry between bosonic states with integer spins and fermionic states with half-odd integer spins. The operators in the algebra (1)-(3) are defined in quantum-mechanical systems and hence the supercharge Q does not possess any spinor index. We call the symmetry which obeys the algebra (1)-(3) quantum-mechanical supersymmetry, or simply QM SUSY in this paper.

Let us first clarify characteristic properties of the algebra (1)-(3), We now show that if the system obeys the algebra, the spectrum has the following properties:

1) The energy eigenvalues are non-negative, i.e. $E \geq 0$.
2) Any positive energy state $|E,+\rangle$ of $(-1)^F = +1$ forms a pair with the state $|E,-\rangle$ of the same energy E and $(-1)^F = -1$, and vice versa. All positive energy states form supermultiplets.
3) Zero energy states (if exist) do not necessarily form pairs of supermultiplets.

Thus, a typical spectrum of QM SUSY systems will be given by Fig.1.

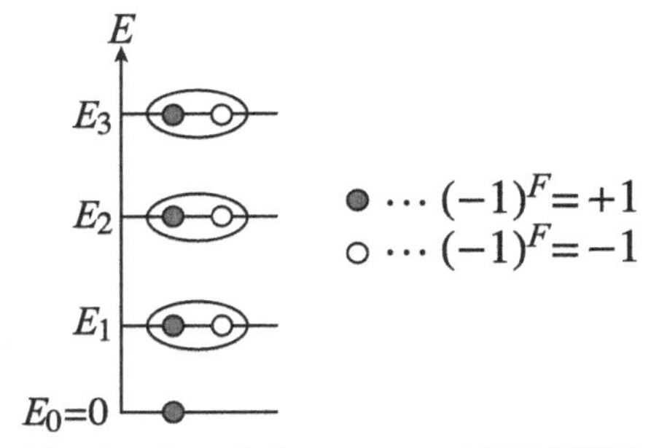

Fig. 1. A typical spectrum of QM SUSY

The first property 1) of $E \geq 0$ is derived from Eq.(1) because

$$E = \langle E|H|E\rangle = \langle E|Q^2|E\rangle = \| \, Q|E\,\rangle \|^2 \geq 0 \tag{4}$$

for any normalized energy eigenstate $|E\rangle$. Here, we have used the facts that the supercharge Q is hermitian and the norm of any state is non-negative. The second property 2) can be shown as follows: Suppose that $|E,+\rangle$ is an energy eigenstate with $(-1)^F = +1$. Then, the state $Q|E,+\rangle$ has the same energy eigenvalue E but the opposite eigenvalue of $(-1)^F$ because $H(Q|E,+\rangle) = Q^2(Q|E,+\rangle) = QH|E,+\rangle = E(Q|E,+\rangle)$ and $(-1)^F(Q|E,+\rangle) = -Q(-1)^F|E,+\rangle = -(Q|E,+\rangle)$. These relations imply that $Q|E,+\rangle \propto |E,-\rangle$. Assuming $Q|E,\pm\rangle = \alpha_\pm|E,\mp\rangle$ with $\| \, |E,\pm\,\rangle\|^2 = 1$, we find

$$E = \langle E, \pm | H | E, \pm \rangle = \langle E, \pm | Q^2 | E, \pm \rangle = \| Q | E, \pm \rangle \|^2 = |\alpha_\pm|^2. \qquad (5)$$

We can then take $\alpha_\pm = \sqrt{E}$ without any loss of generality. Thus, the states $|E, + \rangle$ and $|E, - \rangle$ form a supermultiplet and are related each other through the SUSY relations

$$Q | E, \pm \rangle = \sqrt{E} | E, \mp \rangle. \qquad (6)$$

The above result immediately shows that

$$Q | E, \pm \rangle = 0 \quad \text{for } E = 0. \qquad (7)$$

This implies that zero energy states do not necessarily form supermultiplets.

We have found that characteristic properties of QM SUSY is nicely summarized in Fig.1. Thus, if we encounter such a spectrum, QM SUSY is expected to be hidden in the system. In fact, we will find this type of spectrum again and again in the following sections.

3. 5d Scalar

Let us start with a 5d real massless scalar field theory compactified on a circle S^1.

$$S = \int d^4x \int_0^L dy \left\{ \frac{1}{2} \Phi(x,y) (\partial^\mu \partial_\mu + \partial_y{}^2) \Phi(x,y) - V(\Phi) \right\}, \qquad (8)$$

where x^μ ($\mu = 0, 1, 2, 3$) denotes the 4-dimensional Minkowski space-time coordinate and y is the coordinate of the extra dimension on the circle S^1 of the circumference L. The $V(\Phi)$ denotes a potential term but it will not be concerned in our analysis.

Since the extra dimension is compactified on the circle of the circumference L, we have to specify a boundary condition on the field $\Phi(x, y)$. Let us take a periodic boundary condition, as an example, i.e.

$$\Phi(x, y + L) = \Phi(x, y). \qquad (9)$$

We will make a comment on other boundary conditions at the end of this section.

In order to obtain the 4d mass spectrum, we expand the 5d field $\Phi(x, y)$ into the Kaluza-Klein modes such as

$$\Phi(x,y) = \phi_0^{(+)}(x) f_0^{(+)}(y) + \sum_{n=1}^{\infty} \{ \phi_n^{(+)}(x) f_n^{(+)}(y) + \phi_n^{(-)}(x) f_n^{(-)}(y) \}, \quad (10)$$

where $\phi_n^{(\pm)}(x)$ correspond to 4d scalar fields and $f_n^{(\pm)}(y)$ are the mass eigenfunctions of the differential operator $-\partial_y{}^2$, i.e.

$$\begin{cases} f_0^{(+)}(y) = N_0^{(+)}, \\ f_n^{(+)}(y) = N_n^{(+)} \cos\left(\frac{2\pi y}{L}\right), \\ f_n^{(-)}(y) = N_n^{(-)} \sin\left(\frac{2\pi y}{L}\right), \quad n = 1, 2, 3 \cdots . \end{cases} \tag{11}$$

Here, $N_n^{(\pm)}$ denote normalization constants. We should note that the set of $\{f_n^{(\pm)}\}$ forms a complete set so that the expansion (10) should be regarded as an identity.

Inserting the expansion (10) into the action (8) and using the orthogonal relations of $f_n^{(\pm)}(y)$ with an appropriate normalization, we find

$$\begin{aligned} S = \int d^4x \Big\{ \frac{1}{2}\phi_0^{(+)}(x)\partial^\mu\partial_\mu\phi_0^{(+)}(x) + \frac{1}{2}\sum_{n=1}^{\infty}\Big(\phi_n^{(+)}(x)\big(\partial^\mu\partial_\mu - m_n{}^2\big)\phi_n^{(+)}(x) \\ + \frac{1}{2}\phi_n^{(-)}(x)\big(\partial^\mu\partial_\mu - m_n{}^2\big)\phi_n^{(-)}(x)\Big) - V(\phi)\Big\}, \end{aligned} \tag{12}$$

where m_n is the 4d mass of the field $\phi_n^{(\pm)}$ and is given by

$$m_n = \frac{2\pi n}{L}, \quad n = 0, 1, 2, \cdots . \tag{13}$$

It follows that there appears a single massless mode $\phi_0^{(+)}$ and that all massive modes $\phi_n^{(\pm)}$ $(n = 1, 2, \cdots)$ are doubly degenerate. Thus, the 4d mass spectrum is given by Fig.2. This is nothing but a typical QM SUSY spec-

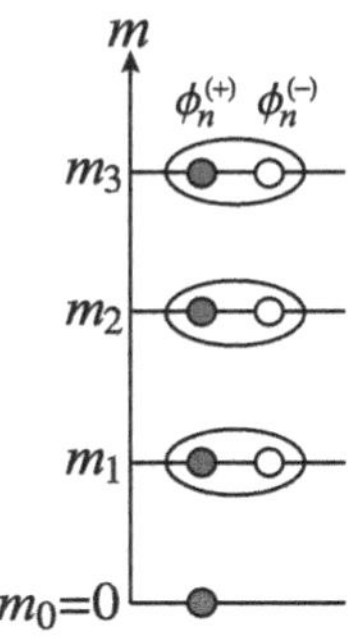

Fig. 2. Spectrum of 5d scalar on $M^4 \times S^1$

trum! We can, in fact, show that the minimal supersymmetry algebra appears in the system and the model will be the simplest higher dimensional field model that possesses QM SUSY.

Now the question is what are the operators H, Q and $(-1)^F$ of QM SUSY in the present system. The answer is

$$H = -\partial_y{}^2, \quad Q = -i\partial_y, \quad (-1)^F = \mathcal{P}. \tag{14}$$

The eigenvalues of the Hamiltonian H correspond to the mass squared $m_n{}^2$. The supercharge Q is just the momentum operator, i.e. $Q = -i\partial_y$ and satisfies the desired relation $H = Q^2$. The operator $(-1)^F$ is given by the parity operator $\mathcal{P}$. It implies that the states of $(-1)^F = +1$ (-1) correspond to even (odd) parity states. It is easy to verify that Q anticommutes with $(-1)^F$, as they should. The parity even (odd) function $f_n^{(+)}(y)$ $(f_n^{(-)}(y))$ has $(-1)^F = +1$ (-1) and they form a supermultiplet because $Qf_n^{(\pm)}(y) = -i\partial_y f_n^{(\pm)}(y) \propto f_n^{(\mp)}(y)$ for $n > 0$. Furthermore, $Qf_0^{(+)} = 0$ because $f_0^{(+)}$ is independent of y. This implies that the zero mode $f_0^{(+)}$ has no superpartner, as expected. Therefore, we have confirmed that the degeneracy for the nonzero modes in the 4d mass spectrum (see Fig.2) can be explained by QM SUSY.

As mentioned before, we make a comment on boundary conditions. In the above analysis, we assumed the periodic boundary condition (9). We can show that the system with the antiperiodic boundary condition $\Phi(y+L) = -\Phi(y)$ possesses QM SUSY as well but without any massless 4d state. Boundary conditions other than periodic and antiperiodic boundary conditions,[a] however, lead to non-degenerate 4d spectrum and no QM SUSY.[b] Thus, we conclude that QM SUSY found in the scalar field theory is accidental and there is no general mechanism to guarantee QM SUSY in any scalar field theories.

4. 5d Spinor

In this section, we consider a 5d spinor field on an interval $(0 \leq y \leq L)$:

$$S = \int d^4x \int_0^L dy\, \bar{\Psi}(x,y)\Big(i\gamma^\mu\partial_\mu + i\gamma^y\partial_y + M + \lambda\varphi(y)\Big)\Psi(x,y), \tag{15}$$

where $\Psi(x,y)$ is a 4-component 5d Dirac spinor field and M is a 5-dimensional (bulk) mass. The γ^y is given by $\gamma^y \equiv \gamma^0\gamma^1\gamma^2\gamma^3 = -i\gamma^5$. Here, we have introduced a coupling to a real scalar field $\varphi(y)$ and allow it to

[a]Examples of other boundary conditions are Dirichlet boundary condition (b.c.) $\Phi(0) = 0$, Neumann b.c. $\partial_y\Phi(0) = 0$, twisted b.c. $\Phi(y+L) = e^{i\theta}\Phi(y)$ for a complex scalar.

[b]The systems with periodic or antiperiodic boundary condition can have an accidental symmetry, i.e. parity symmetry. This is the origin of QM SUSY as well as the degeneracy in the 4d spectrum.

have a nontrivial y-dependence as a background field. The spinor $\Psi(x,y)$ can be expanded as

$$\Psi(x,y) = \sum_n \{\psi_{+,n}(x) f_n(y) + \psi_{-,n}(x) g_n(y)\}, \tag{16}$$

where $\psi_{\pm,n}$ are 4-dimensional chiral spinors defined by $\gamma^5 \psi_{\pm,n} = \pm\psi_{\pm,n}$. The sets of functions $\{f_n(y)\}$ and $\{g_n(y)\}$ are assumed separately to form complete sets and should be chosen for $\psi_{\pm,n}(x)$ to be 4d mass eigenstates. From the representation theory of the Poincaré group, a massive 4d Dirac spinor ψ_n consists of chiral spinors $\psi_{+,n}$ and $\psi_{-,n}$ and they form the mass terms $m_n \bar{\psi}_{\pm,n} \psi_{\mp,n}$. On the other hand, a massless 4d spinor is chiral and hence does not necessarily form a pair of $\psi_{+,0}$ and $\psi_{-,0}$. Therefore, the 4d mass spectrum of (infinitely many) 4d spinors $\{\psi_{\pm,n}\}$ will be schematically given by Fig.3. This is nothing but a typical QM SUSY spectrum, as

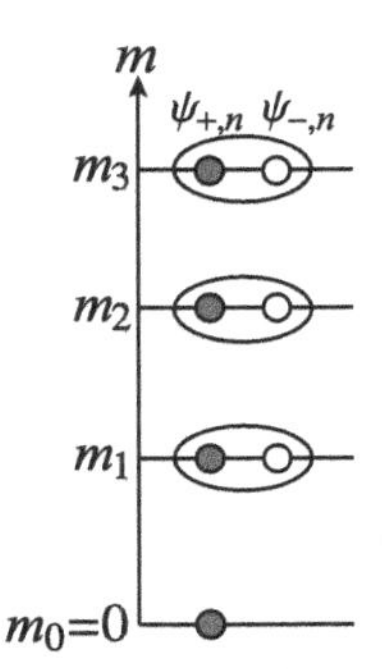

Fig. 3. Spectrum of 5d spinor

discussed in section 2. Thus, we expect that the minimal supersymmetry algebra is hidden in the 4d spectrum. This is indeed the case, as we will see below.

The 5d Dirac equation for $\Psi(x,y)$ is given by

$$\left[i\gamma^\mu \partial_\mu + \gamma^5 \partial_y + M + \lambda\varphi(y)\right] \Psi(x,y) = 0. \tag{17}$$

In terms of $\psi_{\pm,n}$, the above equation can be decomposed as

$$\sum_n \big(i\gamma^\mu \partial_\mu \psi_{+,n}(x)\big) f_n(y) + \sum_n \psi_{-,n}(x) \big(\mathcal{D}^\dagger g_n(y)\big) = 0, \tag{18}$$

$$\sum_n \big(i\gamma^\mu \partial_\mu \psi_{-,n}(x)\big) g_n(y) + \sum_n \psi_{+,n}(x) \big(\mathcal{D} f_n(y)\big) = 0, \tag{19}$$

where

$$\mathcal{D} = \partial_y + M + \lambda\varphi(y), \quad \mathcal{D}^\dagger = -\partial_y + M + \lambda\varphi(y). \tag{20}$$

We then require $f_n(y)$ an $g_n(y)$ to be the eigenfunctions of the differential operators $\mathcal{D}^\dagger\mathcal{D}$ and $\mathcal{D}\mathcal{D}^\dagger$, respectively, i.e.

$$\mathcal{D}^\dagger\mathcal{D} f_n(y) = m_n{}^2 f_n(y), \tag{21}$$

$$\mathcal{D}\mathcal{D}^\dagger g_n(y) = m_n{}^2 g_n(y). \tag{22}$$

Since $\mathcal{D}^\dagger\mathcal{D}$ and $\mathcal{D}\mathcal{D}^\dagger$ are hermitian,[c] the sets of $\{f_n(y)\}$ and $\{g_n(y)\}$ form complete sets, as they should. It follows from Eqs.(21), (22) that $\mathcal{D}f_n$ ($\mathcal{D}^\dagger g_n$) obeys the same eigenequation (22) ((21)) as g_n (f_n), and hence that f_n and g_n are related each other through the SUSY relations

$$m_n g_n(y) = \mathcal{D} f_n(y), \tag{23}$$

$$m_n f_n(y) = \mathcal{D}^\dagger g_n(y), \tag{24}$$

with appropriate normalizations. Thus the eigenvalues of f_n and g_n are doubly degenerate (except for $m_n = 0$), as expected.

The minimal supersymmetry algebra is manifest by introducing the operators as

$$H = Q^2 = \begin{pmatrix} \mathcal{D}^\dagger\mathcal{D} & 0 \\ 0 & \mathcal{D}\mathcal{D}^\dagger \end{pmatrix}, \quad Q = \begin{pmatrix} 0 & \mathcal{D}^\dagger \\ \mathcal{D} & 0 \end{pmatrix}, \quad (-1)^F = \begin{pmatrix} 1 & 0 \\ 0 & -1 \end{pmatrix}. \tag{25}$$

Those operators act on 2-component wavefunctions

$$|\Psi\rangle = \begin{pmatrix} f(y) \\ g(y) \end{pmatrix}. \tag{26}$$

Now, it is not difficult to show that with the SUSY relations (23), (24) and with the orthonormal relations of $\{f_n(y)\}$ and $\{g_n(y)\}$ the action can be written, in terms of the 4d spinors, into the form

$$S = \int d^4x \{\mathcal{L}_{m=0} + \mathcal{L}_{m\neq 0}\}, \tag{27}$$

where $\mathcal{L}_{m=0}$ is the part of the Lagrangian consisting of massless chiral spinors and

$$\mathcal{L}_{m\neq 0} = \sum_{m_n\neq 0} \bar{\psi}_n(x)(i\gamma^\mu\partial_\mu + m_n)\psi_n(x) \tag{28}$$

[c] Boundary conditions for $f_n(y)$ and $g_n(y)$ have to be chosen for $\mathcal{D}^\dagger$ to be hermitian conjugate to $\mathcal{D}$. We will discuss how to determine boundary conditions later.

with $\psi_n = \psi_{+,n} + \psi_{-,n}$ for $m_n \neq 0$. Thus, we have shown that $\psi_{\pm,n}$ are mass eigenstates with m_n, as announced before.

To determine the chiral zero mode part $\mathcal{L}_{m=0}$, we need to specify boundary conditions at $y = 0, L$ for $f_n(y)$ and $g_n(y)$. It turns out that the choice of boundary conditions is crucial for the existence of massless chiral spinors. Allowed boundary conditions compatible with QM SUSY have been classified and are listed below[16,17]:

i) $\mathcal{D}f_n(0) = 0 = \mathcal{D}f_n(L),\ g_n(0) = 0 = g_n(L)$,
ii) $f_n(0) = 0 = f_n(L),\ \mathcal{D}^\dagger g_n(0) = 0 = \mathcal{D}^\dagger g_n(L)$,
iii) $\mathcal{D}f_n(0) = 0 = f_n(L),\ g_n(0) = 0 = \mathcal{D}^\dagger g_n(L)$,
iv) $f_n(0) = 0 = \mathcal{D}f_n(L),\ \mathcal{D}^\dagger g_n(0) = 0 = g_n(L)$.

Since the mode functions $f_n(y)$ and $g_n(y)$ obey the SUSY relations (23) and (24), chiral zero modes (if any) should satisfy

$$\mathcal{D}f_0(y) = 0, \tag{29}$$

$$\mathcal{D}^\dagger g_0(y) = 0, \tag{30}$$

with $m_0 = 0$. These first order differential equations can easily be solved as

$$f_0(y) = N_0 \exp\Big\{-\int_0^y dy' \big(M + \lambda\varphi(y')\big)\Big\}, \tag{31}$$

$$g_0(y) = \bar{N}_0 \exp\Big\{+\int_0^y dy' \big(M + \lambda\varphi(y')\big)\Big\}. \tag{32}$$

We should emphasize that the above solutions do not insure the existence of the massless chiral spinors $\psi_{+,0}$ and $\psi_{-,0}$ because they have to be discarded from the physical spectrum if $f_0(y)$ and/or $g_0(y)$ do not obey the boundary conditions. It is easy to see that $f_0(y)$ given in (31) obeys the boundary conditions only for i) and that $g_0(y)$ in (32) obeys them only for ii). Therefore, we find that

$$\mathcal{L}_{m=0} = \begin{cases} \bar{\psi}_{+,0}(x) i\gamma^\mu \partial_\mu \psi_{+,0}(x) & \text{for i)}, \\ \bar{\psi}_{-,0}(x) i\gamma^\mu \partial_\mu \psi_{-,0}(x) & \text{for ii)}, \\ 0 & \text{for iii) and iv)}. \end{cases} \tag{33}$$

The extension of the above analysis to higher dimensions $M^4 \times K^N$ will be straightforward. The Γ-matrices on $M^4 \times K^N$ may be constructed, in terms of the γ-matrices on M^4 and the $\bar{\gamma}$-matrices on K^N, as

$$\begin{aligned} \Gamma^\mu &= \gamma^\mu \otimes I_{2^{[N/2]}}, & \mu &= 0,1,2,3, \\ \Gamma^i &= \gamma^5 \otimes \bar{\gamma}^i, & i &= 1,2,\cdots,N, \end{aligned} \tag{34}$$

which satisfy

$$
\begin{aligned}
\{\Gamma^\mu, \Gamma^\nu\} &= -2\eta^{\mu\nu} I_4 \otimes I_{2^{[N/2]}}, & \mu, \nu &= 0,1,2,3, \\
\{\Gamma^i, \Gamma^j\} &= -2\delta^{ij} I_4 \otimes I_{2^{[N/2]}}, & i, j &= 1,2,\cdots,N, \\
\{\Gamma^\mu, \Gamma^j\} &= 0.
\end{aligned}
\tag{35}
$$

Here, $[N/2]$ denotes the Gauss symbol and I_n is the $n \times n$ identity matrix. The structure of the Γ-matrices may imply that a $(4+N)$-dimensional spinor $\Psi(x,y)$ can be expanded as

$$
\Psi(x,y) = \sum_n \{\psi_{+,n}(x) \otimes \xi_{+,n}(y) + \psi_{-,n}(x) \otimes \xi_{-,n}(y)\}, \tag{36}
$$

where $\psi_{\pm,n}(x)$ ($\xi_{\pm,n}(y)$) denote 4-dimensional (N-dimensional) spinors and x^μ (y^i) are the coordinates of M^4 (K^N). The 4d mass spectrum of $\psi_{\pm,n}$ will be schematically given just like Fig.3 and the mass eigenfunctions $\xi_{+,n}(y)$ and $\xi_{-,n}(y)$ will form a supermultiplet, though we will not proceed further.

5. 5d Vector

In this section, we consider a (4+1)-dimensional abelian gauge theory on an interval ($0 \le y \le L$):

$$
S = \int d^4x \int_0^L dy \sqrt{-g(y)} \left\{ -\frac{1}{4} F_{MN}(x,y) F^{MN}(x,y) \right\} \tag{37}
$$

with a non-factorizable metric

$$
ds^2 = e^{-4W(y)} \eta_{\mu\nu} dx^\mu dx^\nu + g_{55}(y) dy^2. \tag{38}
$$

The metric reduces to the warped metric discussed by Randall and Sundrum[12] when $g_{55}(y) = 1$ and $W(y) = \frac{1}{2}k|y|$. Another choice of $g_{55}(y) = e^{-4W(y)}$ leads to the model discussed in Ref. 14, in which a hierarchical mass spectrum has been observed.

In order to expand the 5d gauge fields $A_\mu(x,y)$ and $A_y(x,y)$ into 4d mass eigenstates and to make a QM SUSY structure manifest, we introduce the operators H, Q and $(-1)^F$ as follows[18,19]:

$$
H = Q^2 = \begin{pmatrix} -\frac{1}{\sqrt{g_{55}}} \partial_y \frac{e^{-4W}}{\sqrt{g_{55}}} \partial_y & 0 \\ 0 & -\partial_y \frac{1}{\sqrt{g_{55}}} \partial_y \frac{e^{-4W}}{\sqrt{g_{55}}} \end{pmatrix}, \tag{39}
$$

$$
Q = \begin{pmatrix} 0 & -\frac{1}{\sqrt{g_{55}}} \partial_y \frac{e^{-4W}}{\sqrt{g_{55}}} \\ \partial_y & 0 \end{pmatrix}, \tag{40}
$$

$$
(-1)^F = \begin{pmatrix} 1 & 0 \\ 0 & -1 \end{pmatrix}, \tag{41}
$$

which act on two-component vectors

$$|\Psi\rangle = \begin{pmatrix} f(y) \\ g(y) \end{pmatrix}. \tag{42}$$

The inner product of two states $|\Psi_1\rangle$ and $|\Psi_2\rangle$ is defined by

$$\langle\Psi_2|\Psi_1\rangle = \int_0^L dy \sqrt{g_{55}(y)} \left\{ f_2(y) f_1(y) + \frac{e^{-4W(y)}}{g_{55}(y)} g_2(y) g_1(y) \right\}. \tag{43}$$

To obtain consistent boundary conditions for the functions $f(y)$ and $g(y)$ in $|\Psi\rangle$, we first require that the supercharge Q is hermitian with respect to the inner product (43), i.e.

$$\langle\Psi_2|Q\Psi_1\rangle = \langle Q\Psi_2|\Psi_1\rangle. \tag{44}$$

It turns out that the functions $f(y)$ and $g(y)$ have to obey one of the following four types of boundary conditions:

$$\text{i) } g(0) = g(L) = 0, \tag{45}$$

$$\text{ii) } f(0) = f(L) = 0, \tag{46}$$

$$\text{iii) } g(0) = f(L) = 0, \tag{47}$$

$$\text{iv) } f(0) = g(L) = 0. \tag{48}$$

We further require that the state $Q|\Psi\rangle$ obeys the same boundary conditions as $|\Psi\rangle$, otherwise Q is not a well defined operator and "bosonic" and "fermionic" states would not form supermultiplets. The requirement leads to

$$\partial_y f(0) = \partial_y f(L) = 0 \qquad \text{for i)}, \tag{49}$$

$$\partial_y \left(\frac{e^{-4W}}{\sqrt{g_{55}}} g \right)(0) = \partial_y \left(\frac{e^{-4W}}{\sqrt{g_{55}}} g \right)(L) = 0 \qquad \text{for ii)}, \tag{50}$$

$$\partial_y f(0) = \partial_y \left(\frac{e^{-4W}}{\sqrt{g_{55}}} g \right)(L) = 0 \qquad \text{for iii)}, \tag{51}$$

$$\partial_y \left(\frac{e^{-4W}}{\sqrt{g_{55}}} g \right)(0) = \partial_y f(L) = 0 \qquad \text{for iv)}. \tag{52}$$

Combining all the above results, we have found the four types of boundary conditions compatible with supersymmetry[18,19],

$$\text{Type (N,N)}: \begin{cases} \partial_y f(0) = \partial_y f(L) = 0, \\ g(0) = g(L) = 0, \end{cases} \tag{53}$$

$$\text{Type (D,D)}: \begin{cases} f(0) = f(L) = 0, \\ \partial_y \left(\frac{e^{-4W}}{\sqrt{g_{55}}} g \right)(0) = \partial_y \left(\frac{e^{-4W}}{\sqrt{g_{55}}} g \right)(L) = 0, \end{cases} \tag{54}$$

$$\text{Type (N,D)}: \begin{cases} \partial_y f(0) = f(L) = 0, \\ g(0) = \partial_y \left(\frac{e^{-4W}}{\sqrt{g_{55}}} g \right)(L) = 0, \end{cases} \tag{55}$$

$$\text{Type (D,N)}: \begin{cases} f(0) = \partial_y f(L) = 0, \\ \partial_y \left(\frac{e^{-4W}}{\sqrt{g_{55}}} g \right)(0) = g(L) = 0. \end{cases} \tag{56}$$

It follows that the above boundary conditions ensure the hermiticity of the Hamiltonian, i.e.

$$\langle \Psi_2 | H \Psi_1 \rangle = \langle H \Psi_2 | \Psi_1 \rangle. \tag{57}$$

Therefore, we have succeeded to obtain the consistent set of boundary conditions that ensure the hermiticity of the supercharge and the Hamiltonian and also that the action of the supercharge on $|\Psi\rangle$ is well defined. Since the supersymmetry is a direct consequence of higher-dimensional gauge invariance, our requirements on boundary conditions should be, at least, necessary conditions to preserve it. It turns out that the boundary conditions obtained above are consistent with those in Ref. 5, although it is less obvious how the requirement of the least action principle proposed in Ref. 5 is connected to gauge invariance. We should emphasize that the supercharge Q is well defined for all the boundary conditions (53)-(56) and hence that the supersymmetric structure always appears in the spectrum, though the boundary conditions other than the type (N,N) break 4d gauge symmetries, as we will see below.

From the above analysis, the 5d gauge fields $A_\mu(x,y)$ and $A_y(x,y)$ are expanded in the mass eigenstates as follows:

$$A_\mu(x,y) = \sum_n A_{\mu,n}(x) f_n(y), \tag{58}$$

$$A_y(x,y) = \sum_n h_n(x) g_n(y), \tag{59}$$

where $f_n(y)$ and $g_n(y)$ are the eigenstates of the Schrödinger-like equations

$$-\frac{1}{\sqrt{g_{55}}}\partial_y \frac{e^{-4W}}{\sqrt{g_{55}}}\partial_y f_n(y) = m_n^2 f_n(y), \tag{60}$$

$$-\partial_y \frac{1}{\sqrt{g_{55}}}\partial_y \frac{e^{-4W}}{\sqrt{g_{55}}} g_n(y) = m_n^2 g_n(y) \tag{61}$$

with one of the four types of the boundary conditions (53)-(56) and they are actually related each other through the SUSY relations:

$$m_n g_n(y) = \partial_y f_n(y), \tag{62}$$

$$m_n f_n(y) = -\frac{1}{\sqrt{g_{55}}}\partial_y \frac{e^{-4W}}{\sqrt{g_{55}}} g_n. \tag{63}$$

Since the massless states are especially important in phenomenology, let us investigate the massless states of the equations (60) and (61). Thanks to supersymmetry, the massless modes would be the solutions to the first order differential equation $Q|\Psi_0\rangle = 0$, i.e.

$$\partial_y f_0(y) = 0, \tag{64}$$

$$\partial_y \left(\frac{e^{-4W}}{\sqrt{g_{55}}} g_0(y) \right) = 0. \tag{65}$$

The solutions are easily found to be

$$f_0(y) = N_0, \tag{66}$$

$$g_0(y) = \bar{N}_0 \, e^{4W(y)} \sqrt{g_{55}(y)}, \tag{67}$$

where N_0 and $\bar{N}_0$ are some constants. We should emphasize that the above solutions do not necessarily imply physical massless states of $A_{\mu,0}(x)$ and $h_0(x)$ in the spectrum. This is because the boundary conditions exclude some or all of them from the physical spectrum. Indeed, $f_0(y)$ ($g_0(y)$) satisfies only the boundary conditions of the type (N,N) (type (D,D)). Thus, a massless vector $A_{\mu,0}(x)$ (a massless scalar $h_0(x)$) appears only for the type (N,N) (type (D,D)) boundary conditions (see Fig.4). This implies that the 4d gauge symmetry is broken except for the type (N,N) boundary conditions.

It is instructive to discuss the relation between the QM SUSY and the higher dimensional gauge symmetry. The relation becomes apparent by expressing the action, in terms of the 4d mass eigenstates, as

$$S = \int d^4x \{ \mathcal{L}_{m=0} + \mathcal{L}_{m\neq 0} \}, \tag{68}$$

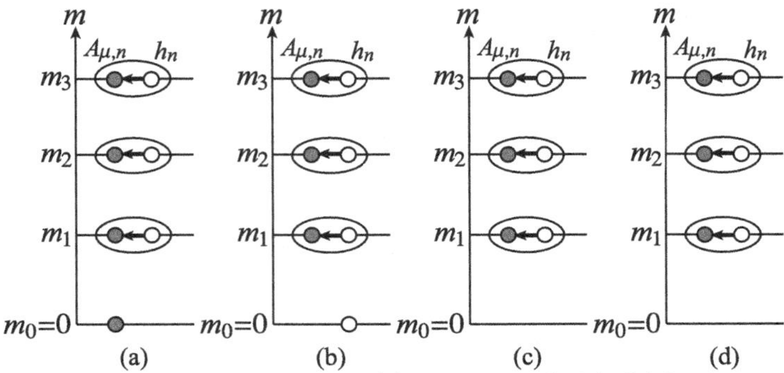

Fig. 4. Spectrum of the 5d gauge theory (a) for the type (N,N), (b) for the type (D,D), (c) for the type (N,D), (d) for the type (D,N) boundary conditions.

where $\mathcal{L}_{m=0}$ is the Lagrangian consisting of the massless fields and

$$\mathcal{L}_{m\neq 0} = \sum_{m_n\neq 0} \left\{ -\frac{1}{4}\left(F_{\mu\nu,n}(x)\right)^2 - \frac{m_n{}^2}{2}\left(A_{\mu,n}(x) - \frac{1}{m_n}\partial_\mu h_n(x)\right)^2 \right\} \quad (69)$$

with $F_{\mu\nu,n} = \partial_\mu A_{\nu,n} - \partial_\nu A_{\mu,n}$. It follows that every nonzero mode $h_n(x)$ for $m_n \neq 0$ can be absorbed into the longitudinal mode of $A_{\mu,n}(x)$ and then $A_{\mu,n}(x)$ becomes massive with three degrees of freedom, as it should be. The choice of $h_n(x) = 0$ ($m_n \neq 0$) is called a unitary gauge. It should be emphasized that the Lagrangian (69) has been derived by use of the SUSY relations (62) and (63). Therefore, the QM SUSY is necessary for $A_{\mu,n}(x)$ ($m_n \neq 0$) to become massive by absorbing the unphysical mode $h_n(x)$, which is a consequence of the higher dimensional gauge symmetry. This observation is summarized in Fig.4.

We have restricted our considerations to a 5d gauge theory. The extension to any higher dimensional gauge theory is possible and QM SUSY is found in the 4d mass spectrum. The details have been given in Refs. 18 and 19.

6. 5d Gravity

In this section, we investigate the 4d mass spectrum of the 5d Randall-Sundrum gravity theory[12] with a warped metric

$$ds^2 = e^{2A(y)}\left(\eta_{\mu\nu}dx^\mu dx^\nu + dy^2\right), \quad (70)$$

where $A(y)$ is the warp factor which turns out to play a role of a superpotential in the $N = 2$ Witten model[20]. For the Randall-Sundrum model, the warp factor is given by

$$A(y) = -\ln\Big(\frac{y}{y_1}\Big). \tag{71}$$

Here, the location of the UV brane is chosen such that the warp factor is set equal to 1 on the UV brane at $y = y_1$.

The metric fluctuations h_{MN} around the background metric (70) are given by

$$ds^2 = e^{2A(y)}\big(\eta_{MN} + h_{MN}(x,y)\big)dx^M dx^N \tag{72}$$

and h_{MN} turn out to be useful with the parameterization[21]

$$h_{MN}(x,y) = \begin{pmatrix} h_{\mu\nu}(x,y) - \frac{1}{2}\eta_{\mu\nu}\phi(x,y) & h_{y\nu}(x,y) \\ h_{\mu y}(x,y) & \phi(x,y) \end{pmatrix}. \tag{73}$$

The action is invariant under infinitesimal general coordinate transformations: $x^M \to x'^M = x^M + \xi^M(x,y)$, which are translated into the field transformations of the metric fluctuations:

$$\begin{aligned} \delta h_{\mu\nu} &= -\partial_\mu \xi_\nu - \partial_\nu \xi_\mu - \eta_{\mu\nu}(\partial_y + 3A')\xi_y, \\ \delta h_{\mu y} &= -\partial_y \xi_\mu - \partial_\mu \xi_y, \\ \delta\phi &= -2(\partial_y + A')\xi_y, \end{aligned} \tag{74}$$

where $A'(y) = dA(y)/dy$. The metric fluctuation fields are expanded, in terms of some complete sets of the functions $\{f_n(y)\}$, $\{g_n(y)\}$, $\{k_n(y)\}$, as

$$\begin{aligned} h_{\mu\nu}(x,y) &= \sum_n h_{\mu\nu,n}(x) f_n(y), \\ h_{\mu y}(x,y) &= \sum_n h_{\mu y,n}(x) g_n(y), \\ \phi(x,y) &= \sum_n \phi_n(x) k_n(y). \end{aligned} \tag{75}$$

It is instructive to examine degrees of freedom for massive modes. Each field of $h_{\mu\nu,n}, h_{\mu y,n}, \phi$ has originally $2, 2, 1$ degrees of freedom, respectively because 5d gravity has no mass term in a 5-dimensional point of view. The vector field $h_{\mu y,n}$ could become massive by "eating" one extra degree of freedom and then has three degrees of freedom as a massive vector. The tensor field $h_{\mu\nu,n}$ could become massive by "eating" three extra degrees of freedom and then has five degrees of freedom as a massive graviton. The above Higgs-like mechanism can actually occur in the 5d gravity system. The vector field $h_{\mu y,n}$ "eats" ϕ_n to become a massive vector with three degrees of freedom, and then the tensor field $h_{\mu\nu,n}$ "eats" $h_{\mu y,n}$ to become a massive graviton with five degrees of freedom:

$$\overbrace{h_{\mu\nu,n} \quad \underbrace{h_{\mu y,n} \quad \phi_n}_{\text{a massive vector}}}^{\text{a massive graviton}} \tag{76}$$

The above observation strongly suggests, on the analogy of the 5d gauge theory, that the 5d gravity theory possesses *two* QM SUSY systems in the 4d spectrum: One is realized between the eigenfunctions $g_n(y)$ and $k_n(y)$. The other is between $f_n(y)$ and $g_n(y)$. To verify it, we have to find the eigenequations for $f_n(y), g_n(y)$ and $k_n(y)$, which should diagonalize the quadratic action for $h_{\mu\nu,n}(x), h_{\mu y,n}(x)$ and $\phi_n(x)$. The eigenequations for $f_n(y), g_n(y)$ and $k_n(y)$ are found to be[21]

$$-\left(\partial_y{}^2 + 3A'(y)\partial_y\right) f_n(y) = m_n{}^2 f_n(y), \tag{77}$$

$$-\left(\partial_y{}^2 + 3A'(y)\partial_y + 3A''(y)\right) g_n(y) = m_n{}^2 g_n(y), \tag{78}$$

$$-\left(\partial_y{}^2 + 3A'(y)\partial_y + 4A''(y)\right) k_n(y) = m_n{}^2 k_n(y). \tag{79}$$

The supersymmetric structure between $f_n(y)$ and $g_n(y)$ will become apparent if we express Eqs.(77) and (78) into the form

$$\mathcal{D}^\dagger \mathcal{D} f_n(y) = m_n{}^2 f_n(y), \quad \mathcal{D}\mathcal{D}^\dagger g_n(y) = m_n{}^2 g_n(y), \tag{80}$$

where

$$\mathcal{D} = \partial_y, \quad \mathcal{D}^\dagger = -\left(\partial_y + 3A'(y)\right). \tag{81}$$

The eigenfunctions $f_n(y)$ and $g_n(y)$ are actually related each other through the SUSY relations

$$m_n g_n(y) = \mathcal{D} f_n(y), \quad m_n f_n(y) = \mathcal{D}^\dagger g_n(y). \tag{82}$$

It seems strange that $\mathcal{D}^\dagger$ is hermitian conjugate to $\mathcal{D}$. This is, however, true because the inner product is defined by[21]

$$\langle \psi | \phi \rangle = \int_{y_2}^{y_2} dy e^{3A(y)} \left(\psi(y)\right)^* \phi(y) \tag{83}$$

with the boundary conditions

$$\partial_y f_n(y) = 0 = g_n(y), \quad \text{at } y = y_1, y_2. \tag{84}$$

The factor $e^{3A(y)}$ in Eq.(83) is required because of the presence of it in the action, whose origin comes from the nontrivial background metric (70). The boundary conditions (84) turn out to be compatible with supersymmetry.

The QM SUSY structure will be manifest if we introduce two component wavefunctions

$$|\Psi\rangle = \begin{pmatrix} f(y) \\ g(y) \end{pmatrix}. \tag{85}$$

Then, H, Q and $(-1)^F$ are found to be the same form as Eq.(25) with $\mathcal{D}$ and $\mathcal{D}^\dagger$ defined in Eqs.(81).

Let us next proceed to the analysis of a pair of the eigenfunctions $g_n(y)$ and $k_n(y)$. The supersymmetric structure between them will be apparent if we express Eqs.(78), (79) into the form[21]

$$\bar{\mathcal{D}}^\dagger \bar{\mathcal{D}} g_n(y) = m_n^2 g_n(y), \quad \bar{\mathcal{D}} \bar{\mathcal{D}}^\dagger k_n(y) = m_n^2 k_n(y), \tag{86}$$

where

$$\bar{\mathcal{D}} = \partial_y + A'(y), \quad \bar{\mathcal{D}}^\dagger = -\big(\partial_y + 2A'(y)\big). \tag{87}$$

Here, we have used the relation $(A')^2 = A''$. The eigenfunctions $g_n(y)$ and $k_n(y)$ are related through the SUSY relations

$$m_n k_n(y) = \bar{\mathcal{D}} g_n(y), \quad m_n g_n(y) = \bar{\mathcal{D}}^\dagger k_n(y). \tag{88}$$

The inner product is defined by Eq.(83). This guarantees that $\bar{\mathcal{D}}$ and $\bar{\mathcal{D}}^\dagger$ are hermitian conjugate each other with the boundary conditions

$$g_n(y) = 0 = \bar{\mathcal{D}}^\dagger k_n(y), \quad \text{at } y = y_1, y_2. \tag{89}$$

The supersymmetric structure is manifest if we introduce two component wavefunctions in a similar manner as Eq.(85). Then, $\bar{H}, \bar{Q}$ and $(-1)^{\bar{F}}$ are given by the same form as Eq.(25) with the replacement of $\mathcal{D}$ and $\mathcal{D}^\dagger$ by Eqs.(87).

We have found two QM SUSY systems, as expected. The eigenfunctions $f_n(y)$ and $g_n(y)$ form a supermultiplet, and $g_n(y)$ and $k_n(y)$ form another supermultiplet. The QM SUSY structure turns out to severely restrict the allowed boundary conditions for $f_n(y), g_n(y)$ and $k_n(y)$. In fact, the boundary conditions for them are unique in order to be compatible with the QM SUSY. This fact is especially important in a low energy effective theory point of view. This is because the boundary conditions for $f_n(y), g_n(y)$ and $k_n(y)$ with the eigenvalue equations (80) and (86) determine uniquely the massless modes. In the present Randall-Sundrum model, there exist one massless graviton and one massless scalar (radion). The 4d spectrum of the Randall-Sundrum model is depicted in Fig.5.

We have investigated the 5d Randall-Sundrum model in which the 4-dimensional space-time is a flat Minkowski. Karch and Randall[22] have extended it to 4d de Sitter (dS_4) and anti de Sitter (AdS_4) space-time. In

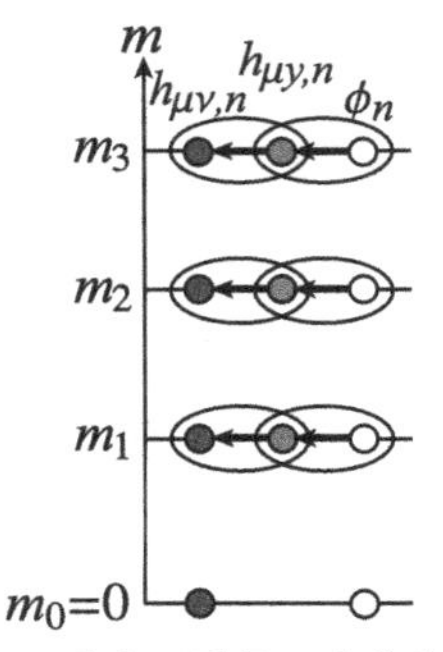

Fig. 5. Spectrum of the 5d Randall-Sundrum gravity

those cases, the warp factor $A(y)$ is different from Eq. (71) with a non-vanishing 4d cosmological constant. The analysis proceeds in a similar way and the results will be reported elsewhere.

Finally, we should make a few comments on interesting observations. The warp factor $A(y)$ cannot be an arbitrary function but has to be a solution of the Einstein equation. This gives a non-trivial constraints on $A(y)$. The differential equations for $f_n(y), g_n(y)$ and $k_n(y)$ are found to be in a class of exactly solvable models with the property of shape invariance[23]. This property holds even for the Karch-Randall models. The second interesting observation is the uniqueness of the boundary conditions which have to be compatible with two QM SUSYs. If we would have a 5d massless theory with a higher spin ($s > 2\hbar$), the 4d mass spectrum of the system could possess more than three QM SUSYs. Our analysis, however, tells us that there are no possible boundary conditions compatible with all QM SUSYs. This may lead to a conclusion that any 5d massless theory on an interval with higher spins ($s > 2\hbar$) has no possible boundary conditions compatible with QM SUSYs. This seems to be consistent with the fact that any non-trivial massless higher spin theory with $s > 2\hbar$ has not been found yet.

7. Conclusions

We have investigated higher dimensional scalar, spinor, gauge and gravity theories from a 4d spectrum point of view. Our analysis has shown that QM SUSY is hidden in 4d mass spectrum of any higher dimensional field theories except for scalars. The origins of QM SUSYs in the 4d mass spectrum are found to be chiral symmetry, higher dimensional gauge symmetry and higher dimensional general covariance symmetry for spinor, gauge and gravity theories, respectively. There is no such symmetry to guarantee QM

SUSY in the 4d mass spectrum for scalar theories. QM SUSY could appear in higher dimensional scalar theories but it is an accidental symmetry.

The higher dimensional gauge invariance guarantees that the nonzero vector mode $A_{\mu,n}(x)$ $(n > 0)$ can absorb the unphysical scalar mode $h_n(x)$ to become massive with three degrees of freedom. This is the origin of QM SUSY between the mass eigenfunctions $f_n(y)$ and $g_n(y)$ for $A_{\mu,n}(x)$ and $h_n(x)$.

The higher dimensional general covariance symmetry similarly guarantees that the nonzero vector mode $h_{\mu y,n}(x)$ $(n > 0)$ can absorb the unphysical scalar mode $\phi_n(x)$ to become massive with three degrees of freedom, and that the nonzero graviton mode $h_{\mu\nu,n}(x)$ $(n > 0)$ can then absorb the massive mode $h_{\mu y,n}(x)$ to become massive with five degrees of freedom. This is the origin of QM SUSY and there appear *two* QM SUSYs in higher dimensional gravity theories: One connects the mass eigenfunction $g_n(y)$ to $k_n(y)$. The other connects the mass eigenfunction $f_n(y)$ to $g_n(y)$.

It is interesting to point out that the chiral, higher dimensional gauge and higher dimensional general covariance symmetries are all related to the symmetries that guarantee the masslessness of spinor, vector and tensor fields, respectively. Since massless particles are crucially important at low energy physics, it would be of great interest to investigate QM SUSY in more details. Our results will be summarized in the Table 1.

Table 1. Summary of our results

higher dim. fields	QM SUSY	origin
scalar	△	accidental
spinor	○	chiral symmetry
vector	○	higher dim. gauge symmetry
tensor	○	higher dim. general covariance symmetry

Acknowledgments

This work has been supported in part by a Grant-in-Aid for Scientific Research (No. 22540281) from the Japanese Ministry of Education, Science, Sports and Culture. The author would like to thank Y. Fujimoto, C.S. Lim, T. Nagasawa, S. Ohya, K. Sakamoto, K. Sekiya, H. Sonoda, K. Takenaga for valuable discussions.

References

1. M. S. Manton, *Nucl. Phys. B* **158** 141 (1979).
2. D. B. Fairlie, *Phys. Lett. B* **82** 97 (1979).
3. J. Sherk and J. Shwarz, *Phys. Lett. B* **82** 60 (1979); *Nucl. Phys. B* **153** 61 (1979).
4. Y. Hosotani, *Phys. Lett. B* **126** 309 (1983); *Ann. Phys. (N.Y)* **190** 233 (1989).
5. C. Csaki, C. Grojean, H. Murayama, L. Pilo and J. Terning, *Phys. Rev. D* **69** 055006 (2004).
6. M. Sakamoto, M. Tachibana and K. Takenaga, *Phys. Lett. B* **458** 231 (1999); *Prog. Theor. Phys.* **104** 633 (2000).
7. M. Sakamoto, M. Tachibana and K. Takenaga, *Phys. Lett. B* **457** 33 (1999).
8. S. Matsumoto, M. Sakamoto and S. Tanimura, *Phys. Lett. B* **518** 163 (2001); M. Sakamoto and S. Tanimura, *Phys. Rev. D* **65** 065004 (2004).
9. H. Hatanaka, K. Ohnishi,M. Sakamoto and K. Takenaga, *Prog. Theor. Phys.* **107** 1191 (2002); *Prog. Theor. Phys.* **110** 791 (2003).
10. K. Ohnishi and M. Sakamoto, *Phys. Lett. B* **486** 179 (2000); H. Hatanaka, S. Matsumoto, K. Ohnishi and M. Sakamoto, *Phys. Rev. D* **63** 105003 (2001).
11. M. Sakamoto and K. Takenaga, *Phys. Rev. D* **80** 085016 (2009).
12. L. Randall and R. Sundrum, *Phys. Rev. Lett.* **83** 4690 (1999).
13. H. Hatanaka, M. Sakamoto, M. Tachibana and K. Takenaga, *Prog. Theor. Phys.* **102** 1213 (1999).
14. T. Nagasawa and M. Sakamoto, *Prog. Theor. Phys.* **112** 629 (2004).
15. M. Sakamoto and K. Takenaga, *Phys. Rev. D* **75** 045015 (2007).
16. T. Nagasawa, M. Sakamoto and K. Takenaga, *Phys. Lett. B* **562** 358 (2003); *Phys. Lett. B* **583** 357 (2004); *J. Phys. A* **38** 8053 (2005).
17. T. Nagasawa, S. Ohya, K. Sakamoto, M. Sakamoto and K Sekiya, *J. Phys. A* **42** 265203 (2009).
18. C.S. Lim, T. Nagasawa, M. Sakamoto and H. Sonoda, *Phys. Rev. D* **72** 064006 (2005).
19. C.S. Lim, T. Nagasawa, S. Ohya, K. Sakamoto and M. Sakamoto, *Phys. Rev. D* **77** 065009 (2008).
20. E. Witten, *Nucl. Phys. B* **188** 513 (1981).
21. C.S. Lim, T. Nagasawa, S. Ohya, K. Sakamoto and M. Sakamoto, *Phys. Rev. D* **77** 045020 (2008).
22. A. Karch and L. Randall, *JHEP* **05** 008 (2001).
23. F. Cooper, A. Khare and U. Sukhatme, *Phys.Rept.* **251** 267 (1995).

ON CHIRAL AND NONCHIRAL 1D SUPERMULTIPLETS

FRANCESCO TOPPAN
CBPF (TEO), Rua Dr. Xavier Sigaud 150,
cep 22290-180, Rio de Janeiro (RJ), Brazil
E-mail: toppan@cbpf.br

In this talk I discuss and clarify some issues concerning chiral and nonchiral properties of the one-dimensional supermultiplets of the $\mathcal{N}$-Extended Supersymmetry. Quaternionic chirality can be defined for $\mathcal{N} = 4, 5, 6, 7, 8$. Octonionic chirality for $\mathcal{N} = 8$ and beyond. Inequivalent chiralities only arise when considering several copies of $\mathcal{N} = 4$ or $\mathcal{N} = 8$ supermultiplets.

Keywords: Supersymmetric quantum mechanics.

1. Introduction

The 1D $\mathcal{N}$-Extended Superalgebra, with $\mathcal{N}$ odd generators Q_I ($I = 1, 2, \ldots, \mathcal{N}$) and a single even generator H satisfying the (anti)-commutation relations

$$\begin{aligned} \{Q_I, Q_J\} &= \delta_{IJ} H, \\ [H, Q_I] &= 0, \end{aligned} \tag{1}$$

is the superalgebra underlying the Supersymmetric Quantum Mechanics[1].

In recent years the structure of its linear representations has been unveiled by a series of works[2−15].

The linear representations under considerations (supermultiplets) contain a finite, equal number of bosonic and fermionic fields depending on a single coordinate (the time). The operators Q_I and H act as differential operators. The linear representations are characterized by a series of properties which, for sake of consistency, are reviewed in the appendix.

The *minimal* linear representations (also called *irreducible supermultiplets*) are given by the minimal number n_{min} of bosonic (fermionic) fields for a given value of $\mathcal{N}$. The value $n_{\min}$ is given[2] by the formula

$$\begin{aligned} \mathcal{N} &= 8l + m, \\ n_{\min} &= 2^{4l} G(m), \end{aligned} \tag{2}$$

where $l = 0, 1, 2, \ldots$ and $m = 1, 2, 3, 4, 5, 6, 7, 8$.

$G(m)$ appearing in (2) is the Radon-Hurwitz function

m	1	2	3	4	5	6	7	8
$G(m)$	1	2	4	4	8	8	8	8

(3)

Non-minimal linear representations have been discussed in Refs. 4,12,13,15.

The construction of off-shell invariant actions (sigma-models) based on these representations has been given in Refs. 4,14,15. In this approach the superfields techniques that can be employed to recover invariant actions for the given supermultiplets[16] are no longer necessary.

In this talk I will address and clarify a specific issue. In Ref. 17 (see also references therein) the $\mathcal{N} = 4$ minimal linear representations (given by the field content $(4,4)$, $(3,4,1)$, $(2,4,2)$ and $(1,4,3)$) are given in two different forms, called "chiral" and "twisted chiral supermultiplets". This double realization of the $\mathcal{N} = 4$ representations can also be extended to the nonlinear realizations (the $(3,4,1)_{\mathrm{nl}}$ and the $(2,4,2)_{\mathrm{nl}}$ which are recovered from the $(4,4)$ root supermultiplet via a specific construction based on the supersymmetrization of the first Hopf fibration, see Ref. 18 and references therein).

For our purposes it is convenient to refer to this doubling as "chirality". The two versions of the supermultiplets will be conveniently denoted as "chiral" and "antichiral", respectively. Chirality seems therefore encoded in all minimal (linear and non-linear) $\mathcal{N} = 4$ representations. On the other hand, the $(4,4)$ root supermultiplet which generates all remaining $\mathcal{N} = 4$ representations (for the linear ones such as $(3,4,1)$, $(2,4,2)$ and $(1,4,3)$ through the dressing, see Ref. 2 and the appendix), is in one-to-one correspondence with the (Weyl-type) realization of the $Cl(4,0)$ Clifford algebra[2]. This realization is unique. Therefore, a natural question to be asked is whether the notion of chirality is truly there. The answer, as we will see, is rather subtle. Since chirality (if there) is inherited from the properties of the root supermultiplets, for our purpose is sufficient to address this problem for the root supermultiplets at a given $\mathcal{N}$; in particular, for $\mathcal{N} = 4$, its $(4,4)$ root supermultiplet and, for $\mathcal{N} = 8$, its $(8,8)$ root supermultiplet.

2. Quaternions, octonions and the $\mathcal{N} = 4, 8$ root supermultiplets: are they chiral?

The $\mathcal{N} = 4$ root supermultiplet of field content $(4, 4)$ admits 4 bosonic fields x, x_i and 4 fermionic fields ψ, ψ_i $(i = 1, 2, 3)$. Its supertransformations are expressed in terms of the quaternionic structure constants (δ_{ij} and the totally antisymmetric tensor ϵ_{ijk}). Without loss of generality we can explicitly realize them as

$$\begin{array}{ll} Q_4 x = \psi, & Q_i x = \psi_i, \\ Q_4 x_j = \psi_j, & Q_i x_j = -\delta_{ij}\psi + s\epsilon_{ijk}\psi_k, \\ Q_4 \psi = \dot{x}, & Q_i \psi = -\dot{x}_i, \\ Q_4 \psi_j = \dot{x}_j, & Q_i \psi_j = \delta_{ij}\dot{x} - s\epsilon_{ijk}\dot{x}_k. \end{array} \tag{4}$$

A sign $s = \pm 1$ has been introduced. It corresponds to the convention of choosing the overall sign of the totally antisymmetric tensor ϵ_{ijk}. It discriminates the chiral $(s = +1)$ from the antichiral $(s = -1)$ $\mathcal{N} = 4$ root supermultiplet. Due to its origin, we refer to this chirality as "quaternionic chirality".

A similar notion of chirality can be introduced for the $\mathcal{N} = 8$ $(8, 8)$ root supermultiplet which, without loss of generality, can be expressed[4] by replacing the quaternionic structure constant ϵ_{ijk} in (4) with the totally antisymmetric octonionic structure constants C_{ijk}, $i, j, k = 1, 2 \ldots, 7$, such that $C_{123} = C_{147} = C_{165} = C_{246} = C_{257} = C_{354} = C_{367} = 1$.

We have the following supertransformations acting on the 8 bosonic fields x, x_i and the 8 fermionic fields ψ, ψ_i:

$$\begin{array}{ll} Q_8 x = \psi, & Q_i x = \psi_i, \\ Q_8 x_j = \psi_j, & Q_i x_j = -\delta_{ij}\psi + sC_{ijk}\psi_k, \\ Q_8 \psi = \dot{x}, & Q_i \psi = -\dot{x}_i, \\ Q_8 \psi_j = \dot{x}_j, & Q_i \psi_j = \delta_{ij}\dot{x} - sC_{ijk}\dot{x}_k. \end{array} \tag{5}$$

The sign $s = \pm 1$ defines the "octonionic chirality" of the $\mathcal{N} = 8$ root supermultiplet.

It is easily realized that the chiral and antichiral supermultiplets are isomorphic and related by a Z_2 transformation. In the $\mathcal{N} = 4$ case this is easily achieved by exchanging $x_2 \leftrightarrow x_3$, $\psi_2 \leftrightarrow \psi_3$ and by relabeling the supersymmetry transformations $Q_2 \leftrightarrow Q_3$. It therefore looks that the chirality is an abusive notion which should be dismissed as useless. In the following we will prove that this is not quite so. It is certainly true that the notion of "quaternionic chirality" turns out to be useless for the $\mathcal{N} = 3$ $(4, 4)$ root supermultiplet (it is obtained from (4) by disregarding the Q_4

supertransformation). The difference w.r.t. the $\mathcal{N}=4$ case lies in the fact that the Z_2 isomorphism in this case can be imposed *without* relabeling the supertransformations (it is sufficient, e.g., to map $\psi \mapsto -\psi$, $\psi_i \mapsto -\psi_i$, while leaving x, x_i unchanged). The relabeling of the supertransformations, which is essential to implement the Z_2 isomorphism for the $\mathcal{N}=4$ quaternionic chirality (and the $\mathcal{N}=8$ octonionic chirality), makes all the difference w.r.t. the $\mathcal{N}=3$ root supermultiplet case.

3. Irreducible $\mathcal{N}=5,6,7$ supermultiplets are nonchiral, reducible supermultiplets are chiral

The $\mathcal{N}=5,6,7,8$ root supermultiplets have field content $(8,8)$. They all admit a decomposition into two minimal $\mathcal{N}=4$ supermultiplets obtained by suitably picking 4 supertransformations out of the $\mathcal{N}$ original ones. For $\mathcal{N}=8$ we have $\begin{pmatrix}8\\4\end{pmatrix}=70$ inequivalent choices of 4 supertransformations. 14 of such choices produce two minimal $\mathcal{N}=4$ supermultiplets (in the remaining 56 cases we end up with a non-minimal, reducible but indecomposable, $\mathcal{N}=4$ representation[15]). This number can be understood as follows: one can pick any 3 supertransformations (necessarily ending up with two $\mathcal{N}=3$ root supermultiplets). Then one is left with 5 possible choices for the fourth supertransformation. In 1 case ($\frac{1}{5}$ of the total) we get two separate minimal $\mathcal{N}=4$ supermultiplets. In the 4 remaining cases we end up instead with an indecomposable non-minimal $\mathcal{N}=4$ supermultiplet.

The $\mathcal{N}=8$ supertransformations in (5) can be associated with the octonions: Q_8 with the octonionic identity and the Q_i's with the 7 imaginary octonions. With this identification, the 14 combinations which produce a decomposition into two separate $\mathcal{N}=4$ supermultiplets are obtained as follows: 7 by picking up Q_8 and 3 supertransformations lying in one of the 7 lines of the Fano plane. The remaining 7 by picking up 4 supertransformations in the Fano plane which are complementary to one of the 7 lines.

Incidentally, this counting proves that for $\mathcal{N}=5$ (and, *a fortiori*, $\mathcal{N}=6,7$) one can always find a decomposition into two minimal $\mathcal{N}=4$ supermultiplets. Indeed, $\mathcal{N}=5$ supertransformations can be obtained in two ways: either, *case a*, with 5 supertransformations lying on the Fano plane or, *case b*, Q_8 and 4 supertransformations lying on the Fano plane. In case a, 4 of the 5 supertransformations are necessarily complementary to a line; in case b we have two possibilities, either the 4 supertransformations lying on the Fano plane are complementary to a line or 3 of them belong to one of the lines. In this case to get the minimal $\mathcal{N}=4$ decomposition we have to pick them together with Q_8.

Well, what all this has to do with chirality and quaternionic chirality? The fact is that, in all cases, no matter which decomposition into two minimal $\mathcal{N}=4$ supermultiplets is taken, one always ends up with two $\mathcal{N}=4$ root supermultiplets of opposite chirality. The chirality, which is irrelevant when a single supermultiplet is concerned, suddenly turns out to be crucial when several copies of them are considered. The $\mathcal{N}=5,6,7$ supermultiplets are essentially non-chiral because in their $\mathcal{N}=4$ minimal decomposition two multiplets, a quaternionic chiral and a quaternionic antichiral one, are produced. The same is true for the $\mathcal{N}=8$ root supermultiplet. It is quaternionic non-chiral in the sense here specified. For $\mathcal{N}=8$ on the other hand, another notion of chirality, the octonionic chirality, can be introduced (no such notion makes sense for $\mathcal{N}=5,6,7$). The $\mathcal{N}=8$ root supermultiplet is quaternionically non-chiral and octonionically chiral.

This result admits the following reformulation: given two minimal $\mathcal{N}=4$ supermultiplets, they can be combined into a single minimal supermultiplet for $\mathcal{N}=5,6,7$ or 8 if and only if the two supermultiplets possess opposite chirality. Indeed, if one tries to link together with an extra supersymmetry transformation two supermultiplets of the same chirality, one ends up in a contradiction.

Two chiral supermultiplets are isomorphic (via the Z_2 isomorphism) with two antichiral supermultiplets, while they are not isomorphic with a chiral and an antichiral supermultiplet, due to the fact that one cannot flip the chirality of the first multiplet without flipping the chirality of the second multiplet. The reason is that the Z_2 isomorphism requires a relabeling of the supertransformations, as discussed in the previous section, not just a transformation of the fields entering the supermultiplets (as it is the case for $\mathcal{N}=3$).

Given a certain number n of $\mathcal{N}=4$ $(4,4)$ root supermultiplets, inequivalent chiralities are discriminated by the modulus (to make it Z_2-invariant) m given by

$$m=|\sum_i s_i|=|n_+-n_-|, \tag{6}$$

where $n=n_++n_-$, with $n_\pm$ denoting the number of chiral (antichiral) supermultiplets.

For $n=1$ we have that $m=1$. The non-chiral case ($m=0$) is only possible for n even.

It is certainly possible and perhaps even convenient to think of m as an energy. In this interpretation the non-chiral state corresponds to the vacuum. For odd n, the vacuum energy is positive. The vacuum is doubly degenerated and spontaneously broken. For the single ($n = 1$) supermultiplet the two (equivalent) choices of the chirality (± 1) can be interpreted as a spontaneous breaking of the Z_2 symmetry.

For a collection of n root supermultiplets of $\mathcal{N} = 8$, the previous steps are repeated in terms of the notion of "octonionic chirality" (each $\mathcal{N} = 8$ supermultiplet is non-chiral for what concerns the quaternionic chirality).

4. Weyl on the Leibniz-Clarke debate on the nature of space

The problem of understanding the nature of chirality for quaternionic (anti)chiral $\mathcal{N} = 4$ root supermultiplets and octonionic (anti)chiral $\mathcal{N} = 8$ root supermultiplets is similar to the problem of understanding the nature of parity (mirror symmetry) of the Euclidean space. A nice framework was provided by Weyl in his popular book on Symmetry[19].

The famous Clarke-Leibniz debate concerning the nature of the space (either absolute, thesis defended by the Newtonians, Clarke was one of them) or relative (thesis defended by Leibniz who anticipated some of the arguments later used by Mach) is well-known. The Clarke-Leibniz debate was expressed in the metaphysical and theological language of the time. Weyl, in his book, reformulates the position of Leibniz (and also Kant) by expressing the relative nature of the space for the specific Z_2 parity transformation associated with the mirror symmetry. Weyl, in his argument, mimicks the theological framework used by Clarke and Leibniz.

The Weyl argument goes as follows. Let us suppose that God at the beginning creates out of nothing, in the empty space, a hand. We have no way to say whether this hand is left or right (in the empty space the hand is so good as its mirror image). Now, let us suppose that after creating the first hand, God creates a second hand. It is only after this second hand has been created that the notion of right or left has been introduced. The second hand can be aligned with the first one or be of opposite type. Right-handedness or left-handedness is a relative notion based on the referential provided by the first hand.

The argument of Weyl clearly applies to the notion of quaternionic (or octonionic) chirality for $\mathcal{N} = 4$ ($\mathcal{N} = 8$) supermultiplets. In the case of supermultiplets on the other hand we can still go further. Let us focus on the quaternionic chirality of the $\mathcal{N} = 4$ supermultiplets. We can go on with theological speculation. Let us suppose now that, after the second hand has been created (so that we have now two hands floating in the empty space) God performs a third act of creation, creating a handless body. This handless body, floating in the empty space, tries to grasp and attach the two floating hands to its handless arms. He/she can only do that if the two hands are of opposite chirality. He/she is unable to do that if they come with the same chirality. Needless to say, the handless body can stay for the full $\mathcal{N}=8$ supersymmetry which can be obtained by linking together two minimal $\mathcal{N}=4$ root supermultiplets. With this baroque image of handless bodies desperately seeking floating hands in the empty space, we can leave the Clarke-Leibniz-Weyl theological speculations.

5. Conclusions

In this paper I clarified the issue of the chirality associated with $\mathcal{N}=4$ and $\mathcal{N}=8$ supermultiplets. I pointed out that there are two notions of chirality which can be introduced: quaternionic chirality which applies to $\mathcal{N}=4,5,6,7,8$ minimal supermultiplets and octonionic chirality which applies to $\mathcal{N}\geq 8$ minimal supermultiplets. For $\mathcal{N}>4$ the minimal supermultiplets are quaternionically nonchiral (every decomposition into two minimal $\mathcal{N}=4$ minimal supermultiplets produces two supermultiplets of opposite chirality). The same is true (concerning octonionic chirality) for the minimal supermultiplets with $\mathcal{N}>8$. The notion of quaternionic chirality applies to the $\mathcal{N}=4$ minimal supermultiplets, while the notion of octonionic chirality to the minimal $\mathcal{N}=8$ supermultiplets.

In both cases a *single* chiral supermultiplet is isomorphic to its antichiral counterpart via a Z_2 transformation. Chirality cannot be detected if we deal with a single $\mathcal{N}=4$ ($\mathcal{N}=8$) supermultiplet.

On the other hand, the chirality issue becomes important when we are dealing with reducible $\mathcal{N}=4$ ($\mathcal{N}=8$) supermultiplets given by several copies of chiral and antichiral supermultiplets. Let us take the example of the reducible $\mathcal{N}=4$ representation given by two separate root supermultiplets (the total field content is $(8,8)=2\times(4,4)$). There are two inequivalent such reducible representations ($(8,8)_{\mathrm{red.,ch.}}$ and $(8,8)_{\mathrm{red.,nc.}}$). One is chiral ($(8,8)_{\mathrm{red.,ch.}}$); it is given by two $\mathcal{N}=4$ root supermultiplets of the same chirality. The other one ($(8,8)_{\mathrm{red.,nc.}}$) is nonchiral and given by two $\mathcal{N}=4$ root supermultiplets of opposite chirality. Only $(8,8)_{\mathrm{red.,nc.}}$ can be "promoted" to a minimal irreducible representation for $\mathcal{N}=5,6,7,8$ by inserting extra supertransformations linking its two $\mathcal{N}=4$ component supermultiplets.

The overall chirality of n $\mathcal{N}=4$ supermultiplets is important when constructing $\mathcal{N}=4$ off-shell invariant actions (sigma-models) for this large set of n supermultiplets.

The notion of quaternionic chirality for root supermultiplets gets extended to the remaining linear supermultiplets which are obtained by dressing (their chirality is encoded in the chirality of their associated root supermultiplets). It is also extended to the two nonlinear $\mathcal{N}=4$ supermultiplets ($(3,4,1)_{\rm nl}$ and $(2,4,2)_{\rm nl}$) which are also produced, see Ref. 18, from the $\mathcal{N}=4$ root supermultiplet. The chirality of the originating root supermultiplet has to be taken into account.

Appendix A

For completeness we report the definitions, applied to the cases used in the text, of the properties characterizing the linear representations of the one-dimensional $\mathcal{N}$-Extended Superalgebra. In particular the notions of *mass-dimension*, *field content*, *dressing transformation*, *connectivity symbol*, *dual supermultiplet* and so on, as well as the association of linear supersymmetry transformations with graphs, will be reviewed following[2–4,9,10,15]. The Reader can consult these papers for broader definitions and more detailed discussions.

Mass-dimension:

A grading, the mass-dimension d, can be assigned to any field entering a linear representation (the hamiltonian H, proportional to the time-derivative operator $\partial \equiv \frac{d}{dt}$, has a mass-dimension 1). Bosonic (fermionic) fields have integer (respectively, half-integer) mass-dimension.

Field content:

Each finite linear representation is characterized by its "field content", i.e. the set of integers $(n_1, n_2, \ldots, n_l)$ specifying the number n_i of fields of mass-dimension d_i ($d_i = d_1 + \frac{i-1}{2}$, with d_1 an arbitrary constant) entering the representation. Physically, the n_l fields of highest dimension are the auxiliary fields which transform as a time-derivative under any supersymmetry generator. The maximal value l (corresponding to the maximal dimensionality d_l) is defined to be the *length* of the representation (a root representation has length $l=2$). Either $n_1, n_3, \ldots$ correspond to the bosonic fields (therefore $n_2, n_4, \ldots$ specify the fermionic fields) or vice versa.
In both cases the equality $n_1 + n_3 + \ldots = n_2 + n_4 + \ldots = n$ is guaranteed.

Dressing transformation:

Higher-length supermultiplets are obtained by applying a dressing transformation to the length-2 root supermultiplet. The root supermultiplet is specified by the $\mathcal{N}$ supersymmetry operators $\widehat{Q}_i$ ($i = 1, \ldots, \mathcal{N}$), expressed in matrix form as

$$\widehat{Q}_j = \frac{1}{\sqrt{2}} \begin{pmatrix} 0 & \gamma_j \\ -\gamma_j \cdot H & 0 \end{pmatrix}, \quad \widehat{Q}_{\mathcal{N}} = \frac{1}{\sqrt{2}} \begin{pmatrix} 0 & \mathbf{1}_n \\ \mathbf{1}_n \cdot H & 0 \end{pmatrix}, \tag{A.1}$$

where the γ_j matrices ($j = 1, \ldots, \mathcal{N} - 1$) satisfy the Euclidean Clifford algebra

$$\{\gamma_i, \gamma_j\} = -2\delta_{ij}\mathbf{1}_n. \tag{A.2}$$

The length-3 supermultiplets are specified by the $\mathcal{N}$ operators Q_i, given by the dressing transformation

$$Q_i = D\widehat{Q}_i D^{-1}, \tag{A.3}$$

where D is a diagonal dressing matrix such that

$$D = \begin{pmatrix} \widetilde{D} & 0 \\ 0 & \mathbf{1}_n \end{pmatrix}, \tag{A.4}$$

with $\widetilde{D}$ an $n \times n$ diagonal matrix whose diagonal entries are either 1 or the derivative operator ∂.

Association with graphs:

The association between linear supersymmetry transformations and $\mathcal{N}$-colored oriented graphs goes as follows. The fields (bosonic and fermionic) entering a representation are expressed as vertices. They can be accommodated into an $X-Y$ plane. The Y coordinate can be chosen to correspond to the mass-dimension d of the fields. Conventionally, the lowest dimensional fields can be associated to vertices lying on the X axis. The higher dimensional fields have positive, integer or half-integer values of Y. A colored edge links two vertices which are connected by a supersymmetry transformation. Each one of the $\mathcal{N}$ Q_i supersymmetry generators is associated to a given color. The edges are oriented. The orientation reflects the sign (positive or negative) of the corresponding supersymmetry transformation connecting the two vertices. Instead of using arrows, alternatively, solid or dashed lines can be associated, respectively, to positive or negative signs. No colored line is drawn for supersymmetry transformations connecting a field with the time-derivative of a lower dimensional field. This is in particular true for the auxiliary fields (the fields of highest dimension in the representation) which are necessarily mapped, under supersymmetry transformations, in the time-derivative of lower-dimensional fields.

Each irreducible supersymmetry transformation can be presented (the identification is not unique) through an oriented $\mathcal{N}$-colored graph with $2n$ vertices. The graph is such that precisely $\mathcal{N}$ edges, one for each color, are linked to any given vertex which represents either a 0-mass dimension or a $\frac{1}{2}$-mass dimension field. An unoriented "color-blind" graph can be associated to the initial graph by disregarding the orientation of the edges and their colors (all edges are painted in black).

Connectivity symbol:

A characterization of length $l = 3$ color-blind, unoriented graphs can be expressed through the connectivity symbol ψ_g, defined as follows

$$\psi_g = (m_1)_{s_1} + (m_2)_{s_2} + \ldots + (m_Z)_{s_Z}. \tag{A.5}$$

The ψ_g symbol encodes the information on the partition of the n $\frac{1}{2}$-mass dimension fields (vertices) into the sets of m_z vertices ($z = 1, \ldots, Z$) with s_z edges connecting them to the $n - k$ 1-mass dimension auxiliary fields. We have

$$m_1 + m_2 + \ldots + m_Z = n, \tag{A.6}$$

while $s_z \neq s_{z'}$ for $z \neq z'$.

Dual supermultiplet:

A dual supermultiplet is obtained by mirror-reversing, upside-down, the graph associated to the original supermultiplet.

Acknowledgments

This work has been supported by CNPq.

References

1. E. Witten, *Nucl. Phys.* B 188, 513 (1981).
2. A. Pashnev and F. Toppan, *J. Math. Phys.* **42**, 5257 (2001) (hep-th/0010135).
3. M. Faux and S. J. Gates Jr., *Phys. Rev.* **D 71**, 065002 (2005) (hep-th/0408004).
4. Z. Kuznetsova, M. Rojas and F. Toppan, *JHEP* **0603**, 098 (2006) (hep-th/0511274).
5. C.F. Doran, M. G. Faux, S. J. Gates Jr., T. Hubsch, K. M. Iga and G. D. Landweber, math-ph/0603012.
6. C.F. Doran, M. G. Faux, S. J. Gates Jr., T. Hubsch, K. M. Iga and G. D. Landweber, hep-th/0611060.
7. F. Toppan, *POS* **IC2006**, 033 (2006) (hep-th/0610180).
8. F. Toppan, in *Quantum, Super and Twistors*, Proc. 22*nd* Max Born Symp., Wrocław 2006. Eds. Kowalski-Glikman and Turko, 143 (2008) (hep-th/0612276).
9. Z. Kuznetsova and F. Toppan, *Mod. Phys. Lett.* **A 23**, 37 (2008) (hep-th/0701225).

10. Z. Kuznetsova and F. Toppan, *Int. J. Mod. Phys.* **A 23**, 3947 (2008) (arXiv:0712.3176).
11. F. Toppan, *Acta Polyt.* **48**, 56 (2008).
12. C.F. Doran, M. G. Faux, S. J. Gates Jr., T. Hubsch, K. M. Iga, G. D. Landweber and R. L. Miller, arXiv:08060050.
13. C.F. Doran, M. G. Faux, S. J. Gates Jr., T. Hubsch, K. M. Iga and G. D. Landweber, arXiv:08060051.
14. M. Gonzales, M. Rojas and F. Toppan, *Int. J. Mod. Phys.* **A 24**, 4317 (2009) (arXiv:0812.3042[hep-th].
15. M. Gonzales, S. Khodaee and F. Toppan, *J. Math. Phys.* **52**, 013514 (2011) (arXiv:1006.4678[hep-th].
16. S. Bellucci, E. Ivanov, S. Krivonos and O. Lechtenfeld, *Nucl. Phys.* **B 699**, 226 (2004) (hep-th/0406015).
17. S.J. Gates Jr. and T. Hubsch, arXiv:1104.0722[hep-th].
18. L. Faria Carvalho, Z. Kuznetsova and F. Toppan *Nucl. Phys.* **B 834**, 237 (2010), arXiv:0912.3279[hep-th].
19. H. Weyl *Symmetry*, (Princeton University Press, 1952) (French edition, Flammarion, 1964).

PART E
Applied Physics

APPLICATION OF PERTURBATION THEORY TO ELASTIC MODELS OF DNA

B. ESLAMI-MOSSALLAM and M. R. EJTEHADI*
Department of Physics, Sharif University of Technology, Teheran, Iran
* *ejtehadi@sharif.edu*

In this paper, we demonstrate the applicability of the perturbation methods to different elastic models of DNA molecule. Two different kinds of perturbation methods are presented to find a first approximation for the force-extension characteristic of DNA in the anisotropic wormlike chain model, and the persistence length of DNA in the asymmetric elastic rod model. In both cases we show that it is meaningful to use the perturbation theory, and a first-order calculation is enough to find the result with an acceptable accuracy.

Keywords: DNA; Elasticity; Perturbation theory; Anisotropic wormlike chain model; Asymmetric elastic rod model.

1. Introduction

Studying the elastic behavior of DNA molecules is important for understanding its biological functions. One of the best theoretical models to explain the elastic behavior of long DNA molecules is the isotropic wormlike chain model[1], which belongs to a larger class of models named elastic rod models[2,3]. In this model it is assumed that the elastic energy is a harmonic function of the deformation, and the bending energy of DNA is isotropic. The isotropic wormlike-chain model can predict very accurately the elastic properties of long DNA molecules[1,4].

The isotropic wormlike chain model has been widely studied. Analytic or semianalytic expressions have been obtained for the distribution function of the bending angle of DNA[5], as well as the force-extension characteristic[1] and the loop formation probability.[4] However, in the more complex models, such as the anisotropic wormlike chain model[6], or the asymmetric elastic rod model[7], exact analytical calculations may become difficult. The purpose of this paper is to introduce the perturbation theory as a useful tool to obtain approximate analytical expressions for the elastic properties of DNA in these models. We study two examples in this paper: Calculation of the force-extension characteristic of DNA in the anisotropic wormlike chain model, and calculation of the persistence length of DNA in the asymmetric elastic rod model. As we will discuss, in both cases finding exact analytical results is difficult. We present perturbation methods which can be used to find approximate analytic results in the first approximation. We show that, even in the first approximation, perturbation theory can reveal some important elastic properties of DNA.

2. Models and Methods

2.1. *General formalism of local elastic rod models*

In the elastic rod models, DNA is represented by a flexible inextensible rod[2,3], which can be deformed in response of the external forces or torques. Here we use the discrete elastic rod model[3], where the rod is divided into discrete segments each representing a DNA base pair. The base pairs are considered as rigid bodies, and a local coordinate system with an orthonormal basis $\{\hat{d_1}, \hat{d_2}, \hat{d_3}\}$ is attached to each base pair. It is assumed that $\hat{d_3}$ is perpendicular to the base pair surface, $\hat{d_1}$ lies in the base pair plane and points toward the major groove, and $\hat{d_2} = \hat{d_3} \times \hat{d_1}$.
Since it is assumed that the DNA is inextensible, each base pair only has three rotational degrees of freedom. The orientation of the $k+1$-th base pair with respect to the k-th base pair is then determined by a rotation transformation $\mathbf{R}(k)$,

$$\hat{d_i}(k+1) = \mathbf{R}(k)\, \hat{d_i}(k) \qquad i = 1, 2, 3\,. \tag{1}$$

The rotation matrix $\mathbf{R}(k)$ can be parametrized by a vector $\vec{\Theta}(k)$ which is perpendicular to the plane of rotation and its magnitude is equal to the rotation angle. The components of $\vec{\Theta}$ in the local coordinate system attached to the k-th base pair are denoted by $\Theta_1(k)$, $\Theta_2(k)$, and $\Theta_3(k)$, and are called tilt, roll, and twist respectively. Tilt corresponds to the bending of the base pair over the sugar-phosphate backbone, roll corresponds to the bending in the groove direction, and twist corresponds to the rotation of the base pair about the axis which is perpendicular to its surface. These three angles can be regarded as the rotational degrees of freedom of the $k+1$-th base pair. The position of the $k+1$-th base pair with respect to the k-th base pair is denoted by the vector $\vec{r}(k)$, which is given by[3]

$$\vec{r}(k) = l_0\, \hat{d_3}(k)\,, \tag{2}$$

where $l_0 = 0.34\,\text{nm}$ is the separation between the base pairs.

For an inextensible DNA with N base pair steps, the elastic energy depends on the $3N$ rotational degrees of freedom. The *spatial angular velocity* $\vec{\Omega}$ is defined as

$$\vec{\Omega}(k) = \frac{\vec{\Theta}(k)}{l_0} \,. \tag{3}$$

The elastic energy can then be written as a function of the $\vec{\Omega}$ components. It is assumed that the interactions between the base pairs are local, i.e. each base pair only interacts with its nearest neighbor. In this case the elastic energy can be written as

$$E = \sum_{k=1}^{N} l_0 \, \mathcal{E}_k[\vec{\Omega}(k)] \,, \tag{4}$$

where $\mathcal{E}[\vec{\Omega}]$ is the elastic energy density. The elastic energy density can be expanded in a Taylor series about the equilibrium configuration. If the equilibrium configuration is denoted by $\vec{\Omega}^0(k)$, then the elastic energy density can be written in the form

$$\mathcal{E}_k[\vec{\Omega}(k)] = \sum_{p=2}^{\infty} \sum_{i_1=1}^{3} \cdots \sum_{i_p=1}^{3} \frac{1}{p!} \mathcal{K}_k^{i_1 \cdots i_p} \prod_{q=1}^{p} \Delta\Omega_{i_q}(k) \,, \tag{5}$$

where

$$\mathcal{K}_k^{i_1 \cdots i_p} = \left. \frac{\partial^p \mathcal{E}_k}{\partial\Omega_{i_1}(k) \cdots \partial\Omega_{i_p}(k)} \right|_{\vec{\Omega}(k)=\vec{\Omega}^0(k)} , \tag{6}$$

and

$$\Delta\Omega_{i_q}(k) = \Omega_{i_q}(k) - \Omega^0_{i_q}(k) \,. \tag{7}$$

The elastic energy must respect the structural symmetries of the DNA molecules. Considering a DNA segment which consists of two adjacent base pairs, it can be shown that rotating this DNA segment by 180 degrees about its local $\hat{d}_2$ axis is equivalent to the transformation $[\,\Omega_1,\, \Omega_2,\, \Omega_3\,] \to [\,-\Omega_1,\, \Omega_2,\, \Omega_3\,]$[1]. The elastic energy of the segment must be invariant under this transformation. For a homogeneous DNA, this invariance implies that odd powers of Ω_1 do not appear in the expansion of the elastic energy density. For an inhomogeneous DNA the terms which contain odd powers of Ω_1 are not necessarily excluded from the elastic energy, but, as observed in all-atom MD simulations[8], these terms are expected to be small and can be neglected.

2.2. *Wormlike chain model*

If we expand the elastic energy density in a Taylor series to the second order, we obtain the wormlike chain model which, in a simplified form, can be written as

$$\frac{\mathcal{E}^{\mathrm{WLC}}}{k_B T_r} = \frac{1}{2}\left[A_1\,\Omega_1^2 + A_2\,\Omega_2^2 + A_3\,(\Omega_3 - \omega_0)^2\right] \tag{8}$$

where k_B is the Boltzmann constant and $T_r \simeq 300°K$ is the room temperature. In equation (8) we have assumed that the DNA is homogeneous, with no intrinsic tilt and roll, and a constant intrinsic twist ω_0. We have also neglected a possible coupling between roll and twist[9]. As discussed above, the structural symmetries of the DNA forbid any twist-tilt or roll-tilt coupling of the second order.

The first two terms in equation (8) correspond to the bending energy of the DNA. The first term is the tilt energy with the bending constant A_1 and the second term is the roll energy with the bending constant A_2. The third term in equation (8) corresponds to the twist energy of DNA, with the twist constant A_3. For practical reasons, it is usually assumed that the bending energy of DNA is isotropic, i.e. it is independent of the bending direction. In this case A_1 is equal to A_2. However, it is known that in a real DNA roll requires less energy than the tilt[8,10,11]. Therefore, one expects that $A_2 < A_1$.

At the continuous limit ($l_0 \to 0$), where DNA is regarded as a continuous rod, the elastic energy in the wormlike chain model can be written as

$$\frac{E^{\mathrm{WLC}}}{k_B T_r} = \frac{1}{2}\int_0^L ds\,\left[A_1\,\Omega_1^2 + A_2\,\Omega_2^2 + A_3\,(\Omega_3 - \omega_0)^2\right] \tag{9}$$

where ds and L are the arc length parameter and the total length of the rod.

2.3. *Asymmetric elastic rod model*

DNA molecule has an asymmetric structure, in the sense that the opposite grooves of the DNA are not equal in size[10]. Thus, one expects that the energy required to bend the DNA over its major groove is not equal to the energy required to bend it over its minor groove. To account for the asymmetric structure of DNA in the bending energy, one needs a term which is an odd function of Ω_2, and does not depend on Ω_1 or Ω_3. As can be seen in equation (8), such a term does not exist in the wormlike chain model, and one must consider the higher-order terms in the expansion of elastic energy to include the effect of asymmetry. A simple local elastic rod model that has the desired properties is the asymmetric elastic rod model, which is given by[7]

$$\mathcal{E}_k^{\mathrm{a}}[\vec{\Omega}] = k_B T_r \Bigg[\frac{1}{2} A_1 \, \Omega_1^2 + \frac{1}{2} A_2 \, \Omega_2^2 + \frac{1}{3!} F^2 \, \Omega_2^3 + \frac{1}{4!} G^3 \left(\Omega_1^4 + \Omega_2^4 \right) + \frac{1}{2} A_3 \left(\Omega_3 - \omega_0 \right)^2 \Bigg] . \tag{10}$$

In equation (10), the term $+1/3! \, F^2 \, \Omega_2^3$ makes the elastic energy asymmetric. This term is included in the model with the positive sign, since it is assumed that negative roll is more favorable than positive roll[12–14]. The term $1/4! \, G^3 \, \Omega_2^4$, with positive G, preserves the stability of the elastic energy. The term $1/4! \, G^3 \, \Omega_1^4$ are added for the consistency of the model. Since there is no coupling term in the model, roll, tilt, and twist can be regarded as independent deformations, and the energy density can be decompose into three separate terms

$$\mathcal{E}^{\mathrm{a}}[\vec{\Omega}] = \mathcal{E}_1^{\mathrm{a}}[\Omega_1] + \mathcal{E}_2^{\mathrm{a}}[\Omega_2] + \mathcal{E}_3^{\mathrm{a}}[\Omega_3] , \tag{11}$$

where

$$\mathcal{E}_1^{\mathrm{a}}[\Omega_1] = k_B T_r \Big[\frac{1}{2} A_1 \, \Omega_1^2 + \frac{1}{4!} \, G^3 \, \Omega_1^4 \Big] , \tag{12}$$

$$\mathcal{E}_2^{\mathrm{a}}[\Omega_2] = k_B T_r \Big[\frac{1}{2} A_2 \, \Omega_2^2 + \frac{1}{3!} \, F^2 \, \Omega_2^3 + \frac{1}{4!} \, G^3 \, \Omega_2^4 \Big] , \tag{13}$$

$$\mathcal{E}_3^{\mathrm{a}}[\Omega_3] = \frac{1}{2} k_B T_r \, A_3 (\Omega_3 - \omega_0)^2 . \tag{14}$$

2.4. *Perturbation theory for the calculation of the force-extension relation in the anisotropic wormlike model*

Consider a DNA molecule of length L, which is stretched by the external force f in $\hat{z}$ direction. The force-extension relation is obtained by calculating $\langle z \rangle$ as a function of f, where z is the projection of the end-to-end vector of DNA along the direction of the force. To calculate $\langle z \rangle$, one must find the partition function of the stretched DNA. For the wormlike chain model in the continuous limit, it can be shown that the partition function is given by[15]

$$Z = \int \Psi(\Phi, L)\, d\Phi\,, \tag{15}$$

where $\Phi = \{\alpha, \beta, \gamma\}$ is the set of three Euler angles, and $\Psi(\Phi, L)$ satisfies a Schrodinger-like equation

$$H\,\Psi(\Phi, L) = -\frac{\partial}{\partial L}\Psi(\Phi, L)\,. \tag{16}$$

The *Hamiltonian* H is given by

$$H = \frac{J_1^2}{2A_1} + \frac{J_2^2}{2A_2} + \frac{J_3^2}{2A_3} + i\,\omega_0\, J_3 - \frac{f}{k_B T_r}\cos\beta\,, \tag{17}$$

where J_1, J_2, and J_3 are the angular momentum operators of a quantum top, which can be written in terms of the Euler angles and their derivatives[16]. Defining $A = 2(1/A_1 + 1/A_2)^{-1}$, $\lambda = (A_1 - A_2)/(A_1 + A_2)$, and $J_\pm = J_1 \pm iJ_2$, the Hamiltonian can be written in the form

$$H = H_0 + \lambda V\,, \tag{18}$$

where

$$H_0 = \frac{1}{2A}J^2 + (\frac{1}{2A_3} - \frac{1}{2A})J_3^2 + i\,\omega_0\, J_3 - \frac{f}{k_B T_r}\cos\beta\,, \tag{19}$$

and

$$V = \frac{1}{4\,A}(J_+^2 + J_-^2)\,. \tag{20}$$

It can be seen that the term $i\omega_0\, J_3$ makes the Hamiltonian non-Hermitian. In fact, the Hamiltonian commutes with the time reversal operator and belongs to a class of Hamiltonians which are called pseudo-Hermitian[17]. In the isotropic wormlike chain model, we have $\lambda = 0$, and the Hamiltonian commutes with

$$H_0^R = \frac{J^2}{2\,A} + (\frac{1}{2\,A_3} - \frac{1}{2\,A})\,J_3^2 - \frac{f}{k_B T_r}\cos\beta\,. \tag{21}$$

Therefore, one can find the simultaneous eigenvectors of H and H_0^R. Since H_0^R is Hermitian, its eigenvectors constitute a compete basis for the Hilbert space. Therefore, one can solve the differential equation (16) by expanding $\Psi(\Phi, L)$ in terms of these eigenvectors[1], as is convenient in quantum mechanics[16]. However, in the anisotropic elastic rod model, there is no guarantee that the eigenvectors of Hamiltonian form a complete basis, thus the usual methods of quantum mechanics do not necessarily work. Nevertheless perturbation theory can be used to solve equation (16), using λ as the perturbation parameter. Since $0 < \lambda < 1$, it is meaningful to use the perturbation theory.

Here we only describe the basis of the perturbation theory. The details are given elsewhere[11]. Since H^0 commutes with H_R^0, it has a complete set of eigenvectors. In Dirac notation, we denote these eigenvectors by $\langle\Phi|\hat{E}_n^0\rangle$, and the corresponding eigenvalues by $\hat{E}_n^0$. We expand $\Psi(\Phi, L)$ in terms of $\langle\Phi|\hat{E}_n^0\rangle$ in the form

$$\Psi(\Phi, L) = \sum_n C_n(L)\, \mathrm{e}^{-\hat{E}_n^0\, L}\, \langle\Phi|\hat{E}_n^0\rangle\,. \tag{22}$$

The expansion coefficient $C_n(L)$ can be expanded in powers of λ as follows

$$C_n(L) = \sum_{p=0}^{\infty} \lambda^p\, C_n^{(p)}(L)\,. \tag{23}$$

From equations (16), (22) and (23), we obtain

$$\frac{\partial}{\partial L}\, C_n^{(0)} = 0\,, \tag{24}$$

and

$$\frac{\partial}{\partial L}\, C_n^{(p)} = -\sum_{n'} \langle\hat{E}_n^0|V|\hat{E}_{n'}^0\rangle\, \mathrm{e}^{-(\hat{E}_{n'}^0 - \hat{E}_n^0)\, L}\, C_{n'}^{(p-1)} \qquad p > 0\,. \tag{25}$$

Solving equations (24) and (25), one can calculate $\Psi(\Phi, L)$ and the partition function in powers of λ. The average end-to-end extension is then obtained by differentiating the partition function with respect to the force

$$\frac{\langle z\rangle}{L} = \frac{k_B T}{L}\, \frac{\partial \ln Z}{\partial f}\,. \tag{26}$$

It can be shown that, in the first approximation, the average end-to-end extension of a long DNA is given by[11]

$$\langle z\rangle = \langle z^{(0)}\rangle + \lambda^2\, \langle z^{(2)}\rangle\,, \tag{27}$$

with

$$\frac{\langle z^{(0)}\rangle}{L} = -k_B T\, \frac{\partial}{\partial f} \hat{E}_0^0\,, \tag{28}$$

and

$$\frac{\langle z^{(2)} \rangle}{L} = -k_B T \frac{\partial}{\partial f} \Delta \hat{E}_0^0 , \tag{29}$$

where $\hat{E}_0^0$ is the smallest eigenvalue of H^0, and $\Delta \hat{E}_0^0$ is defined as

$$\Delta \hat{E}_0^0 = \sum_{n \neq 0} \left[\frac{|\langle \hat{E}_0^0 | V | \hat{E}_n^0 \rangle|^2}{(\hat{E}_0^0 - \hat{E}_n^0)} \right] \tag{30}$$

2.5. *Perturbation theory for the calculation of the persistence length in the asymmetric elastic rod model*

We define the bending angle of the DNA as

$$\cos\theta = \hat{d}_3(1) \cdot \hat{d}_3(N+1) , \tag{31}$$

where N is the number of the base pair steps in DNA. Then it can be shown that in any local elastic rod model $\langle \cos(\theta) \rangle$ decays exponentially with the DNA length[18].

$$\langle \cos(\theta) \rangle = \exp(-L/l_p) , \tag{32}$$

where l_p is the persistence length and $L = N\, l_0$ is the DNA length. In the wormlike chain model it has been shown that[19] $l_p = 2(1/A_1 + 1/A_2)^{-1}$. However, it is difficult to find an exact analytical expression for the persistence length in the asymmetric elastic rod model, because the partition function is non-Gaussian in this model. However, perturbation theory can be used to obtain an approximate analytical expression for the persistence length. The procedure is as follows: For any DNA conformation, the basis vectors of the local coordinate systems at the two ends of DNA are related together by the equation

$$\hat{d}_i(N+1) = \mathcal{R}_{ij}\, \hat{d}_j(1) \tag{33}$$

with the rotation matrix $\mathcal{R}$ given by

$$\mathcal{R} = \prod_{k=1}^{N} \exp[l_0\, \vec{K} \cdot \vec{\Omega}(k)], \tag{34}$$

where the components of the vector $\vec{K} = [K_1,\, K_2,\, K_3]$ are the generators of the $SO(3)$ group

$$[K_i]_{jk} = \varepsilon_{ijk} \,. \tag{35}$$

Then one can write

$$\langle \cos(\theta) \rangle = \langle \hat{d}_3(N+1) \cdot \hat{d}_3(1) \rangle = \langle \mathcal{R}_{33} \rangle \,, \tag{36}$$

and it can be seen from equation (32) that the persistence length is given by

$$l_p = \Big[-\frac{\partial}{\partial L} \langle \mathcal{R}_{33} \rangle \Big|_{L=0} \Big]^{-1} . \tag{37}$$

One can interpret $\mathcal{R}$ as a time evolution operator of a system with the Hamiltonian $\mathcal{H} = i\vec{K} \cdot \vec{\Omega}$. Thus $\mathcal{R}$ can be expanded in a Dyson series in the form[20]

$$\mathcal{R} = \mathbf{1} + \sum_{n=1}^{\infty} \mathcal{R}^{(n)} \,, \tag{38}$$

with

$$\mathcal{R}^{(n)} = \frac{l_0^n}{n!} \Big[\sum_{k_1=1}^{N} \sum_{k_2=1}^{N} \cdots \sum_{k_n=1}^{N} \mathcal{T} \Big\{ \hat{\Omega}(k_1)\, \hat{\Omega}(k_2) \cdots \hat{\Omega}(k_n) \Big\} \Big] , \tag{39}$$

where $\hat{\Omega}(k) \equiv \vec{K} \cdot \vec{\Omega}(k)$ and $\mathcal{T}$ is the *time ordering operator* which operates on the argument of $\hat{\Omega}$. It can be shown that $\mathcal{R}_{33}^{(1)} = 0$. Thus the persistence length is given by

$$l_p = \Big[-\frac{\partial}{\partial L} \sum_{n=2}^{\infty} \langle \mathcal{R}_{33}^{(n)} \rangle \Big|_{L=0} \Big]^{-1} . \tag{40}$$

To calculate $\langle \mathcal{R}^{(n)} \rangle$ in the asymmetric elastic rod model, we define the partition functions $\mathcal{Z}_1[\mathcal{J}_1]$, $\mathcal{Z}_2[\mathcal{J}_2]$ and $\mathcal{Z}_3[\mathcal{J}_3]$ as follows

$$\mathcal{Z}_i[\mathcal{J}_i] = \mathcal{N}_i \int \prod_{j=1}^{N} \mathrm{d}\Omega_i(j) \, \exp \Big[-l_0 \sum_{k=1}^{N} \frac{\mathcal{E}_i^a[\Omega_i(k)]}{k_B T_r} - \mathcal{J}_i(k)\, \Omega_i(k) \Big] , \qquad i = 1, 2, 3 \,, \tag{41}$$

where $\mathcal{E}_i^a$ are given in equations (12), (13) and (14), and the normalization constants $\mathcal{N}_i$ are chosen so that $\mathcal{Z}_i[\mathcal{J}_i]\Big|_{\mathcal{J}_i=0} = 1$. Then one can calculate the statistical average of any function of Ω_is by differentiating $\mathcal{Z}_i$s with respect to $\mathcal{J}_i$s. From field theory we know that the partition functions are given by[21]

$$\mathcal{Z}_i[\mathcal{J}_i] = \mathcal{N}_i \exp\left[l_0 \sum_{k=1}^{N} \mathcal{E}_i^{\text{int}}\left[\frac{\partial}{\partial \mathcal{J}_i(k)}\right]\right] \times$$

$$\exp\left[l_0 \sum_{k'=1}^{N} \left(\frac{\mathcal{J}_i^2(k')}{2A_i} + \tau_i\, \mathcal{J}_i(k')\right)\right], \tag{42}$$

where $\tau_1 = \tau_2 = 0$, $\tau_3 = \omega_0$, and we define

$$\mathcal{E}_1^{\text{int}}\left[\frac{\partial}{\partial \mathcal{J}_1(k)}\right] = \frac{G^3}{4!}\left[\frac{\partial}{\partial \mathcal{J}_1(k)}\right]^4, \tag{43}$$

$$\mathcal{E}_2^{\text{int}}\left[\frac{\partial}{\partial \mathcal{J}_2(k)}\right] = \frac{F^2}{3!}\left[\frac{\partial}{\partial \mathcal{J}_2(k)}\right]^3 + \frac{G^3}{4!}\left[\frac{\partial}{\partial \mathcal{J}_2(k)}\right]^4, \tag{44}$$

and

$$\mathcal{E}_3^{\text{int}}\left[\frac{\partial}{\partial \mathcal{J}_3(k)}\right] = 0. \tag{45}$$

One can obtain approximate analytical expressions for $\mathcal{Z}_1$ and $\mathcal{Z}_2$ using the perturbation method, as is convenient in field theory[21]. This can be done by expanding the first exponential operator on the right side of equation (42) in Taylor series. Since we want to obtain only a first-order approximation for the persistence length, we only keep the terms which are proportional to G^3, F^2 and F^4. By direct calculation it can be shown that, to this order of approximation, the contribution of $n > 2$ terms in equation (40) are smaller than the contribution of $n = 2$ term at least by a factor $\frac{l_0}{\mathcal{K}}$, where $\mathcal{K}$ is one of the harmonic elastic constants of DNA, i.e. A_1, A_2 or A_3. Since $\frac{l_0}{\mathcal{K}} \sim 10^{-3}$, these terms can be safely neglected. Calculating $\langle \mathcal{R}_{33}^{(2)} \rangle$, we obtain the first approximation for the persistence length, which is given by

$$l_p = l_p^0(1 - \Delta), \tag{46}$$

where l_p^0 is the persistence length of the DNA in the wormlike chain model, and Δ is given by

$$\Delta = \frac{1}{2}\left(\frac{F}{A_2}\right)^4\left(\frac{l_p^0}{l_0}\right) - \frac{1}{4}\left(\left(\frac{G}{A_1}\right)^3 + \left(\frac{G}{A_2}\right)^3\right)\left(\frac{l_p^0}{l_0}\right). \tag{47}$$

Perturbation theory can be used to obtain the higher-order correction to the persistence length. But it has been known that the contribution of the anharmonic effect to the persistence length is small for a real DNA[7,22] and one expects that the first approximation is enough.

An important feature of equation (47) is the appearance of the negative powers of the base pair separation l_0 in Δ, which is the consequence of renormalization of the partition function in the continuous limit. This implies that, if the base pair separation changes, the parameters F and G must be renormalized to keep the persistence length unchanged.

3. Results and Discussion

Numerical methods can be employed (see appendix B in Ref. 11) to calculate the first approximation for the average end-to-end extension of DNA in the anisotropic wormlike chain model. Figure 1 shows $\langle z^{(2)} \rangle$ as a function of the external force, for $A = 50\,\text{nm}$, $C = 100\,\text{nm}$[23], and $\omega_0 = 1.8\,\text{nm}^{-1}$. It can be seen that $\langle z^{(2)} \rangle$ is positive. Therefore, to second order in λ, anisotropy increases the average extension of DNA. However, $\langle z^{(2)} \rangle$ is much smaller than $\langle z^{(0)} \rangle$. As can be seen from Figure 1, for $A = 50\,\text{nm}$, where the isotropic wormlike chain model best fits the experimental data[1], one must measure $\frac{\langle z \rangle}{L}$ at least with the accuracy 10^{-4} to detect $\langle z^{(2)} \rangle$. Since $L \sim 10\,\mu\text{m}$ in experiments[24], minimum accuracy of 1 nm is required in measuring $\langle z \rangle$. However, the accuracy of the experiments is by far less than this limit[24], therefore $\langle z^{(2)} \rangle$ can not be detected by stretching experiments.

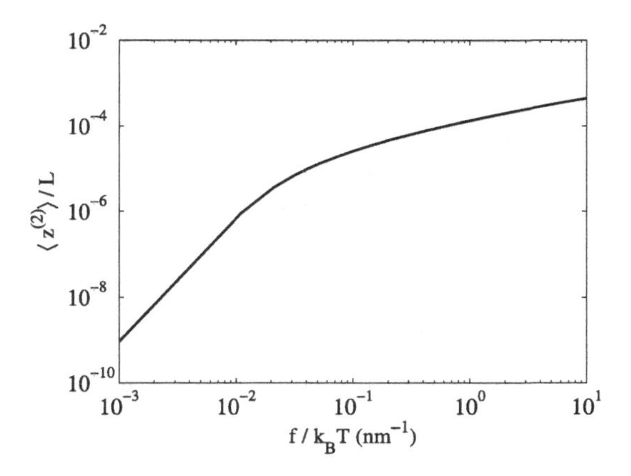

Fig. 1. $\langle z^{(2)} \rangle$ as a function of the external force f, for $A = 50\,\text{nm}$, $C = 100\,\text{nm}$, and $\omega_0 = 1.8\,\text{nm}^{-1}$.

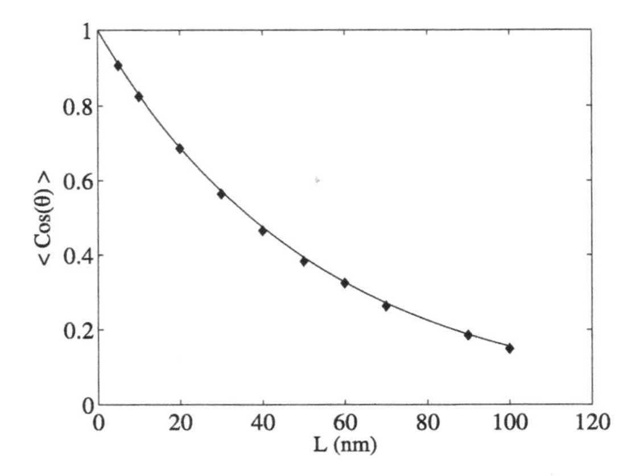

Fig. 2. $\langle\cos(\theta)\rangle$ as a function of DNA length L for the asymmetric model with $A_1 = 87\,\text{nm}$, $A_2 = 43.5\,\text{nm}$, $F = 7.90\,\text{nm}$, $G = 3.20\,\text{nm}$, $C = 100\,\text{nm}$ and $\omega_0 = 1.8\,\text{nm}^{-1}$. Data points show the MC simulation results, and solid curve show the result of perturbation theory.

Figure 2 shows $\langle\cos\theta\rangle$ as a function of DNA length L, in the asymmetric elastic rod model with $A_1 = 87\,\text{nm}$, $A_2 = 43.50\,\text{nm}$, $F = 7.90\,\text{nm}$, $G = 3.20\,\text{nm}$, $C = 100\,\text{nm}$ and $\omega_0 = 1.8\,\text{nm}^{-1}$. These parameters have been obtained from the fitting of the model[7] to the experimental data[25]. The data points correspond to the Monte Carlo simulation of the model[7], and the solid curve corresponds to equation (32), with the persistence length calculated by the perturbation method from equation (46). It can be seen that the perturbation theory gives a good approximation for the persistence length.

4. Conclusion

In this paper, we presented two different kinds of perturbation methods to find a first approximation for the force-extension characteristic of DNA in the anisotropic wormlike chain model, and the persistence length of DNA in the asymmetric elastic rod model. In both cases we showed that it is meaningful to use perturbation theory, and a first-order calculation is enough to find the result with an acceptable accuracy.

We showed that the bending anisotropy has a negligible effect on the force-extension characteristic of long DNA molecules. We also showed that, in the asymmetric elastic rod model, the anharmonic elastic constants affect the persistence length of DNA at all length scales.

References

1. J. F. Marko and E. D. Siggia, *Macromolecules* **28**, 8759 (1995).
2. J. F. Marko and E. D. Siggia, *Macromolecules* **27**, 981 (1994).
3. K. B. Towles, J. F. Beausang, H. G. Garcia, R. Phillips and P. C. Nelson, *Phys. Biol.* **6**, 025001 (2009).
4. J. Shimada and H. Yamakawa, *Macromolecules* **17**, 689 (1984).
5. N. Destainville, M. Manghi and J. Palmeri, *Biophys. J.* **96**, 4464 (2009).
6. F. Mohammad-Rafiee and R. Golestanian, *J. Phys.: Condens. Matter* **17**, 1165 (2005).
7. B. Eslami-Mossallam and M. Ejtehadi, *Phys. Rev. E* **80**, 011919 (2009).
8. F. Lankas, J. Sponer, P. Hobza and J. Langowski, *J. Mol. Biol.* **299**, 695 (2000).
9. W. K. Olson, A. A. Gorin, X. Lu, L. M. Hock and V. B. Zhurkin, *Proc. Natl. Acad. Sci.* **95**, 11163 (1998).
10. C. R. Calladine and Horace R. Drew, *Understanding DNA* (Academic Press, Cambridge, 1999).
11. B. Eslami-Mossallam and M. R. Ejtehadi, *J. Chem. Phys.* **128**, 125106 (2008).
12. F. H. C. Crick and A. Klug, *Nature* **255**, 530 (1975).
13. T. J. Richmond and C. A. Davey, *Nature* **423**, 145 (2003).
14. F. Lankas, R. Lavery and J. H. Maddocks, *Structure* **14**, 1527 (2006).
15. H. Yamakawa and J. Shimada, *J. Phys. Chem.* **68**, 4722 (1978).
16. L. D. Landau and E. M. Lifshitz, *Quantum Mechanics* (Pergamon Press, London, 1977).
17. A. Mostafazadeh, *J. Math. Phys.* **43**, 205 (2002).
18. P. A. Wiggins and P. C. Nelson, *Phys. Rev. E* **73**, 031906 (2006).
19. Davood Norouzi, F. Mohammad-Rafiee and R. Golestanian, *Phys. Rev. Lett.* **101**, 168103 (2008).
20. F. Mandl ang G. Shaw, *Quantum Field Theory* (John Wiley and sons, 1984).
21. L. H. Ryder, *Quantum Field Theory* (Cambridge University Press, Cambridge, 1996).
22. P. A. Wiggins, R. Phillips and P. C. Nelson, *Phys. Rev. E* **71**, 021909 (2005).
23. J. D. Moroz and P. C. Nelson, *Macromolecules* **31**, 6333 (1998).
24. S. B. Smith, L. Finzi and C. Bustamante, *Science* **258**, 1122 (1992).
25. P. A. Wiggins, T. van der Heijden, F. Moreno-Herrero, A. Spakowitz, R. Phillips, J. Widom, C. Dekker and P. C. Nelson, *Nature Nanotechnology* **1**, 137 (2006).

ELECTROMAGNETIC STUDIES OF IONOSPHERIC AND MAGNETOSPHERIC PERTURBATIONS ASSOCIATED WITH THE EARTH, ATMOSPHERIC AND ASTROPHYSICAL PHENOMENA

S. R. TOJIEV*, V. S. MOROZOVA†, B. J. AHMEDOV‡ and H. E. ESHKUVATOV♭

Institute of Nuclear Physics, Ulughbek, Tashkent 100214, Uzbekistan
Ulugh Beg Astronomical Institute, Astronomicheskaya 33, Tashkent 100052, Uzbekistan

*E-mail: sardor@astrin.uz; †E-mail: moroz_vs@yahoo.com
‡E-mail: ahmedov@astrin.uz; ♭E-mail: husan@astrin.uz

Here, we first analyze data from two GPS stations operating in Tashkent and Kitab and new Very Low Frequency (VLF) radio receiver operating in Tashkent for the possible earthquake ionospheric precursors. We find anomalous Total Electron Content (TEC) precursor signals and significant correlation in time between the TEC anomalies and the occurrence of the earthquake in Tashkent on August 22, 2008, $M = 4.4$. The obtained results have revealed a fine agreement with the TEC anomalies observed during the strong earthquake in Tashkent and we demonstrate the capabilities of the GPS technique to detect ionospheric perturbations caused by the local earthquakes. TEC decrease during the solar eclipse on August 1, 2008 is also obtained from the data at GPS stations in Tashkent and Kitab. We used tweek radio atmospherics (originating from lightning discharges) to estimate electron densities in the D-layer of ionosphere. The propagation characteristics of tweek atmospherics observed at the Tashkent station have been studied near the mode cut-off frequencies. It is shown that the height of the Earth-Ionosphere Wave Guide (EIWG) varies from 84 km to 91 km and nighttime electron density varies from 25 el/cm^3 to 27 el/cm^3. We have also studied VLF amplitude anomalies related to the earthquakes occurring on the path from the VLF transmitters to the Tashkent station. For analyzing narrowband data we have used the Nighttime Fluctuation method paying attention to the data obtained during the local nighttime (18:00 LT- 06:00 LT). The deathlines for rotating as well as oscillating magnetars are obtained for the different modes of oscillations and it is shown that the oscillations increase the region in the $P - \dot{P}$ (P is the period) diagram of the magnetars which is allowed for the radio emission.

Keywords: Slant TEC (sTEC); Vertical TEC (vTEC); GPS; Ionosphere; Earthquake; VLF; AWESOME; Magnetar; Oscillation; Deathline.

1. Introduction

The ionosphere is an inhomogeneous (it consists of a number of horizontal layers (D, E, F) in an altitude between 60-1000 km with varying density of charged particles), anisotropic (the refractive index depends on the propagation direction of the wave) and dispersive (the phase velocity of a wave is frequency dependent) medium. The D-layer of ionosphere is the innermost layer (from 50 km to 90 km above the surface of the Earth) where ionization is due to Lyman series-alpha hydrogen radiation. In addition, when the Sun is active with 50 or more sunspots, hard X-rays (wavelength < 1nm) ionize the air. During the night cosmic rays produce a residual amount of ionization. Recombination is high in the D-layer, thus the net ionization effect is very low and as a result high-frequency radio waves are not reflected by the D-layer. The frequency of collisions between electrons and other particles in this region during the day is about 10 MHz. The D-layer is mainly responsible for absorption of radio waves, particularly at 10 MHz and below, with progressively smaller absorption as the frequency gets higher. The absorption is small at night and greatest about midday. The layer reduces greatly after sunset, but remains due to the galactic cosmic rays.

The major source of electromagnetic energy in the lower atmosphere is the lightning discharges radiating electromagnetic waves that are the strongest in the VLF frequency range. A number of naval transmitters also operate in the VLF frequency range. The conductivity increases exponentially with height in the lower and middle atmosphere and causes attenuation and dispersion of the low-frequency waves. At radio frequencies, the lower and middle atmosphere behaves like a vacuum and due to this reason low-frequency waves could control the electrodynamics of the lower and middle atmosphere. Sub-ionospherically propagating VLF waves are uniquely suited for the investigation of the nighttime D-layer, also known as the 'ignorosphere' (40-100 km altitudes) so named due to the difficulty of making systematic measurements[1] because it is too low for satellites to complete an orbit (due to high atmospheric drag) and too high for even the largest balloons ($\sim$ 30 km) and even the highly specialized extremely high altitude aircraft ($\sim$ 20 km) and is thus only accessible by rockets in the course of a single brief traverse through the region. Unfortunately, such measurements do not allow the systematic study of variability of the region and are specifically unsuited for the localized and highly transient type of lightning-induced disturbances that are the subject of our proposed study.

Subionospheric VLF measurements are particularly relevant to the study of this altitude range due to the fact that the nighttime reflection height is in the vicinity of 80-85 km altitude[2]. Measurements of the amplitude and phase of VLF signals propagating in the earth-ionosphere waveguide have long been used effectively for remote sensing of the ionosphere[3–5].

Ionospheric anomalies, such as variations in the electron density several hours before earthquakes, are one of the precursory signals proposed. It is also shown[6–8] that the earthquakes can be preceded by the electromagnetic signals in the ULF, ELF, and VLF bands detectable from ground- and space-based measurements. Earthquakes are the explosions inside the Earth due to movement and interaction of tectonic plates, which can be characterized by the location of epicenter as well as the main parameters of the rupture (magnitude, seismic moment, source mechanism, orientation of the fault plane and direction of motion). Except for mechanical properties, there is sufficient evidence to show the ionospheric perturbations are caused by earthquakes. Even the electromagnetic emissions (ULF, ELF, VLF and HF ranges) emitted during earthquakes modify the ionosphere while propagating through it. Perturbations in VLF phase and amplitude have been reported to occur before large earthquakes, and were associated with phase and amplitude variations. The terminator time (TT), is defined as the time where a minimum occurs in the received phase (or amplitude) during sunrise and sunset. A few days before the earthquake the evening TT deviated significantly from the monthly average one[9]. Simple theory suggested that the observed effect could be explained by decreasing the VLF reflection height by up to few kms.

The Global Positioning System (GPS) derived total electron content (TEC) disturbances before earthquakes were discovered in the last years using global and regional TEC maps, TEC measurements over individual stations as well as measurements along individual GPS satellite passes. Since GPS data can be used to measure the ionospheric TEC, the technique has received attention as a potential tool to detect ionospheric perturbations related to the earthquakes. The GPS based TEC measurements have been discussed in a number of recent papers as ionospheric precursors correlated with the earthquakes[10,11].

Recent observations of quasi-periodic oscillations in the initially rising spike and decaying tail of the spectra of soft gamma-ray repeaters (SGRs) (see Refs. 12–15) let one assume that the neutron stars may be subject to the seismic events, so called glitches, giving rise to the mechanical oscillations of the stellar crust[16–18]. The study of the proper oscillations of isolated neutron stars may provide an opportunity to get important information about the internal structure of these objects what makes it an extremely interesting astrophysical task. Oscillations of the stellar crust, as in the case of the earthquakes, may cause electromagnetic events in the plasma magnetosphere of the neutron stars and have influence on the parameters of this magnetosphere. Investigations of the electrodynamics of the oscillating neutron star magnetosphere were performed in the Refs. 19–22.

Magnetars are neutron stars with very strong surface magnetic field $B_0 \simeq 10^{14} - 10^{15}$G (see Ref. 23) compared to the magnetic field of pulsars $B_0 \simeq 10^{12}$G. At the moment around 15 magnetars were observed, some of them reveal themselves in the forms of SGRs and some of them in the form of anomalous X-ray pulsars (AXPs). Magnetars have larger periods of rotation, $P \simeq 5-10$ s, than ordinary pulsars, and spin down much faster, the time derivative of period for magnetars is of the order of $\dot{P} \simeq 10^{-10} - 10^{-12}$ versus $\dot{P} \simeq 10^{-15}$ for ordinary pulsars, which gives such large values for the magnetic field evaluated as $B_0 \simeq 2(P\dot{P}_{-15})^{1/2}10^{12}$G, where $\dot{P}_{-15}$ is the period time derivative measured in fs/s.

The energetics of magnetars is not clearly understood at the moment, it may be noted that the energy stored in the magnetic field of a magnetar exceeds its rotational energy, which is not true for radio pulsars. There is no periodic radio emission from magnetars in the same range of frequencies as for the ordinary pulsars. Activity of magnetars is observed in the form of bursts, which let one assume that they may be subject to different kind of oscillations with higher probability than ordinary pulsars. It was recently shown[24] that the absence of radio emission from the magnetars is likely to be related to the slow rotation of magnetars and, consequently, low energy of the primary particles, accelerated near the surface of the star. One of the aims of this research is to explore the influence of magnetars oscillations on its magnetosphere and on the conditions for the generation of radio emission from the magnetars.

In section 2 we demonstrate TEC extraction from GPS data using dual frequency GPS signals and study the TEC data obtained at Tashkent and Kitab for the Tashkent earthquake on 22-Aug-2008 and Solar eclipse on 01-Aug-2008. The data are analyzed using mean and standard deviation around the mean approach. To identify the anomalous values of TEC during the earthquakes we calculate differential TEC (dTEC).[25] dTEC is obtained by subtracting 15 days backward running mean of vTEC from the values of observed vTEC at each epoch. This procedure removes normal variations in TEC.

In section 3 we have used tweek radio atmospherics (originating from lightning discharges) to estimate electron densities in the D-layer of ionosphere. The propagation characteristics of tweek atmospherics observed at the Tashkent VLF station have been studied near the mode cut-off frequencies. It is shown that the height of the EIWG varies from 84 km to 91 km and nighttime electron density varies from 25 el/cm^3 to 27 el/cm^3. We have also studied VLF amplitude anomalies related to the earthquakes on paths observed at the Tashkent VLF station. For analyzing the narrowband data we have used the Nighttime Fluctuation method suggested by Hayakawa where we pay much attention to the data received during the local nighttime (18:00 LT- 06:00 LT) and we estimate the mean nighttime amplitude and its nighttime fluctuation[26,27].

In the section 4 the conditions for the radio emission generation from rotating and oscillating magnetars are considered and presented in the form of deathlines separating the regions where the magnetar, in principle, could/could not produce radio emission. We summarize our main scientific results in the section 5.

2. TEC extraction

One of the most important characteristics of the Earth's ionosphere is the Total Electron Content (TEC), however, to date, over the territory of Central Asia its global monitoring was not performed. Analytical models give a good estimate of this parameter provided quiet geomagnetic conditions, but in the case of a perturbed ionosphere TEC assessment becomes less accurate. Radio-raying of the atmosphere by means of signals of satellite navigation systems and a network of ground based stations are readily available and low-cost way to monitor the F-layer of the ionosphere in real time. On the basis of this technique, we created a software to analyze data, which are provided by the international network of International GPS Service (IGS) in the standard `RINEX` format.

There appear two different TEC parameters: the vertical TEC (vTEC) that corresponds to a TEC with the integration path vertical to the Earth's surface and the slant TEC (sTEC) which corresponds to a slant integration path (Figure 1).

2.1. *Total Electron Content*

TEC is a frequently used quantity in ionospheric science. Since the number of electrons approximately equals to the number of positive ions,

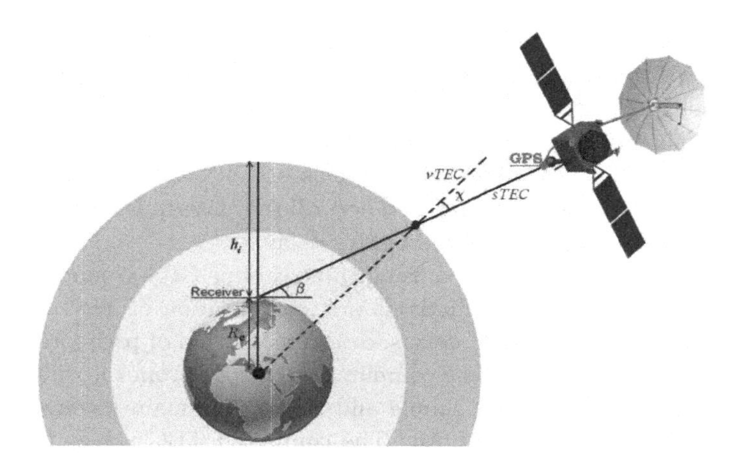

Fig. 1. Geometry of satellite-receiver.

the TEC represents a suitable parameter for the degree of ionization. The TEC is defined as the integral over the electron density distribution N_e along a defined path s:

$$\text{TEC} = \int N_e ds \ . \tag{1}$$

Since N_e is a volumetric density and TEC is defined by the integral over a path, the TEC can be thought as the total number of electrons that is contained in a volume with a cross section area being equal to 1 m^2 and length being equal to the path length. The common unit used for measuring the TEC is called Total Electron Content unit (TECU) and 1 TECU is equivalent to 10^{16} el/m^2. Depending on local time, Solar activity, geomagnetic conditions, region of the Earth, etc., the vTEC can vary from about 1 to 180 TECU.

2.2. *Method of analysis and results*

Space segment of GPS nominally consists of 24 main satellites and four spares. Spacecrafts are moving along six circular orbits at 20200 km with the inclination angle 55° and evenly spaced in the longitude by 60°. This configuration assumes that at any point on the Earth at any time in the zone of radio visibility there are 6-8 satellites which allow the continuous monitoring of the ionosphere. Each GPS satellite emits two high-stable signal at the frequencies $f_1 = 1575.42$ MHz and $f_2 = 1227.60$ MHz. The signals are refracted due to electron density gradients, and since the ionosphere is a dispersive medium, the ray paths of the f_1 and f_2 signals will be slightly different. The obtained phase and pseudo-range measurements contain information about the TEC along the ray paths. Dual-frequency group delay measurements of signals of GPS satellites can provide ionospheric delay of the signal, and accordingly determine the absolute value of TEC, which is proportional to this delay[28,29].

TEC can be computed using three various GPS-ground based measurement methods. These methods are: computation of TEC using the pseudoranges, carrier phase delays, or a combination of pseudorange and phase delay measurements. The computation of TEC from the difference of pseudoranges (P1 − P2) is simple and robust but main inconvenience is that the measurements are noisy. The computed TEC values from the difference of carrier phase delays (L1-L2) is less noisy than previous method and more immune to multipath corruption than those obtained from the pseudoranges but the computation procedure suffers from an initial phase ambiguity due to the hardware used in satellites and receivers. The third computation method aims to combine the advantages of the above methods by using both (L1-L2) and (P1 − P2) data. The (L1-L2) and (P1 − P2) data can be combined in TEC estimation by many different algorithms[30,31].

Currently, there are more than 3000 GPS stations that perform continuous monitoring on a regular basis and freely give their data to the global community. Each individual station monitors the ionosphere within a radius of more than 1000 km, including the hard to reach places.

GPS technology allows us to simultaneously measure the group (P1, P2) and phase (L1, L2) delay signals, which can be written as follows [1]:

$$\mathrm{P}_i = \rho + c(dt^{\mathrm{rec.}} - dt^{\mathrm{sat.}}) + \Delta_i^{\mathrm{iono.}} + \Delta^{\mathrm{trop.}} + \Delta^{\mathrm{instr.}} \,, \tag{2}$$

$$\mathrm{L}_i = \rho + c(dt^{\mathrm{rec.}} - dt^{\mathrm{sat.}}) - \Delta_i^{\mathrm{iono.}} + \Delta^{\mathrm{trop.}} + \Delta^{\mathrm{instr.}} + \lambda_i N_i \,, \tag{3}$$

where the index $i = 1, 2$ corresponds to the carrier frequencies f_1 and f_2; P_i and L_i are the code pseudorange measurements and the carrier phase observations respectively in distance units (to be able to combine equations for f_1 and f_2, we use P1 and P_1, L1 and L_1, etc. interchangeably); ρ is the geometrical range between satellite and receiver; c is the vacuum light speed; $dt^{\mathrm{rec.}}$ and $dt^{\mathrm{sat.}}$ are the receiver and satellites clock offsets from GPS time; $\Delta_i^{\mathrm{iono.}} = \mathrm{TEC} \cdot 40.3/f_i^2$ are the ionospheric delay terms; $\Delta^{\mathrm{trop.}}$ is the tropospheric delay; $\Delta^{\mathrm{instr.}}$ are the receiver and satellite instrumental delay; λ_i are the wavelengths; N_i are the unknown integer carrier phase ambiguities.

Combining the pseudoranges observations P_i , a TEC value is obtained

$$\mathrm{TEC_P} = 9.52 \cdot (\mathrm{P_2} - \mathrm{P_1}) \tag{4}$$

which is very noisy.

And after combination of carrier phase observations L_i we get

$$\mathrm{TEC_L} = 9.5 \cdot [(\mathrm{L_1} - \mathrm{L_2}) - (\lambda_1 N_1 - \lambda_2 N_2)] \tag{5}$$

which is less noisy than $\mathrm{TEC_P}$, but ambiguous. In practice, the calculation of TEC using the pseudorange data can only produce a noisy result. It is desirable to use in addition the relative phase delay between the two carrier frequencies in order to obtain a more precise result. Differential carrier phase provides a precise measurement of relative TEC variations. However the absolute TEC cannot be found unless the pseudorange is also used because the actual number of cycles of phase is unknown. Pseudorange gives the absolute scale for the TEC while the differential phase increases measurement precision.

***Slant and Vertical* TEC**. Slant TEC is a measure of the total electron content of the ionosphere along the ray path from the satellite to the receiver, represented in Figure 1 as the quantity sTEC. It can be calculated by using pseudorange and carrier phase measurements as described above. As sTEC is a quantity which is dependent on the ray path geometry through the ionosphere, it is desirable to calculate an equivalent vertical value of TEC which is independent of the elevation of the ray path. In order to refer the resulting vTEC to a point with specific coordinates, i.e. in order to assign the vTEC value to a specific point in the ionosphere, the so-called single-layer (or thin-shell) model is usually adopted for the ionosphere[31]. Figure 1 shows a schematic representation of this model. In this model all free electrons are contained in a shell of infinitesimal thickness at altitude h_i. This idealized layer is usually set to be at 350, 400 or 450 kilometers, approximately corresponding to the altitude of maximum electron density. Figure 1 depicts the relationship between slant(sTEC) and vertical(vTEC) TEC. Vertical TEC (vTEC) can be regarded as:

$$\mathrm{vTEC} = \mathrm{sTEC} \cdot \cos\chi \ . \tag{6}$$

From Figure 1 a relationship between the zenith angle χ and angle β can be established as:

$$\sin\chi = \frac{R_\oplus}{R_\oplus + h_i} \cos\beta \ , \tag{7}$$

where $R_\oplus$ is the mean Earth radius of 6371 km, h_i is the maximum height of the electron density.

2.3. *Ionospheric perturbations caused by the Tashkent earthquake*

On 22-Aug-2008, at 13 : 27 (LT), the $M = 4.4$ Tashkent earthquake occurred in Tashkent. The epicentre is located at the distance about few kms from the GPS station operating in Ulugh Beg Astronomical Institute. Earthquake was not deep and had a local character. This favorable setting allowed us to test the capability of the GPS system to detect the ionospheric perturbations produced by a moderate size earthquake.

Tashkent GPS station has Geographic Latitude (N) $41°19'5427''$, Longitude (E) $69°17'7671''$ and Geomagnetic Latitude 32.3, Longitude 144.4.

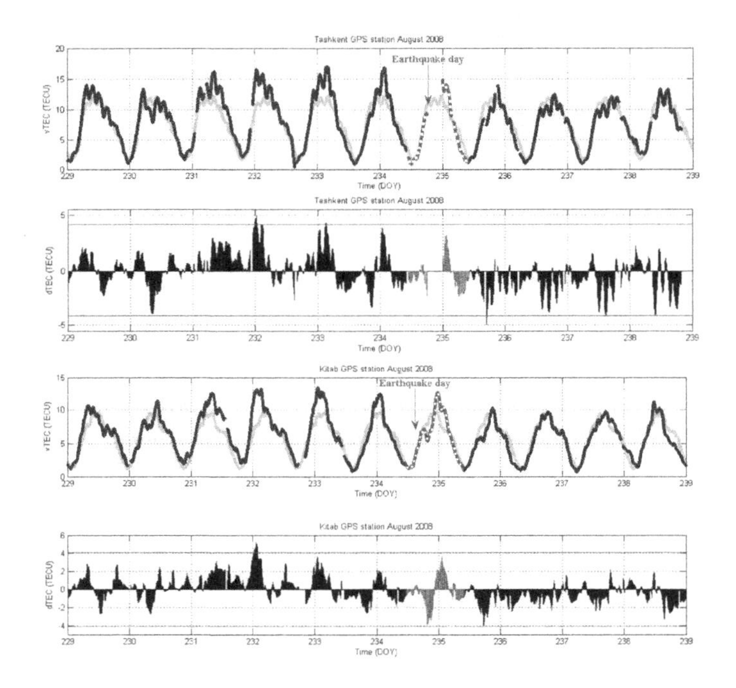

Fig. 2. TEC anomaly for the Tashkent EQ on 22-Aug-2008 [see footnote a].

2.4. *Ionospheric perturbations caused by the Solar eclipse*

The Solar eclipse of 01-Aug-2008 was a total eclipse of the Sun with a magnitude of 1.0394 that was visible from a narrow corridor through northern Canada (Nunavut), Greenland, central Russia, eastern Kazakhstan, western Mongolia and China. It belonged to the so-called midnight Sun eclipses, as it was visible from regions experiencing midnight Sun. The largest city on the path of the eclipse was Novosibirsk in Russia.

A partial eclipse was seen from the much broader path of the Moon's penumbra, including the northeast coast of North America and most of Europe and Asia. In Tashkent, Uzbekistan, the partial eclipse began at 15 h 00 m local time, with a maximum eclipse of $\sim$ 70% at 16 h 06 m, before ending at 17 h 06 m.

Figure[a] 3 illustrates the decrease of TEC during the Solar eclipse of 01-Aug-2008.

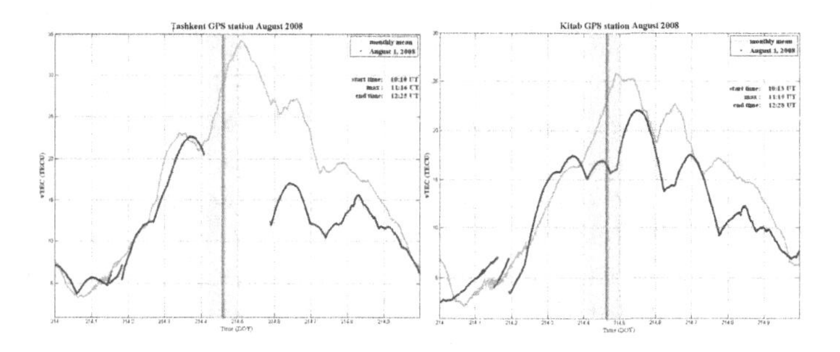

Fig. 3. Variation of TEC amplitude for Tashkent (left panel) and Kitab (right panel) GPS stations during the Solar eclipse on 01-Aug-2008 [see footnote a].

3. VLF study

VLF (3-30 kHz) and Extremely Low Frequency (ELF, 3-3000 Hz) remote sensing measurements can be conducted with Stanford-built equipment in Tashkent. The Stanford ELF/VLF radio receiver[32,33], known as the AWESOME (Atmospheric Weather Electromagnetic System for Observation, Modelling, and Education) Monitor, has been provided to Tashkent by Stanford University under the International Heliophysical Year (IHY). VLF station in Tashkent is the part of AWESOME network being operated globally to study the ionosphere and the magnetosphere with the help of electromagnetic waves in ELF and VLF bands. The results are applicable to climate change, earthquake electromagnetic precursors, lightning, radiation belt dynamics, cosmic rays, via the methods of VLF remote sensing

[a]Color versions of figures are available at http://astrin.uz/vlf/index.html.

of ionospheric perturbations. A less common but newly discovered ionospheric disturbance originates from gamma ray flares of Soft Gamma-Ray Repeaters and provides data on evolution of strongly magnetized neutron stars called magnetars.

3.1. *Experimental Setup*

VLF data are recorded with two orthogonal air core wire loop antennae. Each loop is sensitive to one component of the horizontal magnetic field, enabling directional information to be extracted from the combination. Incoming VLF signals induce voltages in the loop of wire. Using a phase-locked loop, the 1 PPS GPS signal is used to generate a 100 kHz sampling signal, which is fed into a PCI card inside the computer which digitizes and stores the data from both antennae. Software developed by Stanford sets the schedule for recording.

Figure 4 also shows the sample data taken with the AWESOME receiver at the Tashkent VLF station. The top left panel shows the data in the form of a spectrogram, where the strength of the signal as a function of time is divided into frequency bins and indicated with a color scale. The vertical lines are short impulsive radiation from lightning strikes, which could be anywhere in the world, and are known as radio atmospherics, or sferics. These sferics propagate efficiently in the Earth-ionosphere waveguide, to global distances (10 Mm or more from the source), and can reveal properties of the originating lightning stroke, and the ionosphere along the propagation path. An example of a sferic is shown in the lower left panel of the Figure 4. Horizontal lines are VLF transmitter signals, originating from all over the world, and used for long-range communication with submerged submarines. The zoom-in on the top right shows one such transmitter signal, originating from Germany, with the minimum-shift-keying (MSK) modulation signal apparent as up and down frequency shifts. The software written by Stanford determined this MSK pattern, and subtracts out the associated phase shifts in real time, so that a demodulated phase can be extracted and tracked, along with the amplitude of the signal. Because these VLF transmitter signals are guided by the lower ionosphere (being typically reflected between 70 and 85 km during the daytime and nighttime, respectively), they are extremely sensitive to ionospheric disturbances, and are therefore a unique form of ionospheric diagnostics.

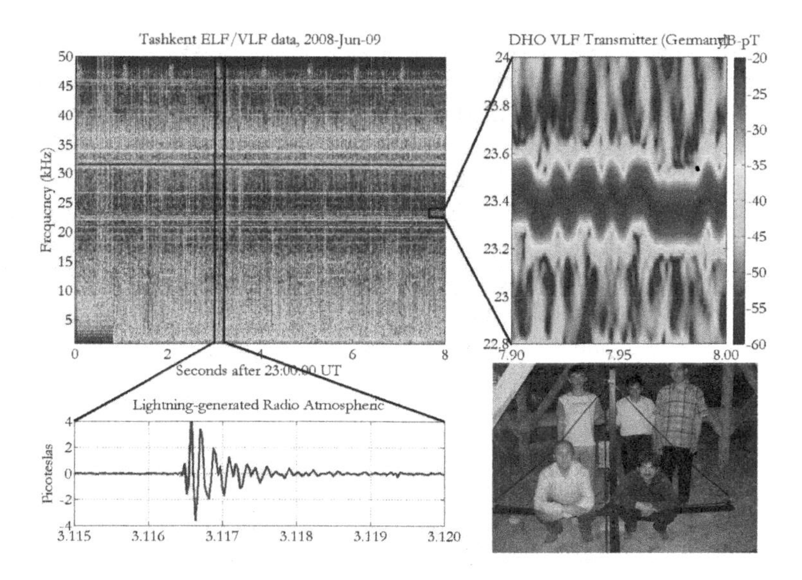

Fig. 4. Data taken with VLF antenna in Tashkent. The top left plot shows a spectrogram, in which the frequency content is divided for individual time bins, and the strength is indicated by the color scale. The bottom plot shows a time-series zoom-in of a radio atmospheric, i.e., short impulsive radiation from a lightning stroke which may be at global distances. The right plot shows a VLF transmitter signal, in this case originating from Germany, with the MSK communication signal evident by the up and down frequency changes. Since both the radio atmospherics and the VLF transmitter signal rely on the D-layer for propagation, the received signals are both extremely sensitive to the various ionospheric disturbances described herein [see footnote a].

3.2. *Tweek radio atmospherics*

Tweek radio atmospherics are electromagnetic pulse signals launched by individual lightning discharges with the certain frequency dispersion characteristics at lower ends. The return strokes of the lightning discharges are powerful transmitters of electromagnetic energy over wide electromagnetic spectrum extended from few Hz[34] to few tens of MHz[35]. The waves in ELF/VLF range propagate by multiple reflection in the natural waveguide formed by Earth and lower boundary of the ionosphere. These waves travel long distances with very low attenuation rate (2-3 dB/1000 km)[36] and are observed around the world from their source lightning discharge. This waveguide mode propagation causes an appreciable dispersion near the cutoff frequency of EIWG around 1.8 kHz[37]. These dispersed sferics are known as “tweeks”. The name tweek is because they sound like “tweek” when the signal is presented to loud speaker. The fact that these waves from lightning discharges are reflected by D-layer makes them important tool for ionospheric studies in this altitude range. Example Spectrograms (frequency vs time) of tweeks recorded at Tashkent VLF station on July 2008 are shown in Figure 5. The diagram of tweeks at about 10.22 s (Fig. 5.a) shows the frequency dispersion at cut-off frequencies near 1.61 kHz and 3.29 kHz. Frequency dispersion is caused by the wave group velocity becoming very small near the cut-off frequencies due to multiple reflections between the D-layer of ionosphere and Earth’s surface. Generally, more frequent tweek atmospherics are received at night, when attenuation is less.

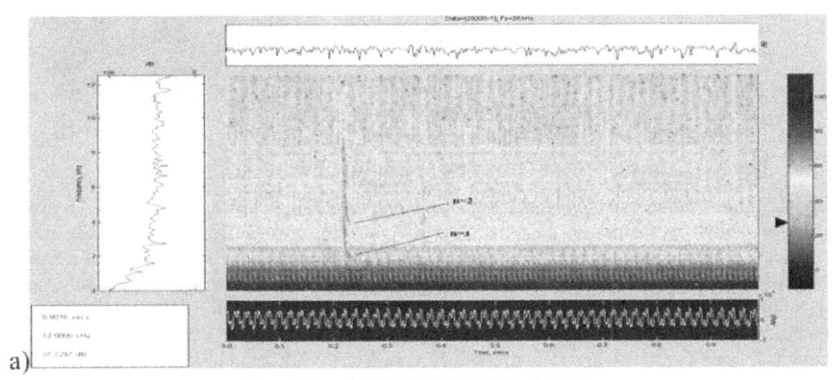

Tashkent 2008.07.02 22:00:10

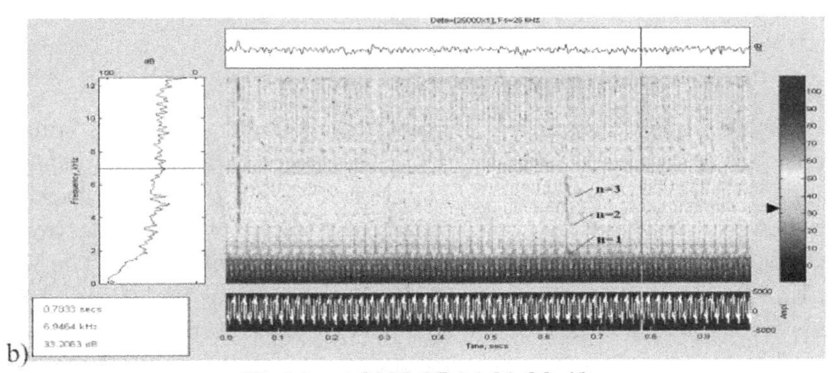

Tashkent 2008.07.14 21:00:40

Fig. 5. Examples of dynamic spectrograms of tweeks observed during the nighttime at Tashkent VLF station on July 2008 [see footnote a].

We have calculated the reflection heights $h = c/(2f_c)$ with help of the accurate reading of the first-order mode cut-off frequency of the tweeks. From this reading, we have estimated the equivalent electron densities $n_e = 1.241 \times 10^{-8} f_c f_H$ at the reflection heights.

In the present study $\sim$ 274 tweeks were recorded during the nights of July 2008 at Tashkent. Dynamic spectrogram of the tweeks observed on 02-Jul-2008 and 14-Jul-2008 is shown in Figure 5. The time duration of the tweek events considered for present study is 12 hours from 06 PM (Tashkent LT) in the evening till 06 AM (LT) next morning. During the observations, tweeks were observed up to the 6th harmonic. More tweeks were observed mainly in post midnight period. Further, tweeks occur mainly during the nighttime as it is difficult to find tweek during the daytime due to the higher attenuation which mainly depends on the collision frequency of ionospheric constituents.

The collision frequency in the daytime is $\sim 10^7$Hz when the ionosphere is at the distance 60-65 km whereas during the night it is about 10^5Hz when the ionosphere evolves up to 90 km.

The variation in ionospheric electron density at different times of the day is different. We separate nighttime to three parts: 1) Dusk (18:00-21:00 LT); 2) Night (21:00-03:00 LT); and 3) Dawn (03:00-06:00 LT). Dusk and Dawn periods of the local night have more attenuation than Night time period. From the table it is also clear that 79.19% number of tweeks observed during the Night period and few tweeks (only 1.46%) observed in the Dusk and 18.97% tweeks in the Dawn period. Higher mode harmonics are observed during the Night and Dawn periods.

Analysis of tweeks observed during the night has shown that the average cutoff frequency for the first mode is 1.6977 kHz. The nighttime ionospheric reflection height estimated from tweeks varies in the range of 84-91 km. Figures 6 a) and b) show the nighttime variation in the reflection height and the electron density estimated from the tweeks observed during July 2008. The heights gradually increase after sunset until midnight when it suddenly decreases conversely to the electron density. During the above days the fundamental mode ($n = 1$) of $\sim$ 93 selected tweeks are analyzed to determine ionospheric reflection height and electron density at the EIWG. Figures 6 a) and b) show that the boundary of the lower ionosphere varies from $\sim$ 84 to 91 km. After the Dusk hours the height of lower ionosphere starts to increase and the behavior of the electron density variation starts. The electron density is found to be in the range of $\sim 25 - 27\,\mathrm{el/cm^3}$.

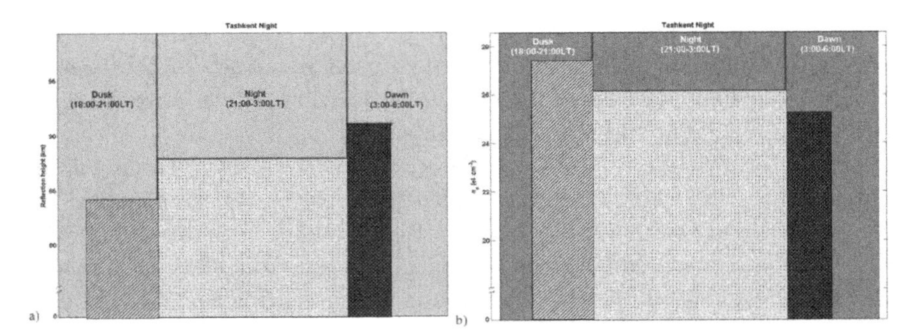

Fig. 6. Figures a and b show the ionospheric reflection height (in km) and electron density at these height (in el/cm^3) at Tashkent VLF site, correspondingly [see footnote a].

3.3. *Studying earthquakes using VLF narrowband data*

VLF subionospheric radio sounding has been discussed in a number of recent papers as a prospective tool for remote detection of perturbations in the lower ionosphere connected with seismic processes[38,39]. The most convincing result on the seismo-ionospheric perturbations with VLF sounding was obtained by Hayakawa (1996) for the Kobe earthquake in year 1995 using method which called TT (Terminator Time) method. They found significant shifts in the terminator times before the earthquake. The morning time shifts to early hours, and evening time shifts to later hours. Here we used another method of VLF data analysis, which is called Nighttime Fluctuation method[26,27] and we pay attention to the data obtained during the local nighttime and we estimate the mean nighttime amplitude and nighttime fluctuation[40,41]. A residual signal of amplitude dA as the difference between the observed signal intensity (amplitude) and the average of several days preceding or following the current day:

$$dA(t) = A(t) - < A(t) >, \tag{8}$$

where $A(t)$ is the amplitude at the time t on a current day and $< A(t) >$ is the corresponding average at the same time t for ± 15 days (15 days before, 15 days after the earthquake and earthquake day). We estimate the following two physical quantities of amplitude: (1) trend (as the average of nighttime amplitude) and (2) nighttime fluctuation ($N.F.$)

$$\text{trend} = \frac{\int_{T_s}^{T_e} dA(t)dt}{T_e - T_s}, \quad N.F. = \int_{T_s}^{T_e} (dA(t))^2 dt \ , \tag{9}$$

where T_s and T_e are the times of starting and ending the nighttime in our analysis and we have selected $T_s = 18$ h LT and $T_e = 06$ h LT. Further we used norm of these two quantities: normalized trend and normalized $N.F.$. When we take an earthquake with a particular date, we estimate the trend on this day and we then calculate the average $<$ trend $>$ over ± 15 days around this date. Then the normalized trend is defined as (trend$-$ $<$ trend $>$)$/\sigma_T$ (σ_T is the standard deviation over ± 15 days around the current date). The same principle is applied to nighttime fluctuation to obtain the normalized $N.F.$

Table 1. The parameters of $M > 6.0$ earthquakes in year 2009.

EQ	YY	MM	DD	Lat.	Lon.	M	Place
1.	2009	08	09	33.167 N	137.941 E	7.1	JAPAN
2.	2009	08	10	34.743 N	138.264 E	6.1	JAPAN
3.	2009	08	12	32.821 N	140.395 E	6.6	JAPAN

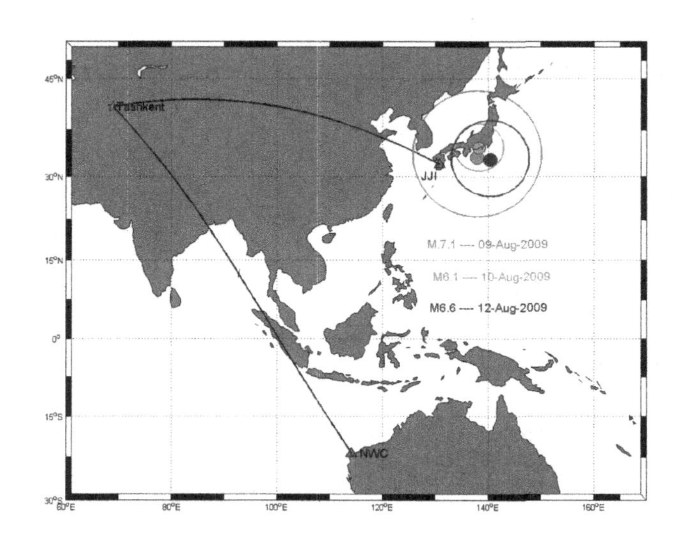

Fig. 7. Map for Tashkent VLF station and Japan earthquakes epicenters [see footnote a].

We have analyzed three Japan earthquakes (EQs) that occurred near Japan transmitter JJI in year 2009 with magnitude more than 6 observed at Tashkent VLF station (Table 1). The amplitude data are analyzed only for the reason that perturbations are identified more clearly in the amplitude data than in phase data. In the Figure 7 locations of VLF transmitters, the receiver and EQ epicenters are presented. It is seen from the figure that the epicenters of EQs on 10-Aug-2009 and 12-Aug-2009 are more distant from the transmitter and preparation zone circle also is not crossed with transmitter-receiver signal path and these EQs may not affect the VLF signal amplitude. But for the EQ on 09-Aug-2009 we have observed significant changes in amplitude parameters (trend, *N.F.*, normalized *N.F.* etc).

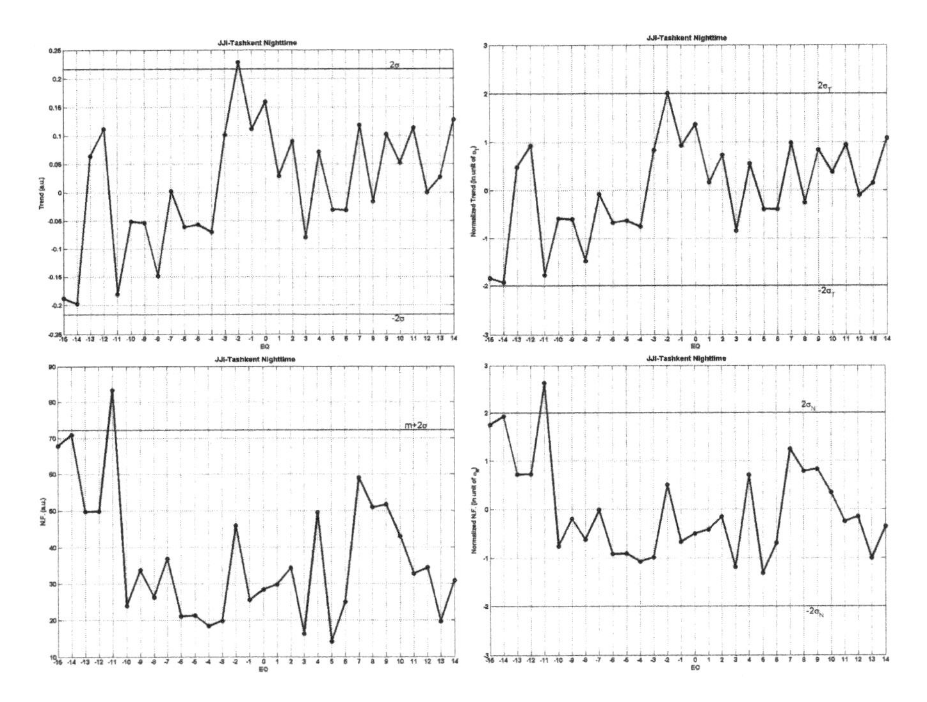

Fig. 8. The nighttime average amplitude (or trend) and nighttime fluctuation variation for Japan EQ on 09-Aug-2009. Top panel indicates trend (left) and normalized trend (right). Bottom panel indicates N.F. (left) and normalized N.F. (right).

4. Deathline for the rotating and oscillating magnetars

In the paper 24 the distribution function of electrons and positrons in the magnetar magnetosphere was investigated and it was shown that the possible Lorentz factor of the particles in electron-positron plasma is restricted by the boundary values γ_{min} and γ_{max}.

The maximum value γ_{max} is determined by the energy k of the photons producing the electron-positron pairs in the neutron star magnetosphere. These photons are emitted by the primary particles, which are pulled out from the surface of the neutron star and accelerated to ultrarelativistic velocities in the close vicinity of the surface of the star due to the presence of the unscreened component of the electric field parallel to the magnetic one in this region. The characteristic energy of these photons, called curvature photons, is given by $k = 3\gamma_0^2/2\rho$ where γ_0 is the Lorentz factor of the primary accelerated particles and ρ is the radius of curvature of the magnetic field lines in the point of emission. The creation of the electron-positron pair by the single photon moving in the magnetic field of the star is possible if the angle χ between the trajectory of the photon and the magnetic field lines reaches some threshold value $\chi_t = 2/k$. As it was shown in Ref. 42 the energy of the photon generating the pair is distributed between the electron and positron almost evenly and the energy of created particles is determined by the expression $\gamma = 1/\chi$. Thus, one may find the maximum value of the energy of the produced particles as $\gamma_{max} = 1/\chi_t \approx 3\gamma_0^2/4\rho$, which, obviously, decreases with increasing radius of curvature of the magnetic field lines and depends on the energy of the primary accelerated particles.

The minimum value of the energy of particles in the electron-positron plasma magnetosphere of the neutron star γ_{min} is determined by the geometry of the magnetic field of the neutron star and increases with the increasing radius of curvature of magnetic field lines. It may be found through the relation $\gamma_{min} = \rho_0/z_0$ (Ref. 24), where ρ_0 is the radius of curvature of the magnetic field lines at the point of curvature photon emission and z_0 is the distance to this point from the center of the star along the dipole axis. Equating γ_{min} and γ_{max} one may get the condition for the minimum value of the magnetic field for which the production of the electron-positron pairs in the magnetosphere of the neutron star is still possible.

It is shown in Ref. 24 that the condition for equality of γ_{min} and γ_{max} is equivalent to the condition $l_{f\ min} = R_s$, where l_f is the mean free path of the photons emitted by primary particles near the stellar surface and R_s is the stellar radius. Using this relation in the paper 24 the minimum value of magnetic field for effective electron-positron plasma production in the magnetosphere of the rotating magnetar was found.

Here we are going to consider a magnetar which is subject to toroidal oscillations, in spherical coordinates (r, θ, ϕ) described by the velocity field

$$\delta v^{\hat{i}} = \left\{0, \frac{1}{\sin\theta}\partial_\phi Y_{lm}(\theta,\phi), -\partial_\theta Y_{lm}(\theta,\phi)\right\} \tilde{\eta}(r) e^{-i\omega t}, \qquad (10)$$

where $\tilde{\eta}$ is the amplitude of the velocity of oscillations, ω is the frequency of oscillations and the spherical orthonormal functions $Y_{lm}(\theta, \phi)$ are the eigenfunctions of the Laplacian in spherical coordinates.

The minimum possible mean free path of curvature photons can be found using the relation $l_f = \rho/\gamma_{max}$ (see Ref. 24) with $\gamma_{max} = 3\gamma_0^3/4\rho$, taking into account that γ_0 is determined by the scalar potential accelerating the first generation of particles $\gamma_0 = |\Psi(\theta, \phi)|$. Considering the distances $(r - R_s)/R_s > 1$ one may get the expression for the scalar potential in the polar cap region of rotating and oscillating magnetar magnetosphere in the following form[22]

$$\Psi(\theta, \phi) = \frac{B_0}{2}\frac{R_s^3}{R_c^2}\kappa\left[1 - \left(\frac{\theta}{\Theta_0}\right)^2\right] - e^{-i\omega t}\tilde{\eta}(R_s)\frac{B_0 R_s}{c}\sum_{l=0}^{\infty}\sum_{m=-l}^{l} Y_{lm}(\theta, \phi) , \tag{11}$$

where Θ_0 is the polar angle of the last open field line at the surface of the star, determining the size of the polar cap, $R_c = c/\Omega$ is the radius of light cylinder, Ω is the angular velocity of neutron star rotation, $\kappa \equiv \varepsilon\beta$, $\varepsilon = 2M/R_s$ is the compactness parameter (in geometrized units) and $\beta = I/I_0$ is the stellar moment of inertia in the units of $I_0 = MR_s^2$. For the particular mode (l, m), using for small polar angles θ the approximation $Y_{lm}(\theta, \phi) \approx A_{lm}(\phi)\theta^m$, dropping the time dependence of the oscillatory term (considering the moment of time $t = 0$), and introducing the new coordinate $\xi = \theta/\Theta_0$, equation (11) may be rewritten as

$$\Psi(\xi, \phi)_{lm} = \frac{B_0}{2}\frac{R_s^3}{R_c^2}\kappa\left(1 - \xi^2\right) - \tilde{\eta}(R_s)\frac{B_0}{c}\frac{R_s^{\frac{m}{2}+1}}{R_c^{\frac{m}{2}}}\xi^m A_{lm}(\phi) , \tag{12}$$

where one takes into account that $\Theta_0 = \sqrt{R_s/R_c}$.

Using equation (12) for the scalar potential in the polar cap region of the neutron star magnetosphere, one may obtain the expression for the mean free path of the curvature photon in the form

$$l_f = \frac{R_s R_c \left(\frac{4}{3}\right)^3 \xi^{-2}}{\left\{\left|\frac{B_0}{2}\frac{R_s^3}{R_c^2}\kappa\left(1 - \xi^2\right) - \tilde{\eta}(R_s)\frac{B_0}{c}\frac{R_s^{\frac{m}{2}+1}}{R_c^{\frac{m}{2}}}\xi^m A_{lm}(\phi)\right|\right\}^3} . \tag{13}$$

The equation describing the minimum of l_f has the form (for $m \neq 0$)

$$8a\xi_{min}^2 + (2b + 3mb)\xi_{min}^m = 2a , \tag{14}$$

where

$$a = \frac{B_0}{2}\frac{R_s^3}{R_c^2}\kappa , \qquad b = \tilde{\eta}(R_s)\frac{B_0}{c}\frac{R_s^{\frac{m}{2}+1}}{R_c^{\frac{m}{2}}}A_{lm}(\phi) . \tag{15}$$

For the modes with $m = 1$ the solution of equation (14) is given by

$$\xi_{min} = \frac{-5b + \sqrt{25b^2 + 64a^2}}{16a} , \tag{16}$$

while for the modes with $m = 2$ the minimum of photon mean free path corresponds to the values of ξ

$$\xi_{min} = \frac{1}{2}\sqrt{\frac{a}{a+b}} . \tag{17}$$

In the absence of oscillations $b = 0$ both cases give the minimum of l_f at the value $\xi_{min} = 1/2$, corresponding to the case of pure rotation[24].

Using obtained minima and performing averaging on the azimuthal angle ϕ one may obtain the analog of the deathline for magnetars, following Ref. 24, in the form (for $m \neq 0$)

$$B_0 > \frac{2^{-\frac{8}{3}} 6\pi}{\left\{\int_0^{2\pi} \xi_{min}^{2/3} \left|\kappa(1-\xi_{min}^2) - 2\frac{\tilde{\eta}(R_s)}{c}\left(\frac{R_s}{R_c}\right)^{\frac{m}{2}-2} \xi_{min}^m A_{lm}(\phi)\right| d\phi\right\}} \times \left(\frac{P}{1s}\right)^{\frac{7}{3}} \left(\frac{R_s}{10km}\right)^{-3} 10^{12}\mathrm{G} . \tag{18}$$

Using the expression for the magnetic field of the neutron star determined from the assumption that the star spins down due to the generation of magnetodipole radiation $B_0 \simeq 2(P\dot{P}_{-15})^{1/2}10^{12}$G, where P is the period of neutron star rotation, $\dot{P}$ is its time derivative and $\dot{P}_{-15}$ is the period time derivative measured in fs/s, one may get (see also Ref. 24) the equation describing the deathline, i.e. the line in the $P - \dot{P}$ diagram separating the regions where the neutron star can/cannot produce radio emission:

$$\log \dot{P}_{-15} = \frac{11}{3}\log P - 0.6 + \log(C^2) , \tag{19}$$

$$C = \frac{2^{-\frac{8}{3}} 6\pi}{\left\{\int_0^{2\pi} \xi_{min}^{2/3} \left|\kappa(1-\xi_{min}^2) - 2\frac{\tilde{\eta}(R_s)}{c}\left(\frac{R_s}{R_c}\right)^{\frac{m}{2}-2} \xi_{min}^m A_{lm}(\phi)\right| d\phi\right\}} . \tag{20}$$

Figure 9 illustrates the position of deathline for the rotating as well as oscillating magnetars for two modes of oscillation and different values of the parameter $K = \tilde{\eta}(R_s)/(\Omega R_s)$ equal to the ratio between the velocity of oscillation and the linear velocity of magnetar rotation. It is seen from the figures that for increasing velocity of oscillation the deathline slightly changes slope and goes down, thus, enlarging the region possible for radio emission. The values of P and $\dot{P}$ for the magnetars are taken from the ATNF catalog[43].

The deathlines for modes with $m = 1$ are presented in the Fig. 9 as one may expect the strongest effect on radio emission for these modes. To see this more clearly one may refer to Fig. 10, which shows the velocity field for the different modes of stellar oscillations (see also Ref. 19). The electron-positron plasma in the neutron star magnetosphere is expected to be continuously generated in the polar cap region of the magnetosphere, where magnetic field lines are open and the plasma may freely escape from the surface of the star to infinity. When the star rotates, the linear velocity of the motion of the stellar surface is proportional to the distance from the axis of rotation to the considered point. In the polar cap region, thus, this velocity is proportional to the polar angle, which is small within the polar cap region. It is seen from the Fig. 10 that the velocity distribution for the oscillatory modes with $m = 0$ in the polar cap region (the top of each sphere) has the same form as for the case of pure rotation. For the modes $m = 1$, instead, the velocity of oscillations remains almost constant across the polar cap region. It was shown in the paper 22 that oscillations with modes $m = 1$ considerably increase the electromagnetic energy losses from the polar cap region of the neutron stars, which may be several times larger than for the case of pure rotation.

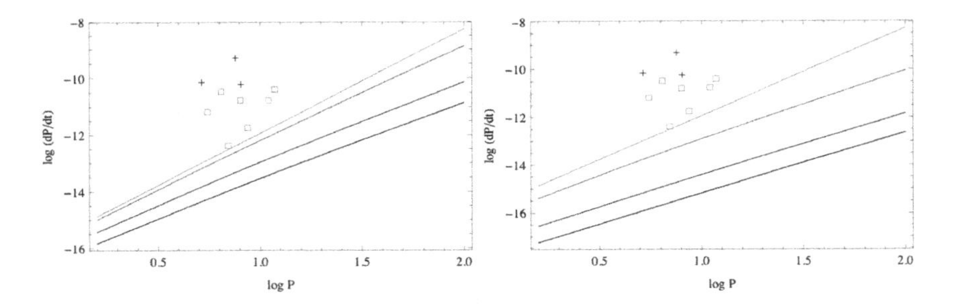

Fig. 9. The picture of deathlines for rotating and oscillating magnetars in the $P - \dot{P}$ diagram. The left panel corresponds to the mode $(1,1)$ and values of $K = 0\ ,0.07\ ,0.1$ (lower lines correspond to greater K). The right panel corresponds to the mode $(2,1)$ and values of $K = 0\ ,0.03\ ,0.07\ ,0.1$ (lower lines correspond to greater K). Other parameters are taken to be $R_s = 10\text{km}$ and $\kappa = 1$, the values of coefficients A_{lm} are equal to $A_{11} = -\sqrt{\frac{3}{8\pi}}\cos\phi$ and $A_{21} = -3\sqrt{\frac{5}{24\pi}}\cos\phi$. Crosses and squares indicate the position of soft gamma-ray repeaters and anomalous X-ray pulsars, correspondingly.

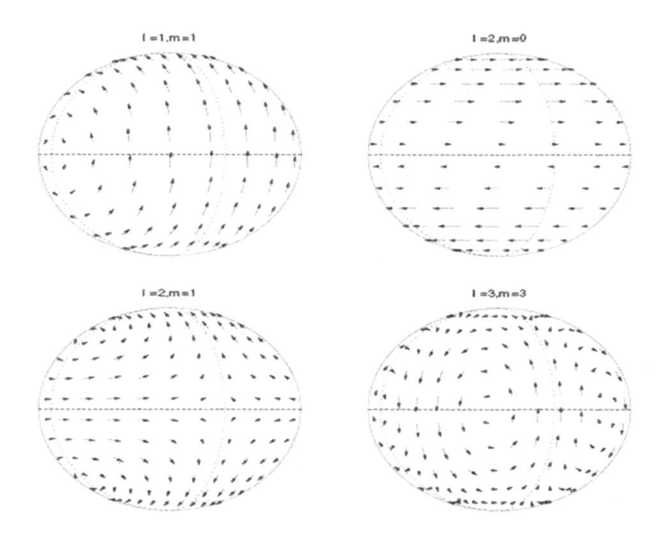

Fig. 10. The velocity distribution for the different modes of stellar oscillations.

5. Conclusion

The results obtained show the principal possibility to detect the earthquake ionospheric precursors before the occurrence of earthquakes (in both GPS and VLF data analysis methods). These results could be useful in giving early warning about the occurrence of earthquakes and the associated risks in Tashkent and Samarkand (near to GPS station in Kitab) could be minimized. The VLF study shows how lightning discharge-generated tweek radio atmospherics find application in diagnostics of lower D-layer of the ionosphere over Central Asia territory. It is found that the propagation anomaly exceeding the 2σ criterion indicating the presence of ionospheric perturbation is significantly correlated with the earthquakes. On the basis of the Tashkent and Kitab GPS data analysis for year 2008 and VLF data analysis obtained at the Tashkent station we can state:

- Ionospheric TEC anomalies have started 3 days before Tashkent, 22-Aug-2008 earthquake. Ionospheric precursor of Tashkent, 22-Aug-2008 earthquake had local character. A similar anomaly was detected in Kitab GPS station which is at the distance of about 300 kms from the epicenter.
- The correctness of applied method of TEC extraction from GPS data is confirmed by the sharp decrease of TEC during the Solar eclipse of 01-Aug-2008.

- Tweeks occur only in the nighttime (19:00 hrs to 05:00 hrs). Tweeks with higher than third harmonics are prominent mostly in the post-midnight period. Observations of higher (above 6th) harmonic tweeks are held. Attenuation is higher in the daytime than in the nighttime, which explains almost no tweek occurrence in the daytime.
- Temporal evolution of lower ionosphere during the night of July, 2008 which is geomagnetically a quiet period is calculated from the observation of tweeks atmospherics. The reflection height was found to vary between 84-91 km and electron density varied in the range of $\sim 25-27\,\mathrm{el/cm^3}$.
- The mean nighttime amplitude (or trend) and normalized trend are found to increase significantly before the EQs with the same tendency as the *N.F.* and normalized *N.F.*. The obtained results have revealed a fine agreement with VLF amplitude anomalies observed in Tashkent VLF station during the strong earthquakes occuring on the path from the transmitters to the receiver.

It is shown that the oscillations of magnetars may have noticeable influence on the conditions of radio emission generation in their magnetospheres. Increasing the scalar potential in the polar cap region of the magnetar magnetosphere increases the energy of primary particles pulled out and accelerated in the vicinity of the stellar surface, thus, increasing the probability of effective electron-positron plasma generation around the magnetar. The largest effect is expected to be for the oscillatory modes with $m = 1$ which has specific velocity field, having no dependence on the small polar angle in the polar cap region.

Acknowledgments

We are grateful to Stanford STAR laboratory members in particular to Umran Inan, Morris Cohen, Ben Cotts, Kevin Graf and Naoshin Haque for involving us in the ionospheric study, providing VLF receiver and software, important comments and useful assistance. Authors would like to thank Francisco Azpilicueta for his invaluable help, fruitful discussions and explanations on the ways of TEC extraction from the GPS data at the early stage of the work. TSR thanks Rajesh Singh and Veenadhari Bhaskara for the hospitality in the IIG, India, where the VLF part of the research has been conducted. This research is supported in part by the UzFFR (projects 1-10 and 11-10) and projects F2-FA-F113, FE2-FA-F134, FA-A17-F077 of the UzAS and by the ICTP through the OEA-PRJ-29 and BIPTUN (NET-53) projects.